The Classic Period of American Toolmaking 1827 - 1930

\

Hand Tools in History Series

- Volume 6: Steel- and Toolmaking Strategies and Techniques before 1870
- Volume 7: Art of the Edge Tool: The Ferrous Metallurgy of New England Shipsmiths and Toolmakers from the Construction of Maine's First Ship, the Pinnace *Virginia* (1607), to 1882
- Volume 8: The Classic Period of American Toolmaking, 1827-1930
- Volume 9: An Archaeology of Tools: A Catalog of the Tool Collection of the Davistown Museum
- Volume 10: Registry of Maine Toolmakers
- Volume 11: Handbook for Ironmongers: A Glossary of Ferrous Metallurgy Terms: A Voyage through the Labyrinth of Steel- and Toolmaking Strategies and Techniques 2000 BC to 1950

The Classic Period of American Toolmaking 1827 - 1930

Including an extensive bibliography and company files on America's most important hand tool manufacturers

H. G. Brack

Davistown Museum Publication Series
Volume 8

© Davistown Museum 2009
ISBN 978-0-9769153-6-2

Copyright © 2009 by Harold G. Brack
ISBN 0-9769153-6-7
ISBN 13 978-0-9769153-6-2
Davistown Museum

Library of Congress Control Number: 2009934186

Cover illustration:

A view of Slater's Mill on the Blackstone River in Pawtucket, RI, America's first partially automated textile mill, built in 1793. The mill is the small wooden building on the right under the bridge. This oil on board painting was probably done between 1810 and 1820 as indicated by more recent masonry factory buildings in the background. Davistown Museum Antiquarian Art Collection, ID# 93004A1.

Cover design by Sett Balise

Frontispiece illustration:

Two views of a plow plane made in 1844 by Elihu Dutcher, Pownal, VT, Rick Floyd collection, Newport, ME. Information on this plane and Elihu Dutcher is in PTAMPIA I & II. Images also provided by Rick Floyd.

This publication was made possible by a donation from Barker Steel LLC.

Pennywheel Press
P.O. Box 144
Hulls Cove, ME 04644

Preface

Davistown Museum *Hand Tools in History*

One of the primary missions of the Davistown Museum is the recovery, preservation, interpretation, and display of the hand tools of the maritime culture of Maine and New England (1607-1900). The *Hand Tools in History* series, sponsored by the museum's Center for the Study of Early Tools, plays a vital role in achieving the museum mission by documenting and interpreting the history, science, and art of toolmaking. The Davistown Museum combines the *Hand Tools in History* publication series, its exhibition of hand tools, and bibliographic, library, and website resources to construct an historical overview of steel- and toolmaking strategies and techniques used by the edge toolmakers of New England's wooden age. Included in this overview are the roots of these strategies and techniques in the early Iron Age, their relationship with modern steelmaking technologies, and their culmination in the florescence of American hand tool manufacturing in the last half of the 19th century.

Background

During 39 years of searching for New England's old woodworking tools for his Jonesport Wood Company stores, curator and series author H. G. Skip Brack collected a wide variety of different tool forms with numerous variations in metallurgical composition, many signed by their makers. The recurrent discovery of forge welded tools made in the 18th and 19th centuries provided the impetus for founding the museum and then researching and writing the *Hand Tools in History* publications. In studying the tools in the museum collection, Brack found that, in many cases, the tools seemed to contradict the popularly held belief that all shipwrights' tools and other edge tools used before the Civil War originated from Sheffield and other English tool-producing centers. In many cases, the tools that he recovered from New England tool chests and collections dating from before 1860 appeared to be American-made rather than imported from English tool-producing centers. Brack's observations and the questions that arose from them led him to research the topic and then to share his findings in the *Hand Tools in History* series.

Hand Tools in History Publications

- Volume 6: *Steel- and Toolmaking Strategies and Techniques before 1870* explores ancient and early modern steel- and toolmaking strategies and techniques, including those of early Iron Age, Roman, medieval, and Renaissance metallurgists and toolmakers. Also reviewed are the technological innovations of the Industrial Revolution, the contributions of the English industrial revolutionaries to the evolution of the factory system of mass production with interchangeable parts, and the development of bulk steelmaking processes and alloy steel technologies in the latter half of the 19th century. Many of these technologies play a role in the florescence of American ironmongers and toolmakers in the 18th and 19th century. Author H. G. Skip Brack cites archaeometallurgists such as Barraclough, Tylecote, Tweedle, Smith, Wertime, Wayman, and many others as useful guides for a journey through the pyrotechnics of ancient and modern metallurgy. Volume 6 includes an extensive bibliography of resources pertaining to steel- and toolmaking techniques from the early Bronze Age to the beginning of bulk-processed steel production after 1870.

- Volume 7: *Art of the Edge Tool: The Ferrous Metallurgy of New England Shipsmiths and Toolmakers* explores the evolution of tool- and steelmaking techniques by New England's shipsmiths and edge toolmakers from 1607-1882. This volume uses the construction of Maine's first ship, the pinnace *Virginia*, at Fort St. George on the Kennebec River in Maine (1607-1608), as the iconic beginning of a critically important component of colonial and early American history. While there were hundreds of small shallops and pinnaces built in North and South America by French, English, Spanish, and other explorers before 1607, the construction of the *Virginia* symbolizes the very beginning of New England's three centuries of wooden shipbuilding. This volume explores the links between the construction of the *Virginia* and the later flowering of the colonial iron industry; the relationship of 17th, 18th, and 19th century edge toolmaking techniques to the steelmaking strategies of the Renaissance; and the roots of America's indigenous iron industry in the bog iron deposits of southeastern Massachusetts and the many forges and furnaces that were built there in the early colonial period. It explores and explains this milieu, which forms the context for the productivity of New England's many shipsmiths and edge toolmakers, including the final flowering of shipbuilding in Maine in the 19th century. Also included is a bibliography of sources cited in the text.

- Volume 8: *The Classic Period of American Toolmaking 1827-1930* considers the wide variety of toolmaking industries that arose after the colonial period and its robust tradition of edge toolmaking. It discusses the origins of the florescence of American toolmaking not only in English and continental traditions, which produced gorgeous hand tools in the 18th and 19th centuries, but also in the poorly documented and often unacknowledged work of New England shipsmiths, blacksmiths, and toolmakers. This volume explicates the success of the innovative American factory system, illustrated by an ever-expanding repertoire of iron- and steelmaking strategies and the widening variety of tools produced by this factory system. It traces the vigorous growth of an American hand toolmaking industry that was based on a rapidly expanding economy, the rich natural resources of North America, and continuous westward expansion until the late 19th century. It also includes a company by company synopsis of America's most important edge toolmakers working before 1900, an extensive bibliography of sources that deal with the Industrial Revolution in America, special topic bibliographies on a variety of trades, and a timeline of the most important developments in this toolmaking florescence.

- Volume 9: *An Archaeology of Tools* contains the ever-expanding list of tools in the Davistown Museum collection, which includes important tools from many sources. The tools in the museum exhibition and school loan program that are listed in Volume 9 serve as a primary resource for information about the diversity of tool- and steelmaking strategies and techniques and the locations of manufacturers of the tools used by American artisans from the colonial period until the late 19th century.

- Volume 10: *Registry of Maine Toolmakers* fulfills an important part of the mission of the Center for the Study of Early Tools, i.e. the documentation of the Maine toolmakers and planemakers working in Maine. It includes an introductory essay on the history and social context of toolmaking in Maine; an annotated list of Maine toolmakers; a bibliography of sources of information on Maine toolmakers; and appendices on shipbuilding in Maine, the metallurgy of edge tools in the museum collection, woodworking tools of the 17th and 18th centuries, and a listing of important New England and Canadian edge toolmakers working outside of Maine. This registry is available on the Davistown Museum website and can be accessed by those wishing to research the history of Maine tools in their possession. The author greatly appreciates receiving information about as yet undocumented Maine toolmakers working before 1900.

- Volume 11: *Handbook for Ironmongers: A Glossary of Ferrous Metallurgy Terms* provides definitions pertinent to the survey of the history of ferrous metallurgy in the preceding five volumes of the *Hand Tools in History* series. The glossary defines terminology relevant to the origins and history of ferrous metallurgy, ranging from ancient metallurgical techniques to the later developments in iron and steel production in America. It also contains definitions of modern steelmaking techniques and recent research on topics such as powdered metallurgy, high resolution electron microscopy, and superplasticity. It also defines terms pertaining to the growth and uncontrolled emissions of a pyrotechnic society that manufactured the hand tools that built the machines that now produce biomass-derived consumer products and their toxic chemical byproducts. It is followed by relevant appendices, a bibliography listing sources used to compile this glossary, and a general bibliography on metallurgy. The author also acknowledges and discusses issues of language and the interpretation of terminology used by ironworkers over a period of centuries. A compilation of the many definitions related to iron and steel and their changing meanings is an important component of our survey of the history of the steel- and toolmaking strategies and techniques and the relationship of these traditions to the accomplishments of New England shipsmiths and their offspring, the edge toolmakers who made shipbuilding tools.

The *Hand Tools in History* series is an ongoing project; new information, citations, and definitions are constantly being added as they are discovered or brought to the author's attention. These updates are posted weekly on the museum website and will appear in future editions. All volumes in the *Hand Tools in History* series are available as bound soft cover editions for sale at the Davistown Museum, Liberty Tool Co., local bookstores and museums, or by order from www.davistownmuseum.org/publications.html, www.Amazon.com, www.BookSurge.com, www.Abebooks.com, and www.Albris.com.

Table of Contents

Introduction

This third volume of essays on the role of hand tools in American history extends the review of toolmaking and tool forms in colonial New England and the early Republic to include a broader consideration of the American industrial florescence, whose roots lie in the shipbuilding industry of 17th and 18th century New England and its intimate relationship with English and continental steel- and toolmaking traditions and tool forms.

The historic role of armor and ordnance production and gunsmithing in European and American history is well documented. Less obvious and not well documented is the underlying significance of earlier metallurgical endeavors and innovative steelmaking strategies and toolmaking techniques, which seem but a footnote to the narration of the exploration and settlement of the New World. Little or no Renaissance era warfare, shipbuilding, world exploration, or industrial innovation would have been possible without a growing and expanding knowledge of how to make steel and steeled edge tools.

The birth and growth of a vigorous colonial ironworking and toolmaking industry and its evolution into the classic period of American toolmaking, 1827 to 1930, is rooted in the social and intellectual history of the Renaissance and the virtually unknown metallurgical magicians who facilitated its physical, industrial, and geographical expansion. Renaissance merchant adventurers, armed with wrought iron guns, cast iron cannons, and sailing ships built with steeled edge tools, explored and conquered a new continent and established new communities whose success depended on the ability to make and use hand tools. The 19th century florescence of American toolmaking grew out of a vigorous and tenacious colonial shipbuilding industry that was an essential component of the American political and industrial revolutions.

The traditions, craftsmanship, and steel- and toolmaking strategies and techniques of this remarkable period of industrial activity are now fading from our memory. Nonetheless, hundreds of thousands of high quality hand tools made by hundreds of American toolmakers and toolmaking factories survive as primary evidence of the vigor of what may be called the classic period of American toolmaking, 1827 - 1930. This publication series attempts to elucidate the transition between ancient steel- and toolmaking strategies and techniques and those available to New England shipsmiths and shipwrights during the early years of colonial and early American shipbuilding along the New England coast and their link to the florescence of American toolmaking, which is the subject of this volume.

The roots of this essentially metallurgical endeavor lie in the earlier prosperity of Elizabethan society and the politics and social values of a rapidly expanding and changing world order. The essential role played by the evolution of direct process smelting furnaces into the more efficient high-shaft blast furnace (1350) in the iron and steel consuming trading economies of the Renaissance was manifested in innovative new strategies for making ordnance and edge tools. The strategies and techniques for making iron and steel tools that characterize the expanding empires of the 16th and 17th centuries were adapted by an indigenous colonial metallurgical industry that played an important, but often unacknowledged, role in the successful settlement of North America.

The forgotten shipsmiths and the ironware and edge tools they made were the lynchpins in the economic triad essential to the success of the New England colonies: forest, fish, and shipbuilding. Those hand tools made of iron and steel smelted in the Forest of Dean and the Weald of Sussex and used to build Maine's first ship, the pinnace *Virginia*, at Fort St. George (1607 – 1608) were also used during the early years of the colonial shipbuilding industry. An examination of some of the hand tools made in the first century of the settlement of New England in the preceding volume of this series helps link English and continental toolmaking techniques with the later florescence of American toolmakers.

American toolmakers did not just suddenly start making tools in 1827, when Samuel Collins completed his first full year of manufacturing axes in his new plant in Collinsville, CT, the year chosen in this publication series to mark the beginning of the classic period of American toolmaking. Many individual artisans, inventors, and entrepreneurs, such as Simon North, Eli Whitney, Oliver Evans, Thomas Blanchard, and John Russell preceded Collins and helped lay the foundation for the birth of the American factory system of manufacturing. The great expansion in American toolmaking that occurred after 1827 was preceded by almost two centuries of forge welded hand tools made by blacksmiths, shipsmiths, and edge toolmakers serving the communities in which they lived. The westward moving population, French and Indian Wars, and the Revolutionary War prompted isolated toolmakers to make tools for use outside their local communities. The growth of the colonial shipbuilding industry and the success of the coasting, West Indies, and North Atlantic maritime trade were important components of the success of the American Revolution and the rapid expansion of the American toolmaking industries that followed. This third volume of the *Hand Tools in History* series traces and explains the creative inventiveness of a robust indigenous toolmaking community that was well established by 1827 and endured for the next century.

The roots of the florescence of American toolmaking during this century of technological innovations, inventive tool designs, and westward industrial expansion lie centuries in the past. A straightforward historical narration of events leading up to the success of the American factory system of hand tool production is an inadequate vehicle for explaining the roots of this florescence. Rather, as noted in the next to last chapter of the previous volume, *Art of the Edge Tool* (Brack 2008a, 144), the phenomenology of history is a series of labyrinths perhaps best described using the dendritic patterns of the microstructure of ferrous metals as a metaphor. Much of this historic labyrinth is beyond retrieval, forgotten events waiting to be re-narrated in the context of the frenetic technological innovations and discoveries of the modern age. A recapitulation of the events leading up to the growth and success of the classic period of American hand tool production involves the often difficult exploration of these historic labyrinths, which, as with the crystalline patterns of ferrous metals, often change over time. The narrative of the many stories of the history of toolmaking derives from none other than the tools themselves.

History Recapitulated

When that shipwright from London constructed the pinnace *Virginia* at Fort St. George, Maine, in 1607-1608, his steel edge tools were "German steel," made by the strategy of decarburizing cast iron. Yet the German Renaissance, which had endured for two centuries, had only a decade of peace remaining. The Thirty Years War (1618-1648,) would destroy its famed watch-making, ornamental iron, lock-smithing, and steelmaking industries. It was this cultural flowering that produced Albrecht Dürer (1471-1528) and ironwork of incredible beauty and complexity. This continental method of producing steel continued to be the principal strategy for steel production until the English technique of carburizing bars of Swedish wrought iron was developed after 1686.

French, and then German ironmongers had immigrated to England beginning in Tudor times, influencing the traditions of English smelters and shipsmiths, many of whom lived along the Thames River and worked at the Royal shipyards. The blast furnaces and ironworks of the Weald, located in Sussex to the south of London, were the principal source of bar iron and German steel in the years before the Forest of Dean, the Severn River, and its upstream Midlands watershed (i.e. Shropshire, which included the communities of Stourbridge and, later, Ironbridge) superseded it as England's most important iron- and steel-producing region. In the late 17th century, the cementation furnace made its first documented appearance in England in the Derwent River Valley, spreading to the Midlands and replacing the continental method for steel production after 1700. Isolated use of the cementation furnace in continental Europe can be dated to the 16th century, suggesting its origins are in the Celtic metallurgical traditions that played such an important role in the German Renaissance. One of many 16th century continental steelmaking strategies, it never superseded German steel production in southern and eastern Europe. It dominated English steel production until the appearance of the Bessemer and Siemens' processes after 1865.

The River Severn on the west coast of England, the location of the important port of Bristol, was the 17th and 18th century equivalent of the interstate highway system connecting the Stour Valley and Birmingham with other Midland ironworks. It also provided access to rapidly expanding European and colonial American economies, which were the most important markets for English iron and ironwares. While iron and steel from the Weald were brought to London by sea from the south coast of England through the British Channel, high quality phosphorus-free iron ore and bar iron were shipped from the Forest of Dean up the River Severn to Midland forges in the Stour Valley even before the English Civil War (1642-1651) resulted in the destruction of Sir Basil Brooke's Forest of Dean blast furnaces. Also supplying the Midland forges, especially in the centuries after the end of the English Civil War, were the ore deposits, blast furnaces, and ironworks of Merthyr Tydfil in south Wales, just northwest of Cardiff at the mouth of the Severn.

Even after the great fire of 1656, London remained England's most important commercial center. The Thames River thus served as a commercial artery second in importance only to the Severn River. The raw materials that made the journey from the Sussex Weald and the Forest of Dean to the Midland forges of Shropshire and the distant steel-producing city of Newcastle were eventually transported to London, often as finely crafted tools to be distributed in a growing world empire.

One important reason for the sudden appearance of the cementation process in England to produce steel was the temporary demise of the German steel industry as a result of the Thirty Years War. England supported the Protestant German nobility in its initial conflict with the Catholic King Ferdinand and his allies within the Holy Roman Empire. The Thirty Years War, essentially an extension of the religious wars that ravaged France in the 16th century (Fisher 2008), raged across Europe from 1618 to 1648 with an erratic pattern of victory and loss on all sides. Eventually, France emerged as the victor (1648) and reigned as the most important European power until the fall of Quebec (1759). One clear fact emerges from the obscurities of this long forgotten conflict: all European nations urgently needed more steel to fight these wars. The Thirty Years War, combined with the later wars of the League of Augsburg and the War of the Spanish Succession, served to dramatically increase the need for iron and steel for ordnance (cannon and cannonballs), guns, swords, and edge tools. The cementation steel furnace, which could only effectively use high quality charcoal-fired Swedish iron bar stock, answered the English need for steel by the end of the 17th century. France, Germany, and Austria continued the production of German steel from fined cast iron. The English method of making steel in the cementation furnace appeared in America after the Treaty of Utrecht ended the War of the Spanish Succession in 1713, and quickly spread from coastal New England southward to Pennsylvania, New York, Maryland, and elsewhere (Bining 1933). No documentation exists that would allow us to evaluate how much steel was produced by the continental method of decarburizing cast iron or by clandestine colonial blister steel furnaces, but the extensive collection of malleable iron hand tools at the Mercer Museum suggests that this continental method was much more well established in the Pennsylvania area than the current literature (Gordon 1996) indicates. By 1750, sporadic domestic production of blister steel began replacing domestic production of German steel (decarburized cast iron) even in Pennsylvania.

The European wars that raged from 1618 to 1763 did more to facilitate colonial shipbuilding and toolmaking than any other factor because of the massive and rapid depletion of timber resources in northwest Europe, a process that was already well under way in England in the 16th century when Elizabeth I began restricting the harvesting of timber in the Royal Forests. Yankee ingenuity, tenacity, and the rich timber resources, bog iron deposits, and shipbuilding skills of New England and other regions along the Atlantic coast were the key components of the flowering of colonial shipbuilding efforts. The massive late 16th and 17th century construction of well armored ships for Spanish, French, English, Dutch, and Portuguese exploration and their conquest of the new-found-lands had a domino effect. Timber resources near European shipbuilding centers (except in the Baltic) were depleted, resulting in rising shipbuilding costs and a lack of trained shipwrights to work in inefficient shipyards. The need to supplement timber supplies, first from Baltic and then North American timber ports, exacerbated the costs of building ships in England and northwestern Europe. As a result, shipbuilding activities exploded in New England after 1645. Warfare in Europe was the prime cause of English and European forest depletion, which then nurtured a robust indigenous shipbuilding industry in the American colonies (Hutchins 1941). This industry, in turn, gave work to New England's shipsmiths and edge toolmakers as well as those in the colonies to the south. The secrets of their finesse at edge tool production and the significance of manganese as a microconstituent in colonial bog iron have yet to be explained. The migration of shipwrights to New England colonies that occurred between 1710-1724, when "half the shipwrights in Great Britain had left the country since 1710" (Goldenberg 1976, 53), pertained to this turmoil and consequent destruction of forest resources in Europe.

The Treaty of Utrecht also signaled a temporary respite in the French and Indian conflicts in North America. Some settlers returned to Maine, only to be harassed again during King William's War, but the dark shadows cast by repeated attacks on frontier settlements, such as Deerfield, Haverhill, and the Piscataqua tidewater towns of New Hampshire, were lifted. The French and Indian wars moved westward to northern New York, western Pennsylvania, and Ohio. Sporadic attacks originating in the missionary communities of the lower St. Lawrence River Valley continued in interior and coastal sections of northern New England (Bourque 1995), as typified by Manwarren Beal's evening shipboard encounter with a native Abenaki at Beal's Harbor on Head Harbor Island in the late 1750s (see Vol. 7 of the *Hand Tools in History* series for a more detailed description of the context of this encounter). Nonetheless, areas suitable for settlement and no longer subject to attacks of the ferocity of the Deerfield massacre greatly expanded after the Treaty of Utrecht, denoting a point in time where English settlement of large areas of North America south of the St. Lawrence River basin was much more certain. The vigorous growth of colonial iron smelting and toolmaking industries followed, much of it esthetically interesting wrought iron hardware and malleable iron hand tools. Blast furnaces, at first located near the bog iron deposits of Saugus and southern Massachusetts, soon appeared in New Jersey, New York, Maryland,

and especially Pennsylvania. The hollowware made at the Carver, MA, blast furnaces for domestic products (1720f.), and probably before that by the Leonard clan (1653f.) at many southeastern Massachusetts furnaces, and by Despard on the North River (1700), soon became the cannon and cannonballs of the Revolutionary War and the War of 1812. The age of the crucible cast steel ax lay far in the future, but the florescence of American toolmaking was well underway in the early 18th century.

After the hiatus of the Revolutionary War, Jefferson's Embargo Act (1807) was the first bump in the road to prosperity for New England's merchants and ship owners. Imports of manufactured ironware to the early Republic were erratic. The War of 1812 continued to interrupt the trading patterns that had been established during the neutral trade. The vast wealth accumulated during the neutral trade as a result of the wars between France and England (1793f.) had already created rows of stately federal-style houses in towns from Wiscasset, ME, to Newburyport and Salem, MA, Newport, RI, and in the communities near the Connecticut shoreline. Philadelphia and then New York replaced Boston as America's most important port. Construction of the steam-powered passenger packets that plied the American coast occurred in New York in the pre-Civil War era. In all locations, the roles of shipsmiths and edge toolmakers, the now invisible enablers of America's maritime florescence, have been forgotten.

By 1856, the brief age of the clipper ship, the last great achievement of the Boston shipyards, had only a few years left. The spread of railroads in southern New England between 1835 and 1850 ended the total dependence of most communities on the coasting trade. Hand tools made in New England's water-powered back country mill towns could now be brought to market by rail. Toolmakers were already moving west. D. A. Barton began his performance as a famous American edge toolmaker in Rochester, New York, in 1832. By 1856, the Buck Brothers were already rolling their famed cast steel edge tools, first at Worcester, MA, a decade before Pittsburg began producing significant quantities of cast steel and then in Millbury, Massachusetts, after 1863. Where did they get their steel? The American factory system was about to leave English toolmakers a distant second in the race to produce high quality hand tools. Were American edge toolmakers working before the Civil War, such as the Buck Brothers, able to produce crucible steel in small quantities by techniques they had learned in England? What untold stories remain to be narrated about English traditional tool forms and their adaptation and production by innovative American toolmakers?

English Traditions; American Resources

The Robert Merchant wantage rule in the collection of the Davistown Museum is among the most important artifacts surviving from the flowering of toolmaking in colonial New England. Meticulously fashioned in its own hardwood slipcase in North Berwick, ME, in 1720, this rule, an icon of American toolmaking, symbolizes the migration of skilled English toolmakers to colonial New England that had been ongoing for almost a century. The skills of American toolmakers are clearly rooted in English traditions and techniques brought in the great migration to Massachusetts Bay and best symbolized by Joseph Jencks, the Leonard clan, and other New England forge masters and toolmakers well versed in the steel- and toolmaking strategies and techniques of the Sussex Weald, the Forest of Dean, and the Midlands forges of the Severn River watershed. That finely made wantage rule by Robert Merchant is just one nonferrous example of the transfer of English and European toolmaking skills to colonial America. That these traditions and techniques are themselves rooted in German, continental, and especially, a Celtic metallurgical legacy is now nearly forgotten. The early 18th century migration of English shipwrights and edge toolmakers cited by Goldenberg (1976) is perhaps the most important component of the transfer of skills and metallurgical traditions to colonial America.

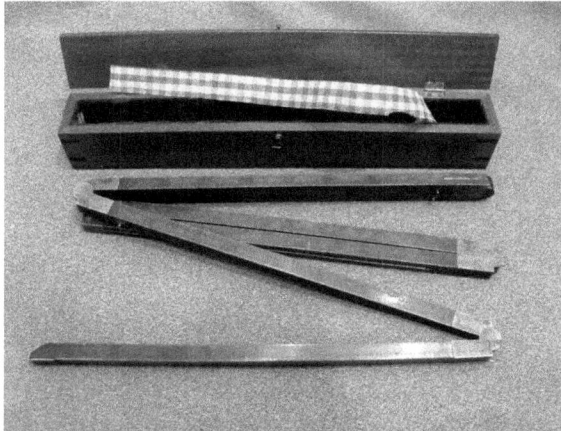

Figure 1. Wantage rule, dated 1720, boxwood and brass, 11 1/8" long box, rule is 10 ¼" long when folded, Davistown Museum MI collection ID# TBW1006.

After the Treaty of Utrecht (1713) and the temporary halt of French-English hostilities in New England, colonial settlements and the many toolmakers and ironmongers who had arrived in the colonies in the great migrations of 1629-43 and 1710-24 moved further inland from

Figure 2. I. Nicholson adjustable plow plane, beech, 1" wide with 1 1/8" wide fence, made in Wrentham, MA around 1733-1740. Photo courtesy of Bob Wheeler.

southern coastal locations. New England shipsmiths and toolmakers had two new and reliable sources of steel, English-made blister steel and steel made in colonial furnaces. A growing domestic capability of producing steel for edge tool production in larger quantities and of a better quality than could usually be forged from a bloom of natural steel accompanied the often clandestine construction of colonial steel furnaces. Making anchors and wrought iron ship fittings from the iron muck bars produced from the bog iron bloom was a long established procedure in which hoists and manpower were the key. After 1713, hardware and anchors weren't the only products of New England forges and furnaces; ironmongers' new blister steel furnaces in New England, Pennsylvania, and elsewhere, in many cases much smaller that the 5 to 10 ton capacity English furnaces, provided shipsmiths, blacksmiths, and edge toolmakers access to small quantities of locally produced steel for forging broad axes, adzes, mast shaves, augers, and timber-framing chisels of quality nearly equal to the best imported English and German steel. While in the early and mid-18th century such production may have only been a trickle, the Revolutionary War unleashed a flood of domestic edge tool production.

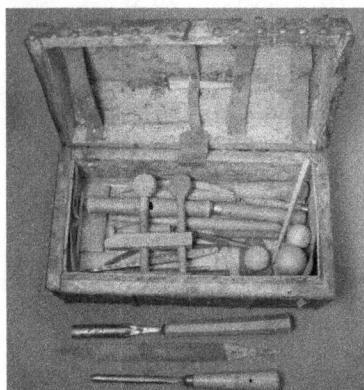

Figure 3. Tool box and files from the Willard family of watch and clockmakers in Boston, 1770 to the mid-1800s, containing edge tools made in Sheffield, England. Davistown Museum MII Collection ID#s 51201T8 and 51201T14. (Also see Figure 52.)

Contrary to the conventional opinion on the incompetence of America's often anonymous steelmakers working after the Revolutionary War, as expressed by Gordon (1996) in his chapter on steel in *American Iron*, and derived from and echoing 19th century sources (Swank 1891, and others,) the numerous domestically made edge tools of exceptional quality and durability that survive from this period tell an alternative tale of the many accomplished toolmakers adept at a variety of steelmaking techniques. Gordon (1996, 171-181) gives several examples of poorly made tools or improperly cast ingots to make the case for two centuries of incompetent steelmaking techniques. However, his index does not contain the word "tool" or "shipsmith," nor does the text contain photographs of any of the thousands of steeled edge tools that are a most important component of the legacy of the American iron industry. American iron 1607 – 1900 indeed! This nation was built in part with tools made with American iron and steel. Sheffield steel is only part of the story. The tools themselves are our primary source of information about the metallurgical finesse of American toolmakers.

In contrast to depleted European resources, colonial America's natural resources were also growing in the sense that westward expansion and exploitation first led to the iron ores of Salisbury, CT, followed by those of Pennsylvania and Maryland. Rock ores mined in these locations could be more easily smelted into high quality malleable iron, either in the charcoal bloomery or by fining blast furnace cast iron, than the more siliceous bloomery bog iron. Nonetheless, natural steel production from bog iron blooms continued despite the introduction of the new technology of the cementation or blister steel furnace. Eric Sloane (1964), in *A Museum of Early American Tools*, has the following intriguing comment, which also applies to so many other tools that survive in old tool chests in workshops long after the ships and houses they built decayed or burned.

> The 1775 gouge in the illustration has an interesting story. It was found in a stone fence. Bright and silverish, its edge is keen; it has no rust. How farm-bound bog iron, privately smelted, hammered together at a farm forge, could be better in any way than today's steel is a mystery. I have compared the best chisels (the most expensive, that is) by leaving them in the rain alongside this ancient tool. The new tool's edge was dulled, and rust appeared within a few days.
>
> The legend is that early surface ore contained much manganese and was purer in iron content. It is also believed that the use of charcoal gave purer carbon content and made a superior iron. (Sloane 1964, 54)

The bog iron used to fashion the edge tool noted by Sloane as being found in a stone fence in Connecticut was probably made near where it was found and from locally smelted bog iron. Maps of major bog-iron-producing regions can be deceiving (see Figures 19 and 20 in *Art of the Edge Tool* (Brack 2008a)). The bog iron deposits used at Saugus or the bog iron swamps of Martha's Vineyard are only several examples of the extent of small bog iron deposits as naturally occurring phenomena. Many local deposits shared the fate of those at Saugus; they were mined, smelted, and soon depleted. The much larger bog iron deposits in southeastern Massachusetts were the primary source of ore smelted in New England's now forgotten network of forges and furnaces between 1653 and 1750.

We have lost the identity of many of the New England shipsmiths who initially made edge tools from iron and steel produced at these domestic forges and furnaces, as well as from imported German and then blister steel. When domestically produced blister steel, often made from imported Swedish bar iron, became available in the first quarter of the 18th century, the colonial ability to make the high quality, heavy duty edge tools of the shipwright was already well established. Imported Sheffield-made light and medium duty crucible (cast) steel edge tools did not become available to American artisans until after 1765. These Sheffield-produced carving tools were soon used by the case furniture artists who created masterpieces out of the hard and soft woods of New England and the tropical forests along the Bay of Campeche, easily accessible as a destination in the West Indies trade. The legacy of New England's ship carvers so well detailed by writers such as Baker (1973) and exhibited in museums, such as the Peabody Essex Museum and the

Maine Maritime Museum, was another art form facilitated by English-made rather than American-made cast steel carving tools. The widespread availability of English cast steel carving and other edge tools after 1765 for American artisans is an integral part of our material cultural history. Of equal significance is the simultaneous domestic production of the heavy duty ship and timber framing tools, the evidence of which is not some written notes in a diary or log, but the tools themselves.

The New England shipwright of the colonial era, as well as those to the south, therefore had access to iron hardware, steel, and steeled edge tools from a wide variety of sources: domestically produced bog iron anchors and ships hardware, manganese and phosphorus laced domestic bog iron, imported Swedish bar iron, imported and domestically made German and blister steel, domestically produced wrought and malleable iron, and, after 1750, Sheffield cast steel and cast steel tools, a most expensive commodity. New England's early colonial bog iron industry was but a pittance in comparison to the vibrant iron ore mining and smelting industries that spread rapidly in all directions from the upper Chesapeake Bay (Principio) and Pennsylvania nucleus in the mid-18th century. By the time of the American Revolution, the American colonies were producing 1/7th of the world's output of smelted iron (Bining 1933). But what percentage of their own shipbuilding tools were they producing and where was the steel used for these tools smelted and forged? A review of the history and context of colonial New England tool production will help answer the first question but not the second.

The Early Toolmakers

When the great migration to Massachusetts Bay brought by the Winthrop Fleet ended in the early 1640s, communities were quickly established along the New England coast at Gloucester, Salem, Ipswich, Portsmouth, and other locations to the north of Boston, and to the south, where settlements dotted the small rivers north of Plymouth. Cape Cod and the southern Massachusetts and Rhode Island coasts were also soon settled, and colonists began moving inland to Concord, Sudbury, Dedham, Natick, and the communities in the Taunton River watershed. While most colonists brought the tools they needed with them from England, the growing need for additional hand tools could not always be satisfied by offerings in colonial shops, such as that of George Corwin in Salem, the contents of which are described in the previous volume in this series (Brack 2008a, 45). Some common hand tools could easily be made by farmers in their backyard bloomery forges. The bog iron hammer (Figure 4) offers an excellent example of how an early colonist could *make do* by forge welding his own hammer from locally smelted bog iron. It simply wasn't practical to journey to Boston to buy an English-made hammer in a merchant shop, nor could the farmer afford its cost and especially the loss of work time that such a journey entailed.

Figure 4. Bog iron hammer, 6" long, 1" square face, southeastern MA, ±1650, Davistown Museum MI collection ID# TAB1003.

Special needs tools, such as the small bowl adz (Figure 5) and the cheese whip (Figure 7), would not have been available as imported tools even if an isolated colonist had the funds to buy such an object. The small bowl adz now in the collection of the Natick Historical Society is a particularly interesting example of a one-of-a-kind, forge welded, malleable iron steeled tool made for a specific use, i.e. wooden-bowl-making. It's possible that this tool was made by an English blacksmith who had immigrated to the colonies, possibly for Native American use, as were many other tools made by early colonists and traded for furs. Or an isolated farmer might have forge welded this tool for his family or for production of wooden bowls, which could have been traded in a community like Natick for products made by other families. From the appearance of the welded cutting edge of this one-of-a-kind adz, it appears that, rather than using bar steel imported from England, the smith who made this forged his own natural steel cutting edge before welding it onto the bowl adz. The smith may have used the ancient technique of submerging a small piece of wrought iron in a charcoal fire for several days to increase its carbon content before welding it onto the body of the adz as its steel cutting edge. It is almost certain that this adz, as well as the bog iron hammer and the cheese whip, were made in Massachusetts in the early colonial period near where they were found. Though not of interest to contemporary tool

13

collectors, this primitive adz from southeastern Massachusetts would have been used in a workshop that surely included a mix of locally made tools, e.g. a shaving horse for making shingles and other woodenware (Figure 10), possibly a crudely fashioned pod auger, the ubiquitous forged malleable iron garden hoe (Figure 8), a French-made trade ax, and a few English tools brought in the great migration.

Figure 5. Bowl adz, 5 ¾" long, handle 6 ½" long, eastern MA, c. 1720. Image used with permission from the Natick Historical Society.

As with most trade axes, the crooked knife (Figure 6,) another common everyday artifact in the toolkits of both colonists and indigenous survivors, was probably forge welded in Europe or possibly Quebec from natural or German steel. It is usually assumed that all tools traded to First Nation communities, such as the ubiquitous trade ax, were made in Europe. Most are probably Spanish, rather than French or English in origin. As with many other edge tools brought from Europe and traded to Native Americans, trade axes appear to have been made out of German steel (decarburized cast iron) and often lack an obvious welded on steel cutting edge. This phenomenon suggests that many of these edge tools were made in one piece from German steel and then had their cutting edges subjected to additional forging (hammering) and heat treatment (quenching and tempering). However, a question arises about the prevailing view of the European origins of both trade axes and English hewing axes, such as the one in The Davistown Museum collection (Fig. 29 in Vol. 7 [Brack 2008a, 77]). Most were probably made in Europe, yet, especially in the case of the English hewing ax, which also might have been forged at the Saugus Ironworks, immigrating English smiths would have brought their knowledge of how to make edge tools with them when they came in the great migration. The same observation can be made about edge toolmakers who immigrated to Quebec, making some of the trade axes encountered in New England today. It's highly unlikely that they would have suddenly changed the style of the hewing axes they traditionally made in England or the trade axes forged in Catalonia.

14

While immigrants participating in the great migration were obviously well prepared and brought many of the tools needed for their first

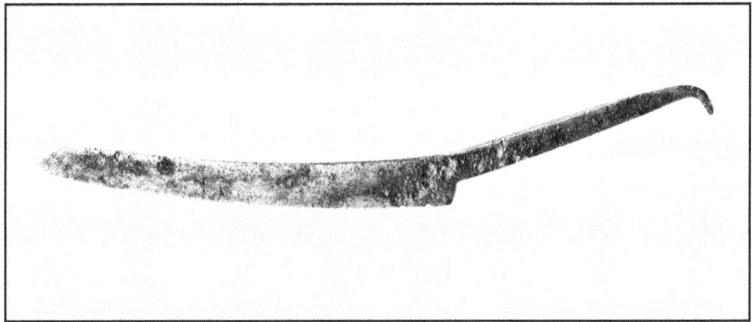

Figure 6. Crooked knife, this one is a specialized basket knife made out of natural (?) steel, 6 7/8" long, ½" wide, Davistown Museum MI collection ID# 011006T1.

years of timber framing and shipbuilding, the rapid expansion of the colonial shipbuilding industry in New England after 1650 must have resulted in a huge demand for edge and other hand tools. How many of these tools were imported from England and elsewhere and how many were produced in the colonies before the appearance of the first steel furnaces remains a mystery that may never be solved. With the establishment of integrated ironworks (blast furnace and finery, chafery, and blacksmith forges) at Braintree (now Furnace Brook, Quincy), Saugus, Taunton, and elsewhere in southeastern New England by the mid-1650s, widespread toolmaking activities by smiths who had already been trained in England were well underway. In addition to these new larger facilities, by the end of the 17th century, hundreds of small bloomery forges had been built throughout New England. Every community had its blacksmith, and most would have worked at forges that were capable of smelting small quantities of roasted bog iron. That same forge would then have been used by the smith to fashion the wrought iron hardware and primitive tools, such as the Natick bowl adz, which have survived in surprisingly large numbers from the first century of colonial settlement in New England.

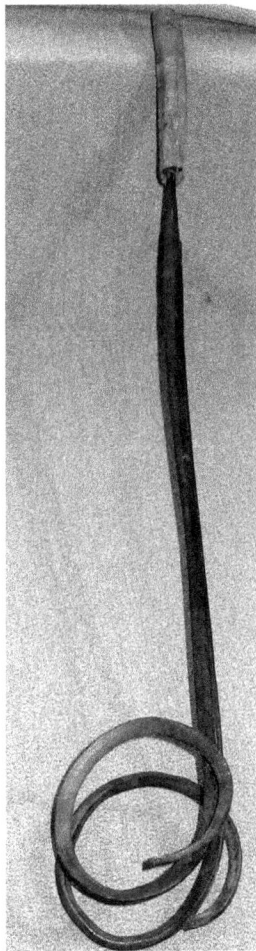

Figure 7. Cheese wisk, 30 ½" long, southeastern MA, ±1680, Davistown Museum MI collection ID# 81801T2.

There were no secret English or German steel- and toolmaking strategies and techniques not known to these early iron mongers. The vast forest resources of New England and the bog iron swamps to the

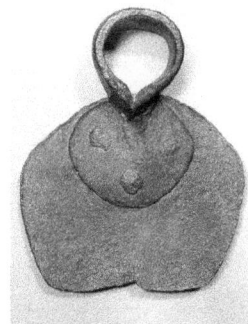

Figure 8. Grub hoe made of forged iron, 4" wide, 6" high, southeastern MA, ±1680, Davistown Museum MI collection ID# TAB1011.

north and especially to the south of Boston were already known to the Massachusetts Bay colonists who legislated explicit regulations, as

15

noted in the Massachusetts Bay Colony records of 1645 and 1646 (Brack 2008a, 17), for the colonial iron industry within a few decades of their arrival in the new-found-lands. These early legislative regulations expressed the consensus of the Massachusetts Bay Colony that ironworks be established as soon as practical. The integrated ironworks at Braintree and Saugus were the first facilities constructed as a result of these regulations. They played an important role in establishing a flourishing indigenous colonial toolmaking industry that has yet to be recognized by many American historians and commentators.

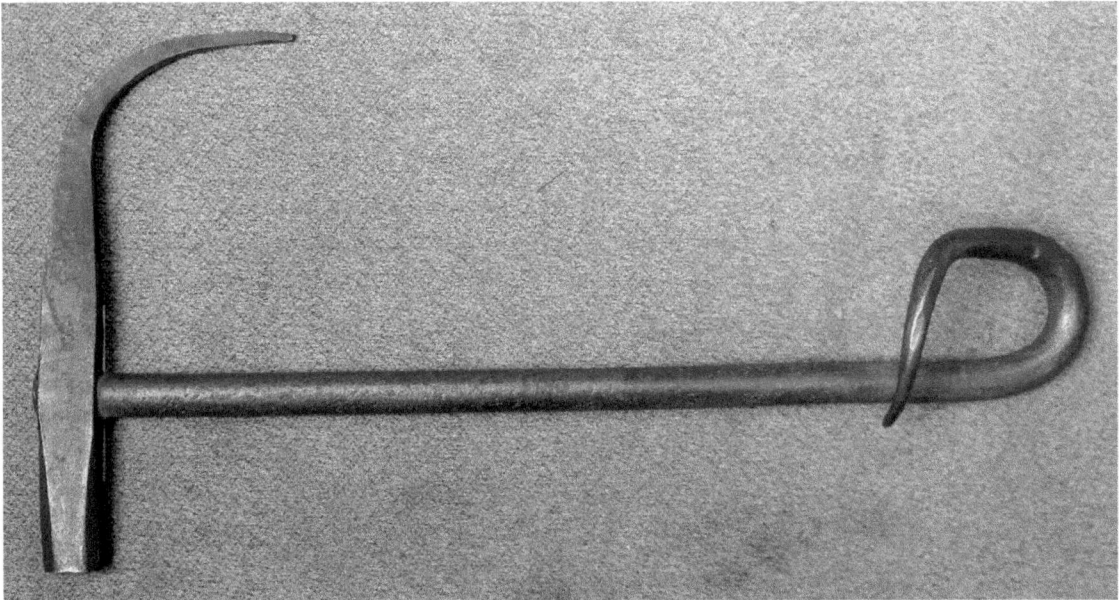

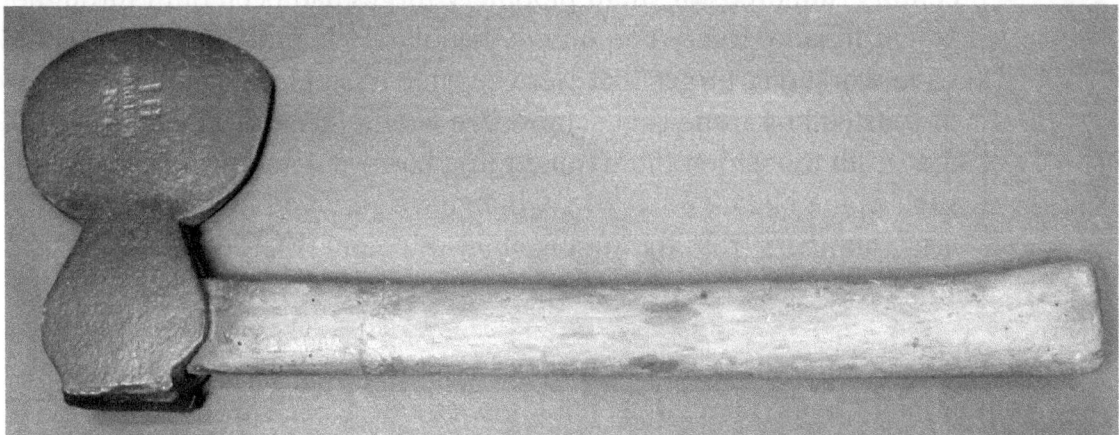

Figure 9. Top: snowshoe hammer, forged iron, 9 ½" long, handle 4 ½" long, ½" round face, c. 1790, Davistown Museum MIII Collection ID# TCM1005. Bottom: hewing ax marked "I H HARRISON N:4", c. 1750, forged iron and steel, 5" long blade, 19" long handle, Davistown Museum MII Collection ID# TBC1003.

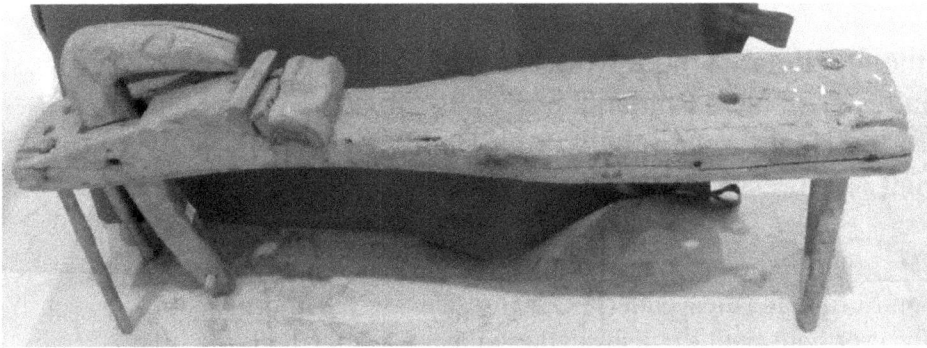

Figure 10. Shaving horse, maple, 57" long, 17" tall, 4" wide clamp, Davistown Museum MI Collection ID# TAB1012.

The Prevailing Viewpoint

Bining (1933) is one of the few lonely voices who has made a case that there was a robust indigenous toolmaking industry in New England, especially after the end of Queen Anne's War and the spread of steel furnaces in New England and elsewhere. In contrast, the conventional viewpoint is expressed by Bolles (1878).

> For two hundred years after the first settlement of the country the inhabitants were really dependent upon Europe for their cutlery. Our forests were felled principally with English axes, the crops cut with English scythes and sickles, the building-arts carried on with chisels and tools from Sheffield, and even the loaf of bread upon the table sliced with an English knife. The quantity and variety of edge-tools made in the New World were extremely small. (Bolles 1878, 221)

Bolles is certainly correct in his observation about the dominance of English cutlery; Sheffield shipped its famed knives and razors by pack horse to other areas of England as early as the Tudor era. But the ubiquitous presence of domestically-made primitive edge tools recovered in New England that date before the rise of the American factory system (1850) tell a different story. Diverse strategies and techniques for making edge tools are reflected in an immense variety of forge welded and steeled tools that have survived in New England from the robust shipbuilding economy of New England's colonial and early republic periods. Writing in 1878, Bolles also extends the hegemony of Sheffield as an edge-tool-producing region to the century preceding Huntsman's adaptation of crucible steel production for his watch spring business (1742). Earlier centers of English edge tool production were Birmingham and Lancashire in the late 17th and early 18th centuries and London prior to the great fire of 1666. The fact remains that we are not sure where the tools used by Boston and other New England shipwrights before 1720 in the construction of vessels for the British merchant fleet, including the East India trade, were forged. We certainly know where they were used: in the shipyards of New England.

Hummel (1968), in *With Hammer in Hand*, has written a comprehensive survey of the tools in the workshop of the East Hampton, NY, Dominy family who were furniture and clockmakers working from 1760 – 1840. The Dominy family continued to work after 1840 but the Hummel study ends there because such shops as the Dominy's "…disappeared in the second quarter of the nineteenth century with the transition from craft to factory production and its resultant need of new types of tools" (Hummel 1968, 31). Hummel then goes on to explain why most tools used before 1840 were imported from England (no mention is made of German tools.)

Craftsmen in both cities and villages whose shops were located near navigable water probably received the cheaper and well-made English imports directly from overseas or by transshipment from major ports. Although there were toolmakers at work in America during the eighteenth and early nineteenth centuries, their products never seemed to rival those of European manufacturers. Transportation costs for the overland movement of goods in the United States prior to 1840 were so high that they almost prohibited the distribution of American-made tools. The report of a United States Senate committee written in 1816 indicated that "a ton of goods could be brought 3,000 miles from Europe to America for about nine dollars, but... for the same sum it could be moved only 30 miles overland in this country." In the same year A. J. Dallas, Secretary of the Treasury, reported to Congress, that hardware, ironmongery, and cutlery were in a class of "manufactures which were so slightly cultivated as to leave the demand of the country wholly, or almost wholly dependent upon foreign sources for a supply." (Hummel 1968, 33)

Gordon (1996) in his classic *American Iron* also comments on the poor quality of American steelmaking efforts. Citing Tweedle (1987), Gordon asserts:

Making steel proved a particularly difficult problem for American ironmasters through the first two-thirds of the nineteenth century. When they converted bar iron to blister steel, manufacturers found it inferior to the Sheffield product. Manufacturers of edge tools and mechanisms such as gunlocks wanted crucible steel, and they used imported metal. English steelmakers retained their American market through vigorous sales efforts. ...The Sheffield steelmakers succeeded through their control of the best Swedish iron, access to coal and the special clay needed to make crucibles, and their experienced, stable work force. Steelmaking remained an art where experience counted for much and formal metallurgical knowledge for little. Others found it difficult to duplicate this success because artisans could not transfer techniques that depended on experience and specific materials unavailable elsewhere. (Gordon 1996, 89)

Later Gordon cites the remains of poor quality axes found at a Canadian site as further evidence for inadequate American toolmaking abilities:

Legends depict colonial or early American smiths as skilled, independent craftsmen making quality products. Artifacts show us, however, that smiths often had to use low-grade steel and sometimes did their work poorly. John Light and Henry Unglik found the remains of twenty axes in their archaeological study of a blacksmith shop used between 1796 and 1812 at Fort St. Joseph, Ontario. Smiths there had folded and welded wrought iron plates to make axe bodies and had then welded-in steel bits. They had made the bits too small and placed them badly. The steel had a variable carbon content and abundant slag inclusions. In none of the blades had the smiths properly hardened the bits. They made poor quality tools that would not have satisfied a demanding user.

In the first years of the republic, the United States had a substantial iron industry and little manufacturing capability. After entrepreneurs like Simeon North (who later became famous as one of the first makers of firearms with interchangeable parts) equipped themselves with tilt hammers in the 1790s to manufacture edge tools, Americans gradually captured a large share of the world market for axes and scythes with factory-made tools. (Gordon 1996, 171-172)

A fourth important source of the prevailing viewpoint of the inadequacy of American toolmaking capabilities can be found in the important Colonial Williamsburg publication *Eighteenth Century Woodworking Tools* (Gaynor 1997), a summary of research papers presented at a 1994 tool symposium. In the second paper in the symposium, noted tool collector Paul B. Kebabian asserts most woodworking tools used in America were of British origin. But then he continues with this qualifying statement.

I shall consider the reasons for that and describe how the tools used in America changed from imported to local production, a trend that was to lay the foundation for a flourishing American tool industry in the nineteenth century. (Kebabian 1997, 23)

Bolles (1878), Hummel (1968), Gordon (1996), and most papers from the tool symposium (Gaynor 1997), summarize the prevailing viewpoint that most hand tools used in American workshops and shipyards were of British origin. Citing a 1975 US Bureau of Census report, Gaynor notes "as late as 1790, the population of the colonies was still almost 79 percent English, Scottish, and Irish" (Gaynor 1997, 24). That the source of most imported tools would therefore be British is understandable, but the tools we find today in New England tool chests, workshops, and collections, some of which date from this era, tell a more complicated story. Kebabian (1997) immediately begins contradicting the prevailing thesis by citing and illustrating a John Nicholson molding plane (second half of the 18th century), a Yankee felling ax dating from before the Revolutionary War, and 18th century plow planes made by Francis Nicholson and John Lindenberger, both working in the Wrentham/Providence area, the center of early American plane making efforts. While the Williamsburg symposium text unfortunately lacks an index, none of the essays on 18th century woodworking tools include any commentary on that most essential category of woodworking trades, those of the shipwright, nor examples of any of the tools they would have used, or information about who made these tools, where, and why. The symposium is, in fact, filled with references to the robust woodworking milieu of the early republic, including the workshop of Samuel Wing in Sandwich, MA and Elbridge Gerry Reed, a chair maker from central Massachusetts (Gaynor 1997).

Another contributor to this symposium, David Hey (1997), professor of local and family history at the University of Sheffield, contends that:

The English toolmaking industry, centered on Sheffield and Birmingham... had already gained a national reputation for some of their products by the sixteenth century and that they had already captured the market for certain tools. During the seventeenth century, they began to export some of their tools to America. (Hey 1997, 9)

In Hey's bibliography, which mostly cites his own publications, there is no mention of Barraclough (1984a, 1984b), which clearly describes the rise of Sheffield steel production capabilities as occurring after canal and road construction in the mid-18th century made Sheffield steel and steel tools more easily transported to other areas of England. Prior to 1725, Birmingham, supplied with iron and steel from the Midlands and earlier from the Forest of Dean, would have been England's principle edge toolmaking center. Before this date, cutlery was Sheffield's most important product. Before the great fire of 1666, London, supplied by the furnaces of the Sussex Weald, would have been England's other principle toolmaking center.

Hey (1997) notes Birmingham and Sheffield as the national center of production of agricultural edge tools and "cheaper knives" by the mid 16th century.

By the 1670s, the Sheffield district had at least six hundred smithies for the manufacture of knives, scissors, sickles, scythes, files, awl blades, nails, and other metal goods. (Hey 1997, 11-12)

But where was the manufacture of edge tools for the shipwright occurring? The central puzzle of the construction of Maine's first ship, the pinnace *Virginia*, is where the tools used for her construction were forged in England. The same question can be asked about the tools used during the first decades of colonial shipbuilding efforts in coastal New England. And when the shipsmiths in New England began making their own adzes, slicks, broad and hewing axes, and mast shaves, where did they obtain their steel? Immigrant blacksmiths from England, who came to New England in the great migration of 1629-1642, many probably from Sheffield as well as from London, brought their knowledge of ferrous metallurgy and their skills at steeling edge tools with them. If they had access to steel suitable for edge tool manufacture, why would they import English tools?

All sources, including the secrets hidden within the contents of those New England tool chests, indicate that England was the source of carving tools, plane blades (though some German blades occasionally appear), and smaller tools, such as gimlets and calipers. The mortising chisels illustrated on page 38 (Gaynor 1997) and the hand vise, screw plate dividers, and nippers illustrated on page 39 (Gaynor 1997) are ubiquitous in American tool collections. The mark of Peter Stubbs, the famed Lancashire file and toolmaker, is among the most commonly encountered English signatures, though not all the tools bearing his name were made in his shops. The gentleman's tool chest and tools, which survive from colonial Williamsburg, and are illustrated on page 43 (Gaynor 1997) also illustrate the frequent appearance in colonial America of gorgeous high quality English-made tools often made in Birmingham and Sheffield. But, with the possible exception of the mortising chisels used to mortise the holes for trunnels in ship construction, none of these would be used by American shipwrights. The question remains as to the origin of their hand tools.

The tools illustrated by Hummel (1968) from the Dominy collection, now housed at the Winterthur Museum, further illustrate this quandary. First, few of the tools in the Dominy workshop would have been used by a shipwright. The Dominys made furniture and clocks, trades which had something in common with most of the tools of the non-shipbuilding trades of Colonial Williamsburg. These carving tools, gimlets, plane blades, and small hand tools were almost always made in England and imported to locations throughout the colonies, including Long Island and Williamsburg. But a closer look at the tools illustrated in Hummel (1968) discloses a very interesting puzzle. Tools noted as "made in England," "America and England," or "probably in England" in the Dominy workshop are fewer in number than we would expect, given that clock and furniture makers in colonial America traditionally used imported English tools. In the Dominy collection, only a few of the clearly stamped touchmarks of the English toolmakers, who always marked their tools with their name and often with its place of manufacture, make an appearance. Hummel divides his text into two sections: first, the woodworking tools and second, the metal working tools used by the Dominy workshop during their early years of operation up until about 1840.

Of the 212 woodworking tools, many incorrectly labeled as of English or probable English origin, only 28 are clearly marked English specimens. One hundred and thirty seven are obviously or probably American-made and are usually the primitive unsigned tools that we often associate with our many early American industries. Another 37 tools seem ambiguous and difficult to identify as to country of origin. Another 10 tools seem distinctly continental in style and either French or German in origin.

Of the 154 metal working tools, 91 have uncertain origins and another 58 are labeled or appear to be American-made, including the primitive clock barrel cutter (fusee engine) which is a classic example of late 18th century American metallurgy. It is not at all as sophisticated as German or English examples, but it got the job done, sometimes more efficiently than if operated by the traditional hide-bound English watchmaker. Only 5 of the metalworking tools in the Dominy collection are clearly identifiable as English.

Another intriguing aspect of the Hummel (1968) text is its use of French illustrations from André-Jacob Roubo (1769-1775) *L'art due Menuisier.* The commonly encountered leg vise illustrated in the plate from Roubo (illustration 169) has a date of 1760-1800 and the attribution "probably England." The form illustrated in plate 169 is ubiquitous even today in farm workshops throughout North America (Figure 11) and it is unlikely that all were imported from England. Most were, in fact, domestically produced. The illustration from Roubo suggests that the design of the leg vise, as with many other tools (such as the hand vises, nippers, and buckle tongs illustrated in the same plate) have continental origins and may, in fact, derive from tools commonly used during the German Renaissance before the Thirty Years War destroyed much of the industrial capacity of the Bavarian toolmakers (1634 and after). Peter Stubbs and the Sheffield toolmakers are famous for their copious production of exactly the same tools illustrated at the bottom of the Roubo plate. Since the illustrations in this plate date from an era prior to the heyday of Stubbs productivity (± 1800, see below), it is highly unlikely that the French derived their tool forms from English sources as Hummel implies. This assertion is further strengthened by the observation that throughout the early modern era of Tudor England, steelmaking innovations originating in Germany and transferred to France were then brought to southern England by French ironmongers trained in the tradition of the manufacture of German steel from decarburized cast iron. During this period, it is highly unlikely that tool forms originated in Sheffield, which was then only accessible by pack horse (Barraclough 1984a), and were then brought to France and Germany. Because these forms, such as the hand vise we still encounter so frequently in American tool chests, were produced in such great numbers by Stubbs and other Sheffield toolmakers, we naturally think of them as English forms when, in fact, they are continental in origin.

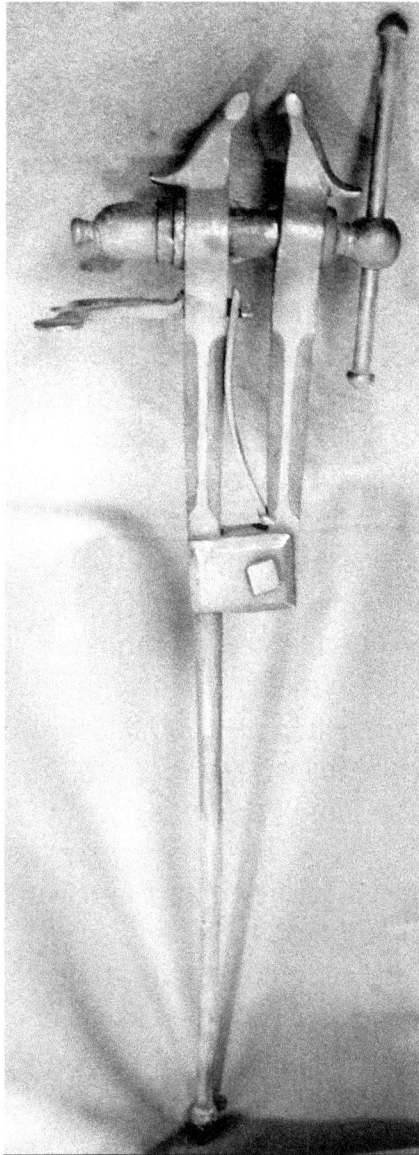

Figure 11. Blacksmiths' leg vise, forged iron, 37 ¾" high, 4 ¼" wide jaw Davistown Museum MIII Collection ID# 4106T10.

Hummel (1968) also asserted that the cost of overland transport was prohibitive. Most American forges, shipsmiths, and toolmakers in the colonial period, especially in New England and in Pennsylvania, were located in close proximity to navigable waterways. It may have been costly to bring iron bar stock or tools to isolated inland locations, but the same ports that were so convenient for imported English tools or Swedish bar iron could easily exchange domestically made iron and tools for other products in colonial America via its robust coasting trade. The edge toolmakers of southern New Hampshire (Garvin 1985) who worked from the late 18th to the late 19th centuries are just one example of a vigorous domestic toolmaking community that shipped edge tools down the rivers of New England to shipbuilders and woodworkers living along the New England coast. The Dominy workshop easily could have been one of their customers, though no Underhill tools, for example, were found in their collection.

Missing from the Hummel index are such words as "shipwright," "edge tools," "iron forge," and "blister steel." This is understandable since the Dominys were furniture and clockmakers, but then can their workshop be used to make a generalized conclusion about our early dependence on English tools? Of the two adzes illustrated on page 44, one is clearly a Yankee pattern lipped adz typical of those used by a shipwright; the second is stamped T. Austin, listed in the *Directory of American Toolmakers* (Nelson 1999) as working circa 1810, location in America unknown. All the augers illustrated in *With Hammer in Hand* (Hummel 1968) are also noted as "probably American." Unsigned and having a rather primitive appearance, these are the ubiquitous nose or pod augers of the late 18th century. The two hand-forged twisted augers were probably made in the early 19th century. Of the three "broad axes" illustrated, "America or England 1800 – 1850," the dates are correct, but aren't these hewing axes, smaller than the mammoth Pennsylvania-style shipwright's broad ax and are not all American makers? "Specimen A" is clearly noted as a stamped Collins "cast steel" ax, probably circa 1840 – 1850 and well used. "B" is a common American form but also nearly exact copy of an English prototype (see comments on pattern books below) with no marks and thus unlikely of English origin. "C" is the I. Conklin ax listed in the *Directory of American Toolmakers* as circa 1825, location unknown. The unusual ax illustrated on page 57 is neither English nor for shipbuilding, but rather continental in origin and an uncommon form. Its suggested use as a wheelwright's ax may be correct. This important collection of tools clearly illustrates the late colonial and early American tendency to manufacture many tools in upstream furnaces and forges with easy access to America's huge fleet of coasting vessels.

Gordon (1996) is certainly correct in noting that Americans were unable to make crucible steel of the quality of the Sheffield manufacturers, but most tools used in America didn't need to be made of this expensive high quality, special purpose cast steel. The Dominy workshop illustrates the wide variety of American-made tools and their uses, as well as the presence of the small but exquisite Sheffield-made carving tools, gimlets, and countersinks that are still found in almost every old New England tool chest. But American political and economic independence was based on industries far more significant and of more consequence than clock- and furniture-making, however important these trades were to America's growing middle class in the post Revolutionary War era. Neither the ubiquitous presence of English carving tools, plane blades, and gentlemen's tool chests, nor a pile of poor quality axes left at an abandoned smithy are sufficient grounds for squelching the celebration of the American tradition of toolmaking. The construction of that first ship, *Virginia*, at Fort St. George in 1607 -1608, is the opening chapter in the rise of indigenous shipbuilding and toolmaking communities that had their roots in the workshops of the anonymous shipsmiths, blacksmiths, and edge toolmakers who followed the ill fated Popham settlement with the successful occupation of North America.

The Iconography of Tools

If we expand our survey of the tool forms, especially woodworking tools that preceded the classic period of American toolmaking, we encounter other narrations and chronicles that help us understand the history of toolmaking. Of particular importance is Mercer's *Ancient Carpenters Tools* (1929). Although few of the tools he illustrates were used by the shipwright, his iconography of tool forms inform us as to what extent the hand tools made in America between 1640 and 1930 derive from English and continental tool forms. Less well known is the ferrous metallurgy of the tools that preceded our modern hand tools.

One of the definitive sources of the iconography of tools is Diderot's *Encyclopedia* ([1751-75] 1959), a pictorial survey of the trades and industries in mid-18th century France. The Diderot encyclopedia is particularly detailed in its depiction of the blast furnace, anchor- and cannon-forging, statue casting, metal mining and smelting, and ornamental iron work manufacture, which was a specialty of the French. Its survey of toolmaking is limited to anvil-making and the threading of screws for machine work. While the pin factory and the making of needles are depicted in detail, there is little mention of woodworking trades or forging of edge tools. The single plate on shipbuilding illustrates the framing of a rather Medieval-looking and bulky hull. Axes, adzes, framing whip saws, and bow saws are illustrated in a shipyard the likes of which no colonist would likely have visited.

Plate XV from Diderot and illustrated in *Hammer in Hand* (Hummel 1968), is actually plate 465 in the Dover edition of Diderot. The axes shown are dissimilar to those found in the Dominy workshop and are tool forms not encountered in colonial New England. Surely a few survive in collections, but they could not have played an important role in New England's shipbuilding industry in the century before the encyclopedia was published. Their continental forms were almost certainly made of German steel. The adz illustrated in figure 5 (Hummel 1968, 52), a block adz, is a much more familiar form and undoubtedly is similar to those used by colonial shipsmiths. Imported from Europe and also probably made from German steel rather than blister steel, such adzes were similar in their metallurgy to the ubiquitous French trading axes so frequently encountered in New England collections. Lacking a welded steel cutting edge, these tools were instead subject to additional forging and heat treatment, which further carburized and tempered their cutting edge.

Five frame saws are illustrated in Diderot's four plates on the cabinetmaker, joiner, and chair-maker. In contrast, carriage-making for the French nobility takes up nine plates. The Diderot plates on shipbuilding and woodworking illustrate the tools of the sawyer, which were nearly universal in all shipbuilding communities of the 17th and 18th century, including those in New England. The most frequently illustrated tool other than European-style hewing axes, the frame saw, would probably be, along with the single whip saw lying on the ground in plate 290 (also see plate 284, construction) and the adz in plate 290, the most commonly encountered tools in the shipyards of colonial New England. The shipsmiths and toolmakers who made the iron fittings and edge tools illustrated in the one plate on shipbuilding remain invisible not only in Diderot but also in other texts.

First printed as etchings between 1631 and 1635, *Jan van Vliet's Book of Crafts and Trades* (Bober 1981), is an even earlier source of information about the tools used in the Netherlands at this time. The trades illustrated in this series of 18 prints by a contemporary and probable student of Rembrandt include blacksmith, locksmith, cooper, sail maker, glazier, and others. The single plate pertaining to woodworking contains excellent illustrations of chisels, planes, augers, calipers, a hammer, a poll adz, and a saw with the old Dutch style bent handle. Any of these tools could have been used by shipwrights of the period in Europe, particularly the huge Medieval-style broad ax. The curved handle planes were already obsolete at this time, but the squares, calipers, brace, and chisel are still encountered in 18th century tool chests in the forms depicted in these plates. The toolmakers who made these tools, particularly the edge tools, remain invisible; the metallurgy of their tools is a lingering mystery. These plates were etched just after the heyday of Dutch exploration and settlement in the New World. The omission of the tools of the shipwright is puzzling.

Even earlier than the iconography of tools in van Vliet's etchings are the surviving tools recovered from the wreck of Henry VIII's *Mary Rose* (1548). The ship carpenter's tools recovered from the *Mary Rose*, now on display at the Portsmouth England Historic Dockyard and reproduced in Goodman (1964), are essentially late medieval forms, none of which may still have been in use at the time of the construction of the pinnace *Virginia* in 1607-1608, the first documented ship construction to have occurred in New England by English settlers (*Art of the Edge Tool*, Brack 2008a). The tools recovered from the *Mary Rose* were those of a carpenter, not of a shipwright and, therefore, tell us little of the tool forms of shipwrights of this period. At this point in time, conflicts with the French were already well underway; the *Mary Rose* was built to defend the south coast of England from privateers and possible invasion. Tudor England and continental Europe were already in an arms race that had begun with the spread of the blast furnace after 1400 (Loewen 1995) and resulted in the rapid growth in cast iron ordnance production. Cannons, made in all sizes, would have had little useful application until somebody built wooden ships to transport these potent new weapons. Again, the presence of shipsmiths and shipwrights is hidden, and the tools they used are poorly documented. The plates in

Diderot (Hummel 1968) illustrating cannon-casting and published three centuries after the *Mary Rose* capsized, after being overloaded with the weight of too many soldiers and their firearms, are among the most interesting in the encyclopedia. The arms race that marks the beginning of the modern era has continued without interruption, always accompanied by the need for improvements in arms manufacturing from foundry casting to hand gun production. Hidden behind this more well documented story are the toolmakers, shipsmiths, and shipwrights who labored in anonymity to build the ships of the navies engaged in this warfare.

William Emmett,

Houſe-Carpenter *and* Joiner,

At the Sign of the *Carpenters-Arms* in the *Market-Street*, PLYMOUTH,

SELLS all Sorts of Ironmongers Wares, *viz.* Axes, Adzes, Hatchets, Chiſſels, Saws, Scythes, Shovels, Spades, Box-Irons, and all Sorts of Braſs Nails, Locks, Scutcheons, Hinges, Powder and Shot, and Gun-Flints: Likewiſe Buckles, Buttons, Combs, Knives and Forks, Sciſſars, all Sorts of Planes, Hammers, Rules, Compaſſes, Sets of Inſtruments, *German* and *Engliſh* Steel, Watchmakers Tools, Emory, Varniſh, Linſeed Oil and Colours, beſt *London* and *Briſtol* Crown Glaſs, and other Glaſs; alſo Looking-Glaſſes of all Sizes, Proſpective Glaſſes, all Sorts of Lines, Hair Brooms and Bruſhes, ready made Coffins, all Sorts of Files, with ſeveral other Sorts of Goods too tedious to mention, at very Reaſonable Rates.

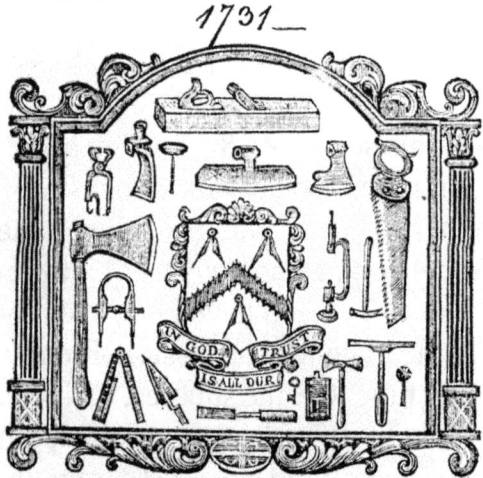

Fig. 1.3. The trade card of William Emmett of Plymouth. Believed to date from 1731, the card lists the tools sold (but not made) by him. Originally published in Devon Notes & Queries *1909.*

Figure 12. Emmett's trade card c. 1731 (Goodman 1993, 4).

Moxon's ([1703] 1989) illustrations of woodworking tools, reproduced in the first volume of the *Hand Tools in History* series, depicts the medieval form of the hand planes recovered from the *Mary Rose*. The modern forms of the jointer plane had already appeared by the time of Moxon's publication, as illustrated by the edge tools in tool dealer William Emmett's trade card 1731, also reproduced in Goodman (1964) (Figure 12). Moxon illustrates the ubiquitous frame saw, whip saw, and buck saw, the one tool still used for cutting coppice and kindling until the mid-20th century. Any of the tools in the Emmett trade card, including the chisels and axes, could have been used in the construction of the *Virginia*, but the ax and adz forms illustrated in Moxon ([1703] 1989) and the Emmett trade card are forms not encountered in the remnants of New England tool collections, shop lots, or tool chests surviving from the colonial period. Perhaps the rendition of the adz in the Emmett trade card is not accurate, but no such forms (note enlarged collared socket) have survived from this (late) date, 1731.

Dramatic changes occurred in tool forms and steelmaking techniques between the time of the publication in 1703 of Moxon's *Mechnick Exercises* and Emmett's trade card (1731) and the appearance, almost 100 years later, of Joseph Smith's ([1816] 1975) *Explanation or Key to the Various Manufactories of Sheffield* and the Timmins pattern book reprinted by Kenneth Roberts (1976) as *Tools for the Trades and Crafts*, discussed in the following chapter.

The well known publication *A Museum of Early American Tools* (Sloane 1964) provides graphic illustrations of how indigenous colonial tool forms, though often based on English prototypes, had evolved by 1800. Sloane only briefly references his lifelong passion for collecting and illustrating the hand tools of colonial Connecticut and nearby states, but his text provides compelling evidence of the degree to which colonists of English, Scotch, and German descent had developed their own unique tool forms. Very few of the hand tools illustrated in the Sloane text were imported from England, and the most common forms of English tools are so labeled. As a whole, the Sloane text, along with his other publications, are profound illustrations of America's comprehensive indigenous production of hand tools for every trade pertaining to rural life and its woodworking based economy. Almost all the tools illustrated by Sloane are wrought, malleable, or steeled hand-forged iron tools. While the numerous tools illustrated in Sloane's text match the forms discussed in the following chapter, few were made in England; most were colonial made. Sloane in particular notes regional forms of hand tools (Pennsylvania versus Connecticut, page 6). Sloane also illustrates European forms – trade, German, and British pattern felling axes, and German goosewing and English poll-less style broad axes. In New England, the American style broad ax (called the New England pattern by Kauffman (1972)), produced in the mid to late 18th century by edge toolmakers such as Faxon at Braintree, and by many later makers, is recovered much more frequently than English or German forms. These were the essential tools used in the woods of both New England and the southern colonies to harvest and shape the white oak so essential for New England's shipwrights. The Sloane text clearly depicts a well established indigenous colonial and early republic toolmaking milieu utilizing domestically produced manganese laced bog ore (hydrated limonite) as well as locally mined rock ores as their principal ingredient. Neither Sloane nor others mention the role of colonial steel furnaces in supplementing imported steel bar stock in the forging of edge tools. The Sloane index does not list topics such as "iron," "bog iron" (discussed in the text, however), "shipwright," "shipsmith," "ship building tools," "steel," "malleable iron," or "wrought iron," even though all played an important role in the creation of colonial and early American woodworking toolkits. Nonetheless, Sloane provides a graphic glimpse of a robust indigenous toolmaking community that provided the basis for the rapid growth of a uniquely American toolmaking industry in the 19th century. One of the more intriguing components of this community was the presence of family clans of toolmakers working over a period of generations.

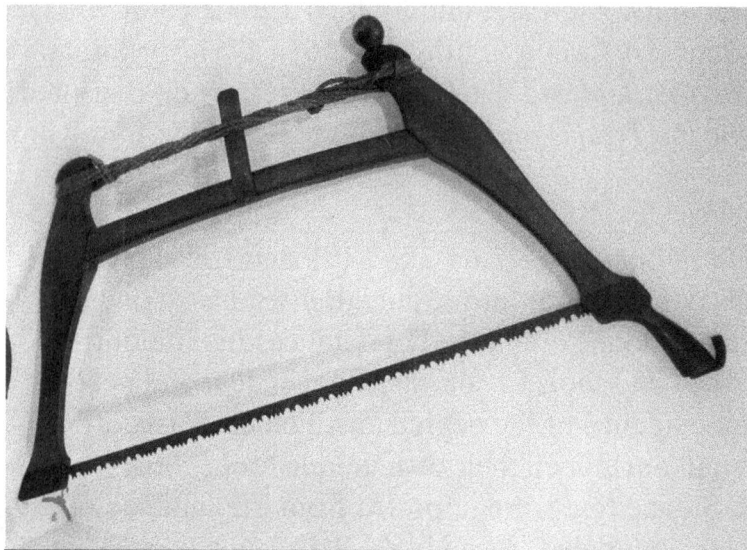

Figure 13. Gentleman's buck saw, c. 1840, steel, rope, leather, wood, 46" wide, 35" long blade, Davistown Museum MIV Collection ID# 7309T6.

The Toolmaking Clans of New England 1652-1930

A wide diversity of tool forms and multiple steelmaking strategies are part of the narrative of the stories told by the thousands of hand tools that have been recovered from New England boat shops, smithies, workshops, and collections by collectors and vendors who were active even before Eric Sloane (1964) began documenting the tools he observed in his Connecticut environs. One of the mysteries of 18th century New England industrial history is the identity of the ironmongers and edge toolmakers who played a critical role in the success of the American Revolution. A major source of information on these and many other American toolmakers is the *Directory of American Toolmakers* (Nelson 1999), now also available on CD. Much of the information in this and other chapters throughout the *Hand Tools in History* series on the working dates of American toolmakers is derived from this source.

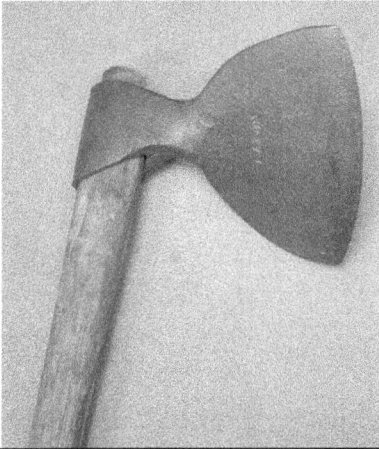

Figure 15. Offset angle hewing ax, marked "FAXON", forged iron, and steel, 8 ½" long, 6 ¾" wide cutting blade, 46" wooden handle, Davistown Museum MIII Collection ID# 7309T2.

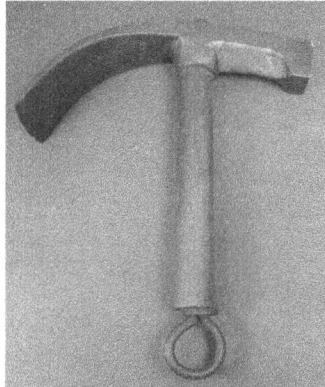

Figure 14. Coopers' adz, marked "FAXON", forged steel and iron, 10" long, 3 ¼" wide blade, 12" long wooden handle, Davistown Museum MIII Collection ID# 7309T1.

The first and possibly the most noteworthy clan of ironmongers working in New England were the Leonards, led by three brothers, Henry, James, and Thomas Leonard. Many other Leonards operated bog iron furnaces and forges throughout southern New England until well into the 19th century. Another well known clan of toolmakers was the Underhills of Chester, NH, Boston, and other nearby communities. The *Directory of American Toolmakers* (Nelson 1999) lists no less than 18 different Underhill family members or companies working for almost a century and a half. Even before the first documented Underhill (Josiah) was working in Chester, NH, the Faxon clan of edge toolmakers was working in Braintree, MA, as well as possibly in the mid-Merrimack River drainage areas of southern New Hampshire. Nelson (1999) reports a Faxon with an unknown first name as dying in 1824 at which time Jessie Underhill purchased his Boston shop. Richard Faxon (died 1821) is recorded as working in Braintree, MA, both before and after 1795. Faxon signed tools are not at all as common as tools made by the Underhill clan, but the broad axes recovered by the Liberty Tool Co. in Quincy, MA, 1973, as well as the two edge tools in Figure 14 and Figure 15 that are clearly marked FAXON, illustrate their productivity in the 18th century as well as early 19th century. The recovery of a vine (?)

ax, a very rare offset form of an ax (Figure 15), and the coopers' adz (Figure 14) from the B. F. Cutter Estate in S. Pelham, NH (see Figure 17) by the Liberty Tool Co. in June of 2009 suggest Faxon clan ironmongers may have also been forging tools in one or more of the Merrimack River watershed communities where the Underhill clan operated after 1760. The history of the Faxon clan is one example of a lost chapter in colonial and early American history waiting to be discovered.

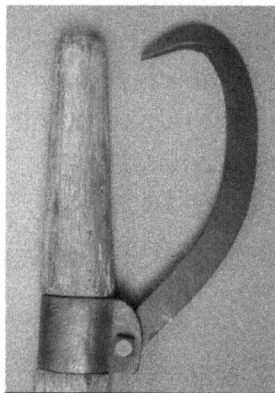

Figure 16. Cant dog, forged malleable iron and wood, 52 ½" long, 14 ¼" long cant dog, Davistown Museum MIII Collection ID# 7309T7.

Many other clans of edge toolmakers can be documented as working in New England in the late 18th and early 19th centuries. In Maine, the Billings clan of edge toolmakers was active in a number of Kennebec River drainage area communities. Numerous examples of their finesse at edge tool forging are on display at the Davistown Museum's *Art of the Edge Tool* exhibition; several are illustrated in

B. F. CUTTER,

——DEALER IN——

FRUIT, FOREST AND ORNAMENTAL TREES,

ORNAMENTAL SHRUBS,

Grape Vines, Currant and Gooseberry Bushes, &c

ALSO, SUPERIOR TOMATO KETCHUP, WHOLESALE AND RETAIL.

MAMMOTH ROAD, FOUR MILES FROM LOWELL. } **PELHAM, N. H.**

Letters and Orders directed to the Lowell Post Office, will receive **PROMPT ATTENTION.**

Figure 17. B. F. Cutter's business card.

volume 7 of the *Hand Tools in History* publication series. The Peavey clan of toolmakers, famous for converting the log rolling cant hook (Figure 16) to the spiked "peavey" still sought and used by woodsmen today, had numerous forging and toolmaking locations throughout central Maine.

In southern New England, one of the most notable families of toolmakers was the North family. Levi North was recorded as working as early as 1782 in Berlin, CT. One of his sons, Jedediah, is reported working from 1810 and a second son, Edmund, was working with him by 1824, manufacturing the tin knockers and sheet metal equipment still recovered today in New England workshops. Also part of the North clan was Simon, an edge toolmaker working before 1800 and famous for the role he played as a principal innovator of the factory system of manufacturing guns with interchangeable parts (see pg. 54).

When Samuel Collins established his ax factory in Collinsville, CT, in 1827, he was part of the Collins family of toolmakers, which included his father, Robert, working by 1805 and Robert's brother (?) David, working by 1809. Johnson Collins Jr. and his son followed Samuel in the family tool business, which continued into the 20th century.

While Samuel was organizing his famous ax company, David Chapin was making planes in New Hartford and Pine Meadows, CT, followed shortly by Hermon in 1828 and Nathaniel by 1840. The Chapin Co. became the Chapin-Stevens Co. and operated until 1929. At the same time, Daniel Copeland was an established planemaker in Hartford, CT, working with his brothers, Melvin and Alfred, as well as the Chapin clan in Pine Meadows from before 1822 into the early 1840s.

Another Connecticut family of toolmakers was the planemakers John and Lester Dennison, working in Saybrook and Winthrop, CT, by 1832 with family members continuing production well into the 1890s. Also working in Enfield, CT, were the Eaton clan of edge toolmakers; Eben, Edward, and then, Edward Jr. are reported as working from the 1840s, with Ephraim Eaton making anvils after 1850. Working in nearby Chester, CT, several decades later (1853) were Charles E. Jennings and Russell Jennings, America's foremost manufacturers of wood bits; the Russell Jennings business continued until 1944.

In Worcester, MA, Loring and Aury Gates Coes were organizing their famous wrench company by 1836. Their sons and grandsons continued operations in Worcester until 1928. In Scituate, MA, the Merritt clan of toolmakers made both planes (Charles H., 1850 and James, after 1860). The edge toolmaker, H. Merritt, was also a member of the Merritt clan; four of his edge tools in the collection of the Scituate Historical Society are illustrated in volume 7 of this publication series, *Art of the Edge Tool* (Brack 2008a, 151). Working outside of New England were many other toolmaking clans and families. The Heller Brothers and family working first in Newark, NJ, and then, Philadelphia, PA (1866f.), were America's most famous and prolific manufacturer in the 19th century of farriers' tools. Shortly after D. A. Barton began manufacturing edge tools in Rochester, NY (1832), the L. & I. J. White family began making coopers' tools (1837). As America's rapidly growing hand tool industry spread west, clan-dominated tool manufacturing that characterized some New England communities was replaced by individual companies that eventually grew into toolmaking factories with hundreds of employees.

The toolmakers of New England, as well as those in New York, Pennsylvania, Ohio, and other states, did not suddenly start making tools in America's rapidly expanding landscape in the late 18th or first four decades of the 19th century. Their finesse and expertise at making tools was based on centuries of making edge tools and planes by rule of thumb techniques passed down from generation to generation of toolmakers who had their origins in the Celtic metallurgical traditions of south central Europe (see volume 6, *Steel- and Toolmaking Strategies and Techniques before 1870* [Brack 2008b] for a review of the early history of toolmaking.) The origins of the iconography of tool forms used by these clans of New England ironmongers and toolmakers derived from these earlier toolmaking communities. American toolmakers went on to invent many distinctive new forms of tools, including new variations of long established edge tool forms. They also reproduced the many functional tool forms that were being produced by Birmingham and Sheffield, England, toolmakers in the first decades of the 19th century, often with minimal changes in their basic designs. The traditions of English and also continental European toolmakers played a key role in the amazing early 19th century florescence of New England toolmaking clans. That these clans were accompanied by thousands of individual edge toolmakers, shipsmiths, and blacksmiths making tools in every New England community (as well as in all other states) is reflected in the thousands of entries in the *Directory of American Toolmakers* (Nelson 1999). As noted, only a few of the most important and commonly encountered toolmakers are listed in the company files of this volume.

The classic period of American toolmaking is hopefully more accurately explicated by considering not only the iconography of tool forms that were their heritage but also the contemporary tool forms that were being produced and, in some cases, imported to the United States by the famed tool manufacturers of Birmingham and Sheffield in the same decades a massive tool manufacturing industry was blossoming in New England.

The Pattern Books: Smith and Timmins

The tool forms illustrated by Sloane (1964), often found in the Connecticut countryside where he lived, were relatively unchanged during the 18th century. His *Diary of an Early American Boy* (Sloane 1965) was about the typical tool kit of a multitasking, nearly self-sufficient Connecticut farm family in the early republic (±1810), well before the sudden appearance of factory-made hand tools in the 1840s. When New England's anonymous edge toolmakers began making tools for colonial shipwrights, a century before the era of Sloane's *Diary*, they copied existing forms, such as the Kent ax and the broad ax illustrated on the Emmett trade card (Figure 12). Other than those few illustrated in Diderot, Moxon, and Goodman, woodworking tools used in Europe between 1600 and 1750 are poorly documented. This lack of documentation suddenly ends with the appearance of two important information sources about tools in the era of crucible steel, steam engines, puddling furnaces, and rolling mills, Smith's ([1816] 1975) *Explanation or Key to the Various Manufactories of Sheffield* and Robert's (1976) reproduction of the pattern book of R. Timmins and Sons of Birmingham, *Tools for the Trades and Crafts*. In his introduction, Roberts reviews the known history of pattern books, the first of which was John Wyke's catalog of tools for clock- and watchmakers. These key documents contain illustrations of tools produced in Birmingham and Sheffield, England's most important late 18th century industrial centers other than London. The pattern books were issued as advertisements for both British and American hardware and tool vendors and retailers and provide an invaluable record of early modern hand tools. The tools found in American shop lots and collections, as well as those found in these pattern books fall into three categories:

1.) Tools imported to the American colonies in the early republic and frequently encountered in the remains of tool collections and tool chests found in New England in the last 39 years by the Liberty Tool Co. Many of these forms were soon copied, but the tools produced in England have two notable characteristics. They were usually signed by their English makers with company names or touchmarks and often with a place of manufacture, especially "Sheffield". Secondly, these tools are more finely made in comparison to more primitive American copies. Eventually, especially after 1840, American makers achieved the capacity to produce forms as finished and sophisticated as any English product.

2.) Tools that appear in New England tool collections that are similar in appearance and design to those in the pattern books, but are neither signed by English toolmakers, nor have the finished look of fine Sheffield tools, and are obviously copies of the English originals, just as a signed Stubbs hand vise is a copy of very similar continental, probably German, prototypes. In some cases, their tools have the signatures and place of manufacture of domestic toolmakers.

3.) Tool forms which were neither imported nor copied in sufficient quantities to appear frequently in New England tool collections.

Since the extensive tool collection of the Davistown Museum is being used as a database, it should be noted that the accumulations of tools from which the museum collections derive not only originated within 50 miles of the shipbuilding areas of coastal New England from eastern Maine to the Narragansett Bay, but also from the Blackstone River valley, Merrimack River valley of eastern New Hampshire, and western Massachusetts along the Route 2 corridor as far west as Greenfield, MA. These hand tools have been purchased in estate lots by the Jonesport Wood Co. (Liberty Tool Co.) at average rates of 2-4 tons per month since 1971. Though only a tiny percentage of the tools recovered date before 1840, this quantity of hand tools recovered over a period of decades provides a representative sampling of tools and artifacts that have survived for decades and tell us about the tool forms used long ago in the New England colonies and in the early republic. The majority of early tools thus recovered were not made in England or Germany. However, the woodworking and other tool forms frequently encountered that lack the touchmark and/or signatures of the English makers often have very similar forms to those made in Sheffield or Birmingham.

Tool forms that have appeared frequently in New England collections and workshops and appear to have been made in England or have English touchmarks in the Timmins pattern book include those in the following list. "L" means that in the Timmins pattern book these tools are noted as Lancashire in origin and, thus, probably made by Stubbs or his subcontractors in or near his factory in Warrington. The pattern book is also somewhat confusing since calipers and compasses made in Lancashire are clearly labeled as such, but, at the beginning of the text, the Birmingham compasses and calipers are unlabeled as to place of origin. Roberts also notes that Stubbs may have made the various shoemakers' files at Warrington. The listing of commonly occurring tools in the Timmins pattern book is followed by a similar listing from Smith ([1816] 1975).

Category 1: Timmins pattern book: (Roberts 1976) commonly encountered imported tool forms:
> Plate 75, 76 and 161, pg. 219: clock and watchmaker hammers (L), "plyers" (L)
> Plate 95 and 160: hand vises (L)
> Plate 159: clock screw plates (L)
> Plate 158: calipers and compasses
> Plate 156: firmer and mortising chisels and gouges
> Plate 143: flat and center bits
> Plate 29, 141, and 142: gentleman's braces with a nut adjustment and associated bits

Plate 37 and 38: turnscrews
Plate 129: bed keys
Plate 98: common bench vise
Plate 85: timber scribe
Plate 81: Kent and boat builder's axes
Plate 80: bung bore
Plate 78: bill hooks, cooper's round shave, egg handled draw knives (B)
Plate 74: box head and long shelled gimlets
Plate 71 and 72: nippers
Plate 66: upholsterer's hammers
Plate 64: saddler's punch
Plate 60: pinking irons
Plate 52: gentleman's bow (frame) saw
Plate 50 - 52: saw sets
Plate 29: saw pad, spoke shave
Plate 28: carpenter's squares
Plate 27: marking gauge
Plate 32: carpenter's pinchers (L)
Plate 25-26: shoemaker's tools
Plate 23: files
Plate 22: carpenter's mallets

Category 1: Smith's *Key to Sheffield Manufactories* (Smith [1816] 1975)
The plates in the Early American Industry Association's reproduction of Smith's *Key* are not labeled, even though every pen knife, shovel, and file has its own number. Smith ([1816] 1975) recapitulates many of the tool forms illustrated in the Timmins pattern book. The following tools must have been imported to America in significant quantities because they often appear in American tool collections.

Sickles, files, parallel rules, hand planes, hand saws (dovetail, sash, and tenon), clamps, ship carpenter's caulking mallets, caulking irons, large mast shave, cooper's ax, cooper's adz, draw knives, fluting gouges, socket chisels and gouges, tanged chisels, gouges, mortising chisels, and carving tools.

Smith's *Key* also contains extensive illustrations of razors, scissors, pen knives, and snuffers, which are the most famous products of Sheffield; their production predates the rise of Sheffield as an edge toolmaking center and may, according to Hey (1997), date from late medieval times. They also appear in America in large quantities but can't be used as evidence that all shipwrights' tools originated in Sheffield.

Category 2:
Numerous tools dating before 1840 found in New England tool collections closely resemble their English prototypes. Their lack of touchmarks and manufacturer's signatures, their slightly varied and often simplified designs and styles, and their subtle lack of the look and quality of English-made tools suggest these tool forms are early copies dating from the late 18th century or early 19th century. Most notable are the following common tool forms, which appear in both Smith's *Key* and the Timmins pattern book:

Tools associated with shipbuilding

Calipers and compasses: Some specimens in New England tool collections seem to be exact duplicates of the English patterns, but they have American marks and signatures. The earlier English designs obviously had a far reaching impact as prototypes for the later proliferation of American-made machinist's measuring tools.

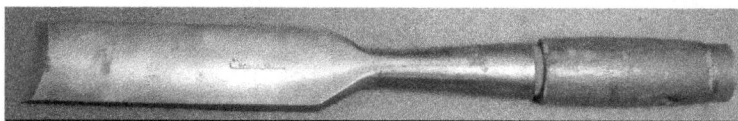

Figure 18. Timber framing chisel, marked "BUCK BROTHERS CAST STEEL", wooden handle, 16 ½" long with an 8" long and 2" wide blade, Davistown Museum MIV Collection ID# 31908T20.

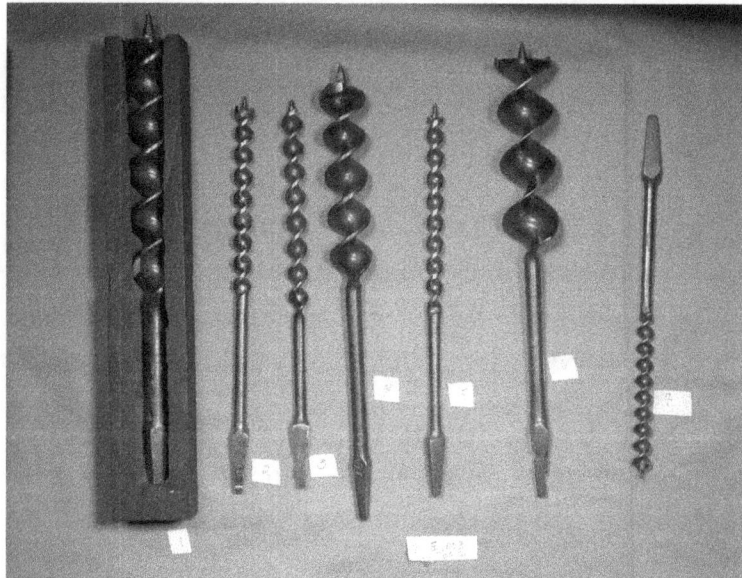

Figure 19. Auger bits, forged iron, ½" to 1 ¼" diameter, Davistown Museum MIII Collection ID#s: TCE1003A1, B2, C3, D4, E5, F6, and G7.

Socket chisels and gouges and tanged chisels and gouges: The same problem as discussed above applies to these tools. Along with numerous signed English examples, signed edged tools often look exactly like those produced in Sheffield, except for those with the mark of an American maker (Figure 18). The one exception is the mortising chisel, which does not seem to have been copied in America to any appreciable extent. Almost all mortising chisels recovered from New England tool collections have English marks. This tool may be the exception to our observation that most specimens of essential tools used by the shipwright in colonial America were made in this country.

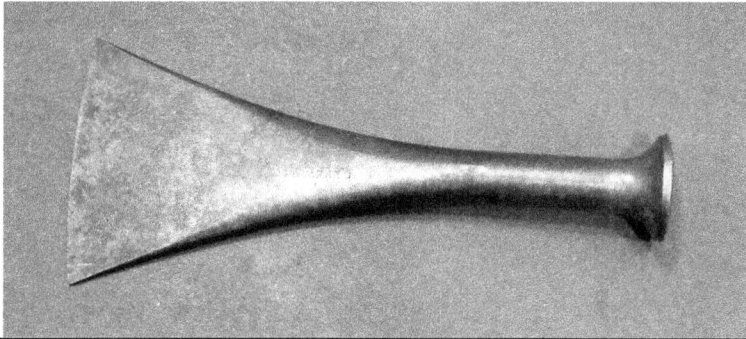

Figure 20. C. Drew caulking iron, cast steel, 6" long, 2 ½" wide, Davistown Museum MIV Collection ID# TCX1002.

Augers: Augers appear everywhere in American workshops and have the same basic designs as most of those shown in the pattern books, but these often unsigned tools, absolutely essential not only for every shipyard but also for many other woodworking tasks, are more primitive looking than the English examples and are probably among the first tools domestically produced by New England shipsmiths (Figure 19). The pod augers illustrated in the Dominy collection are typical of those domestically produced tools essential for every boatyard and workshop and predate the appearance of screw augers, which were produced during and after the first decade of the 19th century.

Caulking irons: The irons illustrated in Smith ([1816] 1975) are very similar to those frequently found in American tool collections, but they have subtle stylistic differences from the American-made irons, which are usually signed by their makers (Figure 20). In contrast to signed American C. Drew irons, English-style caulking irons appear infrequently. Caulking irons with forms obviously different from those illustrated in Smith ([1816] 1975) are also commonly found in New England tool collections and may be continental (German?) in origin, illustrating the diversity of the sources of 18th and early 19th century hand tools.

Hatchets and axes: The look of the hatchets illustrated in the English pattern books is different from most American hatchets (see, in particular, illustrations 232, 234, and 236 in Smith.) North America produced a wide variety of hatchets and axes. Only a few English prototypes are present in the pattern books. American toolmakers soon invented their own regional designs, as illustrated in the excellent surveys by Kauffman (1972), Klenman (1990), and Heavrin (1998). The same comment applies to most of the broad and hewing axes illustrated in both texts. The Kent pattern ax seems to have been closely copied by American makers; many other English patterns are infrequently encountered.

Drawknives: Signed specimens of English "drawing knives" are not uncommon. Many English specimens have distinctive bulbous handles. Much more common, however, are the primitive American shaves, which appear in every tool chest and are precursors of the fine American cast steel knives made by Kimball, Crossman,

Witherby, Wilkinson, and others (See *Appendix C. 18th and 19th Century American Toolmaker Company Files*). Forge welded and steeled examples dating from before the era of drop-forged tools are still commonplace in American workshops and flea markets.

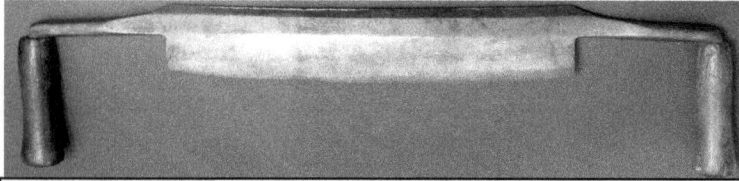

Figure 21. L. & I. J. White mast shave, cast steel, wooden handles, 24" long, 14" blade, 4 ½" wooden handles, Davistown Museum MIV Collection ID# 51100T2.

Adzes: The appearance of signed English adzes in New England tool collections is fairly rare, with the exception of those made by James Cam. A wide variety of American designs seems to have grown out of a small number of English prototypes, which often appear more elongated and curved than those used by New England shipwrights and woodworkers. The obviously steeled poll of the block adz illustrated in the pattern books is either rare on American-made tools or so well hidden by forge welding techniques that it is no longer visible. Steeled polls were soon obsolete after all cast steel adzes began appearing after 1850.

Ship scrapers: These tools appear frequently but seem to be mostly domestically produced, having a slightly different look than the scrapers illustrated in the pattern books. Such tools are generally unmarked, making it difficult to differentiate English from American specimens.

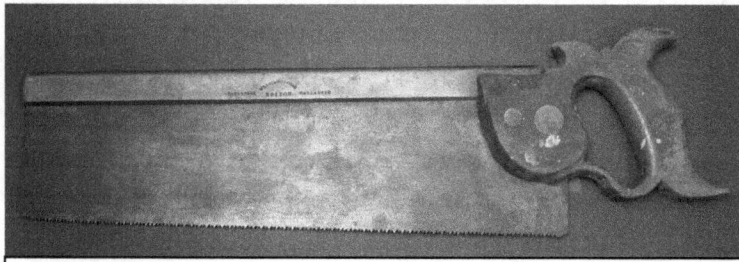

Figure 22. Welch & Griffiths back saw, cast steel, brass, and wood, 13 7/8" long blade, 6 ½" long handle, Davistown Museum MIII Collection ID# TCW1301.

Saws: Only a few saws are illustrated in the pattern books, including the gentleman's bow saw already noted and hand saws with the characteristic flat-bottomed handle, the latter of which makes an occasional appearance in New England tool collections and was also probably copied. On unsigned hand saws, it would be difficult to determine if these "early" looking handles were imported or copied by American handle-makers. All the saws noted as common English imports were also soon copied by American companies, such as Disston and Simmons. Hand saws with characteristic English stamps (e.g. Greaves and Groves), that are so different from American marks are still frequently encountered. Pit saws and cross cut saws must have been widely imported; no known 18ᵗʰ century American makers have yet been documented. As with most wood plane blades, most saw (spring) steel used to manufacture America's domestically produced saws was made in England until at least 1840. Handsaws and carving tools illustrate the continuing high regard for English steel, a phenomenon which makes it easy to overlook America's growing production of domestically produced hand tools.

Figure 23. Coopers' hoop driver, steel, iron, and wood, 7 3/4" long, 2 5/8" wide, Davistown Museum MIV Collection ID# 51201T3.

Coopers' tools: The basic form of the coopers' broad ax remains unchanged, but the specimens that turn up in New England collections, including those in the Davistown Museum collection usually have American marks. Many unsigned specimens of coopers' adzes look just like the English examples, but there are signed American-made coopers' adzes that are also similar to the English prototypes. The same comment could be made about crozes, shaves, and froes, i.e. English forms are obviously the prototype, but most surviving specimens appear to be American-made. In the case of the coopers' Nantucket hoop driver, differentiating unmarked English and American made specimens is difficult.

Other pattern book tools

Bick irons, such as that illustrated in Timmins (Roberts 1976), make an occasional appearance, but the makers would be difficult to determine on unmarked tools. Hand tools, such as pliers, nippers, pinchers, tongs, hammers of all kinds, and many of the

Figure 24. Coopers' broad ax, marked "H. A. W. KING LEWIS STNY", forged iron and weld steel with wood handle, 17 ½" long, 9 ¾" blade, Davistown Museum MIV Collection ID# 7602T2.

Figure 25. Stubbs screw plate, sheaf steel, 5" long, 1 1/8" wide, Davistown Museum MII Collection ID# 913108T17.

other tools in the pattern books, have been widely copied. Some forms may have originated in Germany, but others were made in America. These more commonplace tools don't jump out of a tool chest as possibly being English in origin as do finely made English edge tools.

The English die stock illustrated in Smith ([1816] 1975) is certainly the prototype for American models, but few signed English specimens appear. English wire gauges and screw plates were the prototypes for the huge production of American-made tools. The saddler's heading knife and other leather working tools were soon copied in America. The Osborne Company's finely made leatherworking tools are easily the equal of any earlier Birmingham-made specimens.

English coach wrenches are a controversial tool. Signed English wrenches are not uncommon, but three unsigned, more primitive forge welded coach wrenches have been found in the Boston area and are now in The Davistown Museum collection, suggesting the possibility that these coach wrenches were already being copied in late 18th century Boston. They are the prototypes for production of the later American monkey wrench, which was first manufactured in MA, beginning in the late 1830s (Page 2004) and which gave birth to an American wrench industry that produced a remarkable variety of wrenches of every conceivable design.

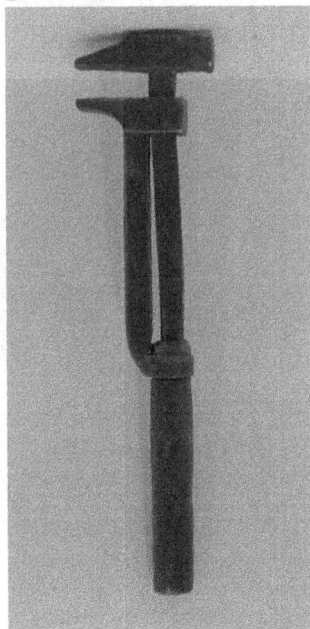

Figure 26. English-made adjustable coach wrench, forged iron, 13 ¼" long, 3 ¼" wide, Davistown Museum MII Collection ID# 32103T4.

Many of the shoemakers' tools in the pattern books would be difficult to differentiate from American products if they are not signed. They look familiar and were soon produced in America in large quantities more or less simultaneously with the appearance of these pattern books. But New England shipwrights didn't use cobblers' hammers, except possibly to make their own shoes, and many of the tools they used don't make an appearance in these English pattern books. These pattern books are a reflection of the rapid growth, both in England and America, of a middle class with a hunger for consumer goods. Upon close inspection, a large majority of the images in these pattern books are not of tools used by the "lower classes" of artisans. Instead, many are illustrations of gentlemen's tools for the workshops of the English aristocracy and the middle class, typified by those "gentlemen's tool chests" depicted at the beginning of the pattern books. If not made for the growing middle class, many

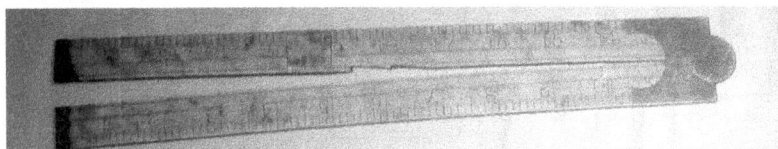

Figure 27. Folding rule, marked "J. WATTS BOSTON", brass and wood, 12" long, Davistown Museum MIII Collection ID# TCP1002A.

of the other tools in the pattern books were manufactured for the specialized trades of the coach-maker, shoemaker, mason, barber, and machinist.

PARTIAL LISTING OF BIRMINGAHM & SHEFFIELD TOOL FIRMS, 1770 - 1849

BIRMINGHAM

	1770	1800	1825	1849*
Awl Blade Makers	6	5	5	6
Brace & Bit Makers	-	-	10	21
Edge Tool Makers	7	6	9	12
File Makers	11	-	24	49
Gimblet Makers	12	14	15	23
Hammer Makers	3	-	-	17
Planemakers	5	6	10	14
Sawmakers	7	-	9	4

SHEFFIELD

	1787	1797	1821	1828	1849****
Anvil Makers	4	4	3	4	4
Auger Makers	-	-	2	3	3
Awl-Blade Makers	-	4	7	9**	-
Brace & Bit Makers	-	-	5	11	20
Edge Tool Makers	12	13	40	47 **	63
File Makers	30	40	47	80	-
Joiner Tool Makers	-	-	10	20	40
Sawmakers	10	14	43	60	-

Figure 28. Roberts, Kenneth D. 1976. *Tools for the trades and crafts: An eighteenth century pattern book: R. Timmins & Sons, Birmingham.* Fitzwilliam, NH: Ken Roberts Publishing Co. Table II. pg. 19.

Just before these pattern books were issued, England and France entered a period of warfare (1793) that ended in the Napoleonic Wars and was soon followed by the War of 1812. There was a huge demand for both new warships to fight these wars and for firearms and ordnance. The pattern books illustrate only a tiny slice of the hand tools needed in this larger social context. The tools of the shipwright, gunsmith, and cannon founder remain invisible in the context of these advertisements for a growing English consumer society, which was the prototype for the rapid growth of an American consumer society in the late 19th and 20th centuries, now recapitulated as the resource-devouring phenomenon of a global consumer society.

When considering the sources of tools imported to North America, the physical locations of toolmaking centers and their access to the transatlantic trade are of interest. The Timmins pattern book (Roberts 1976, 19) illustrates (Figure 28) the dominance of Sheffield over Birmingham as a center of edge tool production, as well as the rapid growth of Sheffield as an edge toolmaking center in the 19th century. Ironically, the Smith pattern book contains the letters from Joseph Smith to Peter Stubbs, the most important Lancashire toolmaker, while illustrating only a few of his tools (Figure 25).

The Timmins pattern book, in contrast, contains excellent illustrations of Lancashire-produced screw plates, calipers, compasses, hack saws, and vises. Birmingham is located in south central England but had easy Atlantic Ocean access via the River Severn. The Lancashire toolmakers, such as Stubbs, were centered at Warrington on the river Mersey, just upstream from Liverpool, a major English port also having easy access to the Atlantic Ocean. In contrast, Sheffield is in a more isolated location in central England. Transport of bar iron from Sweden to the Sheffield edge toolmakers, which Barraclough (1984a) notes occurring as early as 1717, was initially via Birmingham and the Mersey River to Stockworth, then by river craft to Bawtry, and then overland to Sheffield. Canals were built on the river Don, providing access to the North Sea from Rotherham in 1734, and were extended to Tinsley in 1751. The Turnpike Act of 1756 also aided access to Sheffield. The real growth of the Sheffield steel industry had to await the coming of the railroad in 1838, at which time the Sheffield industries spread from the ancient city center, the location of its ancient cutlery trade, throughout the Don Valley (Barraclough 1984a, 103). It was unlikely that any significant quantity of colonial era woodworking tools were made in Sheffield and transported to North America, in contrast to Sheffield's vast production of edge tools in the 19th century. The Timmins listing (Figure 28; Roberts 1976, 19) of Birmingham and Sheffield tool forms illustrates the relative growth and importance of both Birmingham and Sheffield as toolmaking centers after 1756 and the introduction of cast "crucible" steel.

The mystery remains of where New England's shipsmiths and shipwrights obtained the steel for the larger edge tools they were already producing in domestic forges by the late 17th century. Were our domestic ironmongers of the first decades of the 18th century able to smelt kilogram quantities of steel in the now forgotten furnaces of bog iron New England in an era where the forge of the edge toolmaker or shipsmith was an everyday component of the viability of most shipbuilding communities?

A Lost Chapter in Ferrous Metallurgy

Though discussed in detail in the two previous volumes of the *Hand Tools in History* series, a review of steelmaking strategies provides information essential to understanding the evolution of the classic period of American toolmaking. When Benjamin Huntsman rediscovered the lost art of crucible steel production (cast steel, 1742,) he was able to produce the finest steel available in Europe, albeit in very small quantities (± 6 to 8 kg). It was surpassed in quality only by the Wootz steel of the Damascus sword, an earlier form of cast steel, and was adapted by Huntsman for his watch spring business. Cast steel was characterized by a totally homogenous carbon distribution, a lack of most slag contamination, and a steel surface free from the blisters characteristic of cementation steel. But crucible steel, made in clay ingots containing broken up pieces of cementation and charcoal dust, only supplied a tiny percentage of the growing market for steel between 1750 and 1870. The production of two other types of steel was well established and supplied most of European and colonial demand for steel. Cementation steel was made from wrought iron packed in layers of charcoal dust in airtight sandstone furnaces taking 5 to 12 days for carburization. By the early 19th century the cementation process could not produce enough to supply the growing need for steel. The continuing production of high quality German steel by the decarburization of cast iron in finery furnaces preceded both cementation and crucible steel production and was centered in areas in Germany and Austria, where cast iron high in manganese (spiegeleisen) facilitated sulfur removal and ease of production. In the 16th century, manganese-laced carbonate ores had been used for iron and some German steel production in the Weald of Sussex, but, after 1650, these deposits had been depleted. Steel-producing areas without access to iron ores containing manganese, such as England after 1650 and America, could not easily use this process. It remained entrenched in France and other areas in Europe, which, as a result, did not manufacture significant quantities of blister steel.

In this context, the production of large quantities of cast iron machinery and equipment characterized industrial development after 1785. Puddled wrought iron only supplied part of these needs. Construction of bridges, water systems, tunnels, cranes, and, later, locomotives and steamships required huge amounts of cast iron. But just as cast iron could not be used for railroad tracks (wrought iron was used until the era of bulk steel production), brittle cast iron was not the only constituent of the iron machinery designed and produced by the English industrial revolutionaries. Many of these early machines have a steely look to them, and they often feel more like steel than cast iron. The question lingers as to what the early forge masters did to produce special purpose cast iron and, particularly, malleable cast iron for the machine builders before the era of bulk processed steel. What were the alternative steelmaking strategies at the beginning of the 19th century?

Two other steel-producing processes, one ancient and one a modern innovation, may have played a major but undocumented role in providing some of the steel and cast iron with steely characteristics to build the machines designed by the innovative English engineers of the early 19th century. The Brescian method of carburizing wrought iron in molten cast iron may have originated in China at an undetermined time. The date of the first production of cast iron in China is unknown but may coincide with the construction of the first bloomery furnace that could produce heterogeneous blooms that included cast iron, steel nodules, and wrought iron in one firing. Research by Needham (1958) dates the use of cast iron in China at least as early as 700 BC. Tools made of cast iron were known in Egypt in the 6th century BC. During the Italian Renaissance, knowledge of the Brescian method of steel production was likely passed from generation to generation of ironmongers. This technology may have been known and utilized both in the Roman era and in the migration period on a small scale prior to the rise of modern steelmaking technologies following the development of the blast furnace after 1350. Any Roman shaft furnaces could have been operated at a hotter temperature with an altered fuel ore ratio, producing cast iron in sufficient quantities to be fined (decarburized) into steel. However, no written documentation exists to suggest that this alternative to making natural steel was used in Roman era forges, despite the fact that some variation of the Brescian process was used to make steel in China 600 years prior to the formation of the Roman Empire.

There is no reason why English and American foundries and forges could not have used variations of the Brescian method to produce small quantities of steel in the 18th and early 19th centuries, but no records survive of the use of this technique. A variation of the Brescian method consisted of layering wrought iron interspersed with fragments of broken up cast iron, and then heating, piling, folding, and hammering these constituents into steel bar stock. This is reminiscent of the ancient tradition of pattern welding layers of sheet iron and case hardened steel or layers of sheet iron and thin pieces of crucible steel to produce the wide variety of swords and edge tools made in the early Iron Age. With the rapid increase in demand for steel, including edge tools, in both America and Europe in the early and mid-19th century, these obscure early technologies may explain the survival of so many functional and high quality forge welded steel tools that are not made of crucible cast steel. The source of the steel in these hand and edge tools is a lingering mystery that may never be solved.

A second alternative to the use of crucible, cementation, shear, Brescian, and German steel arose when Henry Cort invented the reverbatory furnace. Barraclough (1984a) notes that Cort intended to produce not only high quality wrought iron from decarburized cast iron but also steel. Mass-production of puddled steel from decarburized cast iron in the refractory furnace was not achieved until 1835, after which time it played a major role in filling the gap in steel production needs before the bulk steel processes were perfected.

Before 1835, especially in London and Manchester, major centers of industrial and machine production, puddled steel, or at least a high quality steely cast iron, could have been produced locally and in small quantities at any forge or foundry equipped with the now common-place refractory puddling furnace. The same may be said for any forge or foundry in the vast network of industrial communities, which spread from New England across the continent beginning in the late 18th century. Again, written documentation is lacking. The evidence for the use of these two later steelmaking techniques lies in the survival of both machinery, such as that on exhibit at the Royal Victoria and Albert Museum in London, and hand tools not made from crucible steel (too expensive, not marked "cast steel"), cementation steel (too many blister imperfections in the heterogeneous austenized steel), or German steel (Spathic ores not available). The massive castings of the early machinery made before 1840 in England and in America raise the possibility of a lost chapter in metallurgy. It is not known to what extent these ancient and obscure steelmaking processes were used to produce not only edge tools and other hand tools but also the machinery of the Industrial Revolution between 1785 and 1835, when puddled steel became widely available. Puddled steel or malleable cast iron could have been made both in England and America in relatively small quantities by knowledgeable founders and smelters before it was widely produced after 1835. We don't know to what extent variations of malleable cast iron, cast iron as semi-steel, and annealed white cast iron were utilized to make the machinery used in the early stages of the modern Industrial Revolution. Nor do we know what else pre-1840 English and American machine and hand toolmakers used other than puddled, Brescian, German, or other steels to produce their machinery and tools. Are most of the machines manufactured in this period only made from cast iron, or will archaeometallurgical analyses show, in fact, that both early machine- and toolmakers working in this period utilized a wide variety of steel- and iron-making technologies? It is possible that the early 19th century strategies used to make the machinery of the coming Industrial Revolution constitute another lost chapter in ferrous metallurgy, along with the mystery of Damascus steel and the early use of steel made from fined cast iron by Roman armorers.

England versus America: Resources, Markets, and Ideology

When the English industrial revolutionaries designed and built their lathes, planers, nut cutters, and proto-milling machines, they probably had little awareness of the widespread consequences of their inventions, especially in the distant land of America. London needed water supply systems, cranes, and heavy equipment for infrastructure expansion, and Manchester needed machinery and steam engines for its cotton-spinning textile factories. Hand tools were still made the old-fashioned way in England. Artisans followed long-established craft traditions to make files by hand or gouges and carving tools, which they carefully forged one at a time from hot rolled crucible steel. Even sawyers still used hand tools, such as frame and whip saws, as late as 1880, for pit sawing done by water-powered sawmills in America as early as the mid-18th century. To be fair to these hide-bound English toolmakers and tool users, even in water-powered America some stubborn American shipwrights also continued to utilize pit saws for framing out wooden ships until the late 19th century, but they were the exception to the rapid adoption of machines as prime movers in America's glowing industrial landscape.

When Darby, Huntsman, Watt, Wilkinson, Cort, and many others created that interrelated synchronicity of steam engines, smelting and melting furnaces, whirling rolling mills, and textile machinery, they created a pyrotechnic smoke belching behemoth, i.e. industrial society. In a few decades, late 18th century England increased the efficient production of iron, steel, and factory-made textiles by an order of magnitude. The German Renaissance and its legacy of steel production was already a fading memory. English dependence on high quality Swedish charcoal iron to produce crucible steel would be repeated in a clandestine colonial blister steel industry that shared the English fondness for Swedish bar iron. When interior Pennsylvania, Maryland, and New York ironmongers began producing iron of quality equal to Swedish bar iron, only the coming of the railroad could provide market access for American wrought and malleable iron produced in remote hill country locations not previously accessible to America's vigorous late colonial and early 19th century coasting trade. The remarkable tale of the evolution of the American iron industry is well told by Gordon (1996) in *American Iron, 1607 - 1900*. It began in early colonial New England in the 1640s, spread after 1720 to western Connecticut, the bog iron swamps of New Jersey, and then westward to the Mid-Atlantic States, helping to insure the success of the American Revolution. The success of early American ironmongers was based, in part, on their roots in the robust iron industry of England. The many blacksmiths who came to America in the great migration (1629-42) and manned the Saugus Ironworks (1646f.) represented the beginning of two centuries of the rapid transfer of the latest technological innovations from England to America that culminated in the classic period of American Toolmaking.

In England, machines made machines (e.g. steam engines), which used heated water to do the work formerly performed by human hands. Mass production was what the spinner did with the newly invented textile machinery of Hargraves, Arkwright, and Crompton. The revolutionary block-making machinery Henry Maudslay designed for the British Navy was not made for a market economy but for empire-building, i.e. for the specific purpose of supplying sailing ships with the necessary equipment to insure the dominance of a British Empire that reached its pinnacle of economic influence and global trading during the last half of the 19th century. At the rate of 150,000 blocks per year, the British Navy (1802-1807) needed a little bit of help from the machinery built by Henry Maudslay. Those 45 machines were the essence of labor-saving devices and facilitated worldwide consolidation of a British imperial trading economy on a grand scale.

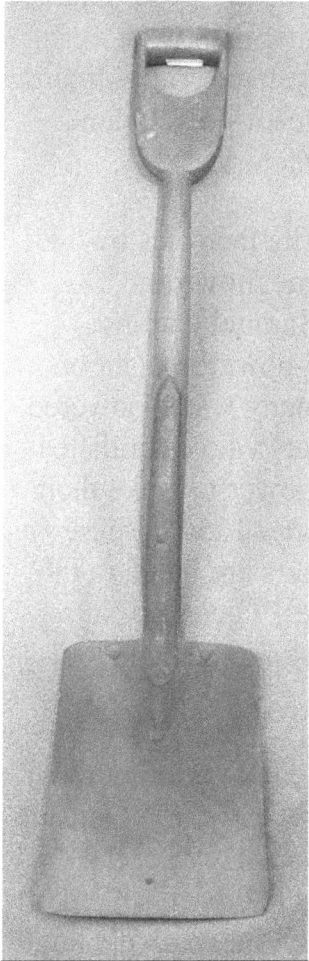

Figure 29. O. Ames shovel, c. 1820, cast iron and wood, 30 ½" high, 7" wide, Davistown Museum MII Collection ID# TCK1002.

America had a different vision, not world empire-building but expansion into wilderness areas with nearly unlimited resources, which soon created a demand for practical, useful, cheaply-made tools, which England could not supply in the quantities needed for westward expansion. In fact, as it had done with textiles, England had provided the colonies and early Republic with an immense variety of tools, especially small hand tools. Copies of most were soon made in America, but England's carving tools, plane blades, and a selection of other small steel tools dominated the American edge tool market until the Civil War. The expanding American frontier required large quantities of axes, brush hooks, picks, shovels, and horticultural tools, which were expensive and time-consuming to import from Great Britain. By the time of the American Revolution, or just after, America was making at least a small majority of its hand tools. Many lacked the finished look of Birmingham and Sheffield tools, but they executed work more efficiently than English tools used in tradition-bound English crafts-based industries. By the beginning of the 19th century, westward expansion became closely intertwined with an expanding American hand tool industry, which continued to grow throughout the century. Often living on the edge of the wilderness, American tool wielders worked their hand tools from dawn to dusk. England built the machinery of the Industrial Revolution but beginning in the 1830s, Americans used the basic design of English inventors to build machines that made hand tools. In turn, these tools harvested and processed the natural resources of a rapidly growing American economy. This is the techno-historical context for the later success of an indigenous American industrial florescence.

Key differences characterize the industrial milieu of England and America. England needed steam-power for its factories and cities because water power from its relatively small rivers was in short supply. Coal and coke were readily available as steam engine and reverbatory furnace fuel. America, and particularly New England, was characterized by a maze of streams and rivers that, with high rainfall and spring snow melt in hilly terrain with narrow valleys and many waterfalls, provided ideal sites for water-powered mills and trip hammers (Hunter 1979). Until George Corliss invented the automatic variable cutoff steam engine, the irregular stroke of the older model steam engines, the lack of coal in New England, and especially the availability of water power postponed the widespread use of the steam engine in the United States for sixty years, with two exceptions: the unique steam boats of America's Midwestern river systems and railroads, which soon spread their networks across the American landscape. Steam-power came only slowly to New England's sawmills, shipyards and toolmaking factories.

Steam engines were already driving the machinery of England's textile factories by the last quarter of the 18th century. Steam power continued to be the prime mover of industrial England throughout the 19th century. In Collinsville, CT, Samuel Collins kicked off the classic period of American toolmaking with his water-powered plant on the Farmington River. His fellow toolmakers quickly occupied the many water-powered privileges on New England's labyrinth of rivers. The many toolmakers who established their factories in CT, MA, VT, NH, and RI initially relied on water power to make their tools. Steam-powered toolmaking only supplemented water power after 1850 in these river valley towns. Hundreds of the most important of these toolmakers are listed in the appendix to this volume. All participated until the last decades of the 19th century in either supplying the toolkits of America's westward migration or forging the edge tools that built the wooden ships of maritime New England.

Another key difference between England and America during the second period of the Industrial Revolution was the existence of wilderness. There was no westward expansion in England; Ireland was not a friendly land for English settlers. Particularly after the Maine, interior New England, New York, and Pennsylvania frontiers had been settled, the westward migration of settlers, which followed the American Revolution, created a huge market for portable, practical, cheaply-made consumer goods. Few settlers could afford expensive hand-forged, Bowie-type, belt-mounted, hunting knives or expensive hand-forged, English, silvered, gentlemen's knives. John Russell solved that problem in Greenfield, MA; he expanded his chisel-making business by using the newly-designed water-powered trip hammer to make drop-forged, punched-out skinning, beaver, and hunting knives for the booming market of westward expansion. The steel that he initially
50

used for both his wood chisels and hunting knives was the best quality, imported, English cast steel. The design of his trip hammers derived from the creative accomplishments of both English and continental industrial revolutionaries. The markets, marketing strategies, and production techniques for his tools were uniquely American. Russell was one of the first to manufacture hand tools using the factory system and its die forging machinery, but he had many American predecessors, especially in the clock- and gun-manufacturing industries.

A third difference between England and America was resources. Equally as important as America's water power resources were its vast forest resources for charcoal production and wooden shipbuilding that made the early 19th century the era of the direct process bloomery furnace, but only in America, not in Europe. The early 19th century also saw America's blacksmiths mastering the art of edge tool manufacture, often, but not always, using imported high quality English crucible steel. America had vast iron deposits. The high quality iron ore deposits of Salisbury, CT, and the wide availability of bog iron in southeast New England and the New Jersey Pine Barrens were soon supplemented by low sulfur iron ore from the Juanita deposits in the Adirondacks and the rich iron deposits in Pennsylvania. The opening of the Erie Canal sent factory system consumer goods west and brought coal and wheat east to the factories and towns of New England, New York, and Pennsylvania. England had no such wealth of natural resources. Gordon (1994) summarizes the essential elements in the growth and success of the American factory system:

> Components of industry necessary to utilize water-power resources for manufacturing included artisans who were willing and able to learn new methods of working, an agricultural surplus, producers of the primary materials used, a transportation system capable of delivering raw materials and distributing products at acceptable cost, sufficient capital to make the initial investments, and a minimum of restrictive trade and labor practices. A conjunction of all these factors in the late eighteenth century helped entrepreneurs start the new American manufacturing technology. (Gordon 1994, 88)

The most important difference between England and America in the Industrial Revolution was cultural. England was traditional, conformist, hide-bound, with afternoon tea and every worker in his niche. English workmen learned a trade, mastered it, and worked at it with regularity and consistency, symbolized not only by the tea break but also a rigid educational system, which may have fostered individual excellence in the educated upper classes but was, in essence, a closed, not open society. America was the land of the liberty men: scubbers who were veterans of the Revolutionary War and went north and east to Vermont, Maine, and west to the Appalachian frontier for a new life, nearly free land, and the opportunity to forge their own social compact. They had no king, royalty, or proprietary landowners to obey (Henry Knox was an anomaly, but soon died). The American industrial tycoons of the late 19th century had not yet made their fortunes.

The podzol soils of New England soon proved inadequate for the agricultural needs of a rapidly expanding regional economy. Westward expansion was the timely progeny of a nontraditional society with an open educational system. Opportunity was there; anyone could become an Oliver Evans or an Eli Whitney and adapt the machinery of the inventive English engineers to new uses, or, in the case of Eli Whitney's cotton gin, invent entirely new devices. The mental attitudes of the educational system and the admittedly primitive milieu of the one room schoolhouse were components of an open society that fostered innovation, experimentation, and freedom of information about new inventions, machine designs, and manufacturing technologies. New frontiers were located not just to the west over the Appalachians or the Alleghenies but just on the other side of the schoolhouse wall. Perhaps it was the claustrophobia of that one room schoolhouse that fostered the realization in America that educational opportunities, including that of adapting already existing machine designs to a new system of manufacturing, could be successfully implemented by any enterprising artisan of any social class. In the newly minted land of America, there were no more walls to be encountered, at least for Caucasians of European descent, in the sense that America did not share the class-based conservative social and manufacturing systems already in place in England since the Enlightenment. Slavery, the American Achilles heel, was not a factor in the efficient functioning and rapid growth of American's northeastern quadrant of watermills, blast furnaces, and bloom smelters.

The luminescent landscapes of 19th century American painting, so radically different from the dark tones of continental painters of the Barbizon school or the wild Impressionism of J. W. Turner, are symbolic of the more well-lit American educational landscape. Less confining, more open, perhaps with a lower horizon line, America's educational system was more conducive to traveling to distant unexplored landscapes, both physical and intellectual. America's physical landscape had been ethnically cleansed of most indigenous communities, inadvertently complementing the impact of an open educational system and a growing free market economy. The luminescent paintings of the 19th century eerily foretell of glowing landscapes to come. America used revolutionary English machine designs for the invention of a factory system of mass production using interchangeable parts that was only an unrealized dream of a few English inventors. Slaves harvested cotton and tobacco in a southern economy that was much more colonial than the early northern republic of shipwrights, shipsmiths, iron mongers, inventors, toolmakers, machinists, and machine operators. No shoe factories ever sprouted in Williamsburg, Virginia. In America, with an open educational system where any male with white skin could advance, new innovative toolmaking technologies were part of the landscape and horizon of opportunities later symbolized by the luminous painters.

The Reinvention of Malleable Cast Iron

In American folklore and in its early industrial history, many inventors inhabit the cultural landscape, including Samuel Slater, Eli Whitney, Oliver Evans, Elisha Root, and many others. Seth Boyden is one of these historical figures, alleged to have discovered the art of making malleable cast iron in 1826 and then to have implemented his discoveries by manufacturing a wide variety of malleable cast iron products after 1831. This rediscovery of malleable cast iron greatly broadened the market for the often brittle products previously made from gray cast iron. But prior to Boyden's alleged discovery, other cultures and other communities knew the secret of malleable cast iron. Citing the famous English Chinese historian Joseph Needham, Barraclough (1984a), notes the widespread production of malleable cast iron tools in the 3rd and 4th century BC in China. R. A. F. de Réaumur, writing in France in 1722, also describes this process (Barraclough 1984a). It wasn't a new idea, but in the early 19th century, Boyden was the American pioneer of the innovative industrial application of the ancient process for producing malleable cast iron by rapid cooling followed by lengthy annealing of cast iron. Numerous variations of this strategy for producing durable machinable cast iron were immediately (within a decade) adapted for the manufacture of a wide variety of tools, machinery, and consumer products for the demands of a growing market economy. No product better illustrates the practical application of this technology than the wide variety of patented malleable iron planes documented by Roger Smith in his two volume treatise on American planemakers (Smith 1981; Smith 1992).

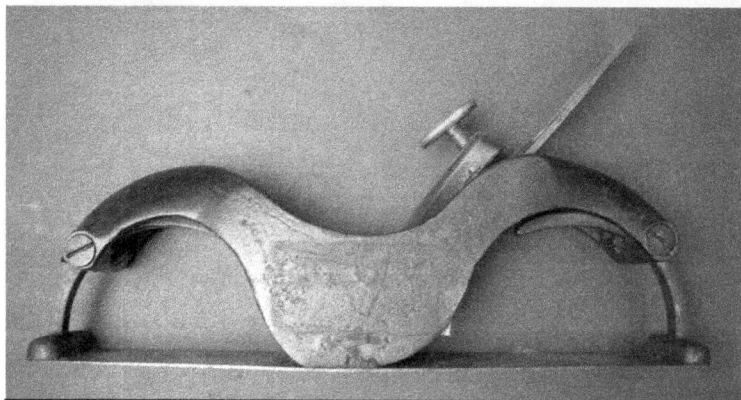

Figure 30. Evan's circular plane, cast iron, 10 3/8" long, 2 3/16" wide, with a 1 5/8" wide blade, Davistown Museum IR Collection ID# TJE1001.

The first step in the production of malleable cast iron is rapidly cooling cast iron in iron molds. Cooling cast iron slowly in sand molds produces the traditional gray cast iron, high in silicon, but also high in uncombined carbon, i.e. loose flakes of graphite, which weaken the iron, producing the traditional gray cast iron that is relatively soft and brittle, allowing some machining but limiting its usefulness. Cooling cast iron rapidly in iron molds rather than slowly in sand molds prevents the precipitation of carbon into its graphite form, retaining it in its combined form, which results in white cast iron, very hard and strong, not machinable, and not very useful for many purposes (Spring 1917). Some clever forge master or founder in ancient times discovered the secret of malleable cast iron, i.e. taking

Figure 31. Birmingham Plane Co. No. 1 smooth plane, cast iron, 6 3/8" long, 1 ¾" wide, blade 1 ½" wide, Davistown Museum IR Collection ID# TJE1002.

very hard white cast iron and packing it in iron ore or mill scale. For reasons unknown, after heating for one or two weeks, much of the carbon is removed from the casting resulting in a white, steely fracture. This is known as the Réaumur process of annealing, named after the 18th century French philosopher and metallurgist who knew and wrote about this ancient process. The Réaumur process is also very different from the Boyden method of producing malleable cast iron, in which the heating time is shorter and more carbon remains in the center of the casting, which is therefore called "blackheart," rather than "whiteheart," due to its higher carbon content. Two forms of blackheart have been traditionally produced in America, one with and one without packing in iron oxide prior to annealing. Both the American and the French methods lend themselves to the production of tools and implements that are extremely durable. Hundreds of variations in heat treatment, production methodology, and alloy variations were used in the 19th century to produce cast iron articles of every description. The Griswold Company, which began operation in Chester, CT, in 1845, is famous for producing the ultimate in malleable cast iron. Their frying pans will literally bounce down the stairs and walk out the door themselves. The legacy of hundreds of American forge masters and founders who produced a wide variety of cast iron products in the age of iron is well-known and well-remembered. An obscure, lingering question remains: in England in 1800, was the secret of Réaumur's malleable cast iron, with its long annealing time in iron oxide packing, used as another strategy for producing steel, i.e. a steely low-carbon malleable cast iron for the designers and machinists who constructed the equipment of the Industrial Revolution before the era of bulk steel production? This question especially applies to the machinery designed by the English industrial revolutionaries, some of which is on display at the Victoria and Albert Museum in London. This steely-looking cast iron machinery was adapted by American entrepreneurs for use in the American factory system.

Few examples of American machine tools made before the Civil War survive outside of specimens in the Smithsonian collection, Charles River Museum of Industry in Waltham, MA, and American Precision Museum in Windsor, VT; interest in the metallurgy of this early industrial equipment is minimal. Given the rapid decline in interest in American

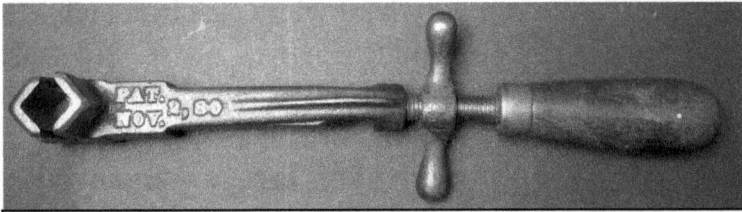

Figure 32. Wagon wrench, marked "PAT. NOV. 2, 80", steel, brass trim, and wood, 10 1/2" long, takes a 5/8" square nut, Davistown Museum IR Collection ID# 31908T31.

history and the decline in the availability of public resources to support museums and historical societies, future archaeometallurgical analyses of surviving examples of machinery used in the early years of manufacturing are, unfortunately, unlikely.

Roots of the American Factory System

French engineers made important contributions to the American factory system, although the invention of the machinery and equipment by the English industrial revolutionaries (1770-1840) discussed in Volume 6 (Brack 2008b) of the *Hand Tools in History* series obscures their contribution to the American factory system. The impact of the introduction of Watt's steam engine in the 1770s and its expeditious improvement resulted in the rapid spread of the textile industry, first in England and then in America (see cover illustration of Samuel Slater's first automated textile mill in Pawtucket, Rhode Island.) The rapid evolution of the technology necessary for the mechanization of textile production paved the way for the innovative adaptation of the block-making machinery of Henry Maudslay by numerous American industries and manufacturing companies. American entrepreneurs were quick to adapt the many machines made by Maudslay for the British Navy for the manufacture of machinery for the production of useful consumer goods, such as sewing machines and steam-powered woodworking tools. There is, however, an earlier chapter in the story of the roots of the American factory system.

Well before Eli Whitney made his contributions to the manufacture of guns with interchangeable parts at Harper's Ferry, innovations had been made in English textile factories to transmit the power generated by watermills to spinning jennies and looms with flying shuttles (John Kay, 1738), spinning jennies for weft spinning (James Hargrave, 1764), and warp spinning frames (John Kay and Richard Arkwright, 1764). The first water-powered spinning mill, which combined weft and warp spinning, was established in Derbyshire by Richard Arkwright in 1771. Improvements in carding, roving, and spinning soon followed. After 1775 steam-powered textile mills began appearing, and by 1793, these innovations in textile manufacturing were transferred to America in the form of Samuel Slater's textile mill on the Blackstone River in Pawtucket, RI.

We may think of the success of the American factory system, and the notable sale of famed Enfield rifles by the Robbins and Lawrence Armory of Windsor, VT to the British army in the 1850s, as signaling America's first totally successful production of guns with interchangeable parts. In reality, American industrial success was not based upon the inventive design of machinery or guns with interchangeable parts but on the tedious step-by-step creation of a factory system based on the use of interchangeable parts that put to practical use the innovative designs of the English industrial revolutionaries and the unique contribution of one obscure French gunsmith, Honoré Blanc.

Honoré Blanc was appointed to the post of controller at the Saint Etienne Royal Arms Manufacture in 1785. Blanc supervised the manufacture of the rifle model known as the

Charleville, which equipped many American soldiers in the American Revolution and which Lafayette possessed while on board the frigate *Hermaine* in 1780 (Allen 1983). Blanc's task was to make muskets at the St. Etienne Armory for the king's regiments. In 1777, Blanc initiated the production of a new model rifle, the "1777", characterized by the use of interchangeable parts. This innovation was executed three decades before Eli Terry began using interchangeable parts in his Connecticut clock manufactory.

Under the supervision of the French chief of ordnance J. B. Vaquette, Blanc built a specially designed workshop in 1783 in the dungeon of the Vincennes Castle and began work in 1786. On November 20th, 1790, Blanc assembled 1,000 gun locks at the Hotel des Invalides and demonstrated the interchangeability of their parts (Allen 1983). Eli Whitney was well aware of Blanc's work. The French artillerist and aid to Lafayette, Louis de Tousard (1809), wrote about Blanc and the issue of interchangeability in his three volume *American Artillerist Companion*. George Washington ordered one or more copies of these volumes and also had de Tousard redesign part of the West Point garrison into a military academy.

In 1815, Eli Whitney (1765-1825) attended a meeting in New Haven called by Col. Decius Wadsworth to begin the manufacture of muskets with interchangeable parts at both the Harper's Ferry, VA, and Springfield, MA armories. In fact, the American rifle model M1816 was based on the designs of the older French M1777 (Smith 1985, 511) and represented the first stage in the domestic manufacturing of guns with interchangeable parts.

In the context of the rise of the American factory system, water power (not steam power) did not supplant the still common use of pedal-driven lathes in the Connecticut clock-making industry until the early 19th century, occurring a few years before the 1815 New Haven conference of Whitney, Wadsworth, and others about the use of interchangeable parts for gun manufacture. The complex system of pulleys and belts that drove these water-powered lathes soon characterized most 19th century factories where the newly designed machines of the English industrial revolutionaries were making their appearance. Interchangeable wooden parts for mass produced clocks soon followed the innovations in transmuting water power into work. Between 1807 and 1814, Eli Terry perfected the mass production of pinions and wheels for his clocks using belt-driven machinery, not the hand work still the tradition in English industries. The American factory system was fast off the starting line.

The American Factory System

Among the first and most important applications of innovative English machinery design in America were those for constructing woodworking tools and machinery. The first sawmill was based on Dutch design and built in South Berwick, ME, in 1634 on John Mason's plantation on the Great Works River (Carroll 1975). Noted by Charles Carroll as the invention of Samuel Miller in England in 1777, the circular saw was adopted in sawmills in America, particularly after 1814. The first American steam sawmill was operating by 1802 (Carroll 1975). The invention of the band saw in London in 1801 was quickly copied in America, cutting down on the waste from the larger kerf of circular saws. Between 1820 and 1850, planing machines, lathes, carding machines, reciprocating mortising machines, and a wide variety of other forms of wood and metalworking machinery were invented or improvised based on English designs. As early as 1795, James Parker, working in Newburyport, MA, patented, designed, and built a water power nail-cutting and nail-heading machine (Rosenberg 1975). By 1797, Amos Whittemore had revolutionized the production of wool cards by automatic machinery at his Boston factory (Winsor 1881, 80).

This was also the dawn of the use of the power loom in textile manufacturing. Samuel Slater had finally constructed successfully operating cotton-spinning machinery for Moses Brown at Pawtucket in 1793. (See cover illustration.) A totally mechanized factory for manufacturing cotton cloth out of raw cotton had never existed before Francis Cabot Lowell built the first such factory on the Charles River in Waltham, MA, in 1813. Spinning with water- and steam-powered mules was widespread in England after Crompton improved Arkwright's spinning machines, but Crompton's spinning mules still required hand weaving on looms to produce the final product. The Lowell mill at Waltham completed the mechanization of the textile industry and remains an icon of the Industrial Revolution. Lowell's system was soon copied and became the basis for all subsequent textile manufacturing centers, including the mammoth complex built by Lowell's (d. 1817) associates on the Merrimac River at the new town of Lowell (1824).

An important early stimulus for mechanization and factory mass production techniques, other than the textile industry, was the need for arms production. The recently improved lathes of the London engineers under Maudslay were the starting point for Thomas Blanchard's creation of the irregular or copy lathe. First used for gun stock production at the Springfield and Harper's Ferry armories in 1818 (Muir 2000), this lathe could turn curves and ovals for making gun stocks. Blanchard soon made further improvements in his lathe design, which could mass produce shoe lasts, ax handles, and other wooden artifacts with irregular turned surfaces. Meanwhile, Simon North, whose family is still remembered for their production of bench plates, tin knocker stakes, and shears, had built a proto-milling machine at his gun factory in Middleton, CT. Improved and redesigned by the Maine gunsmith John Hall, this milling machine used rotary cutters and could cut iron and low carbon puddled steel in various sizes and shapes. Not yet qualifying as a precision milling machine, later developed by J. R. Brown at the Darling, Brown, and Sharpe plant in Providence, RI, and with high speed steel cutting tools still not perfected, the North-Hall milling machine was an essential first step in the process of producing guns with interchangeable parts. Then, after 1827, what had been a rising tide of industrial change turned into a tsunami of industrial innovation.

Among the most important and well documented American manufacturing companies was the Collins Axe Company of Collinsville, CT, whose first full year of operation was 1827, the date used in this publication to signify the beginning of the classic period of American toolmaking. When Elisha K. Root joined the Collins Company in 1832, shaping and forming machines using punches, dies, and patterns may already have begun replacing the tilt hammer for some steps in ax production. Rather than one dramatic change from hand-forged to machine-made axes, the Collins company innovations were a series of improvements in the efficiency and excellence of ax production that culminated in the adoption of shaping and forming machinery using dies and rollers in 1846, accompanied by the use of shaving machines that replaced the always hazardous and tedious task of shaping and finishing axes by grinding. Gordon (1994) provides the following synopsis of the significance and evolution of what was the quintessential American factory of the early years of the classic period of American tool manufacturing.

> Some forging tasks could be done with one hammer blow by placing the metal between dies containing a cavity that was the shape of the desired part and striking the dies with a sledge to cause the metal to fill out the cavity. French mechanician Honoré Blanc used die forging in 1778, and, beginning with John Hall at the Harpers Ferry Armory in the early 1820s, many Americans experimented with the application of mechanical power to Blanc's technique. In the drop hammer, power from overhead shafting turned rollers at the top of the machine to lift a wooden plank with an attached hammerhead; the operator could disengage the rollers at any time to drop the hammer on the dies with the desired force. Guides on the side of the frame held the dies in alignment. Because of the accurate alignment and the high force applied (which caused the metal to flow into the finest detail of the die cavities), artisans could make very precise forgings with a drop hammer. The forge operator had to heat the metal to the right temperature, place it accurately between the dies, and know the correct height from which to drop the hammer. The level of skill needed was comparable to that in hand-forging with a sledge. The operator retained control of the pace of work, but worked alone rather than with a striker; this procedure may have increased the risk of injury from inattention, particularly when a long run of forgings of a given part was to be made at a high production rate. The violence of the blows struck by a drop hammer was hard on the machinery and dies and made a noisy, dangerous work environment. As we have seen, tool designers at the Collins axe

works in the 1840s overcame these problems by substituting pressing and rolling technology for forging. (Gordon 1994, 358-9)

The roots of the growing factory system of mass production of hand tools, guns, and other equipment with interchangeable parts lie in the labyrinths of New England's river systems. The Underhill Edge Tool Co. of Nashua, NH, was the largest of hundreds of tool manufacturers along the Merrimack River, which had many tributaries, each with multiple water-powered mill sites that often produced only a few horsepower to run trip hammers and bellows for blacksmiths who still made hand-forged hand tools. The Connecticut River was another major power source, and its eastern tributary, the Millers River, in northern Massachusetts was the site of major tool-producing centers in the mid-19th century. At the confluence of the Greenfield and Connecticut rivers, John Russell was the first New England manufacturer to use trip hammers to make cutlery (1834). Soon after, he was using a punching machine to make tang holes in knives for their handles. To the north, Windsor, VT, was the location of one of the most important and innovative tool and arms production factories, the Robbins and Lawrence Armory and Machine Shop, located on one of the many outfalls along the Vermont / New Hampshire section of the Connecticut River.

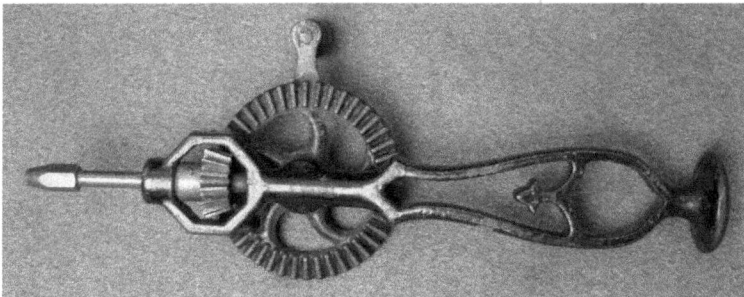

Figure 33. Millers Falls hand drill, steel and wood, 14" long, Davistown Museum IR Collection ID# 112400T1.

The Blackstone River running south from Worcester, MA, to Pawtucket and Providence, RI, gave rise to numerous tool and textile factories. Slater's Mill was only the most famous and was a harbinger of what was to evolve in the coming decades. To the east of the Blackstone River lies the Taunton River, to the west the Quinebaug, Willimantic, Connecticut, Naugatuck, and Housatonic rivers. New England was the center of hand tool production, which supplied the needs of a rapidly expanding and westward moving population. Major tool-producing centers simultaneously arose to the west on the Hudson and Mohawk rivers. In 1832, D. R. Barton, located on the Genesee River, began his cooper and edge tool manufacturing business in Rochester, NY, becoming a major supplier for New England coopers to the east, as well as for the growing tool market to the west, since tools could be shipped easily on the great lakes. Ancient D. R. Barton cooper's jiggers make surprisingly frequent late 20th century appearances in New England workshops and tool collections even though they haven't seen use for almost 125 years. See Figure 37 of L. & I. J White's jigger.

Figure 34. D. R. Barton socket chisel, marked "1832", steel, 6" long, 1 ¾" wide, Davistown Museum MIV Collection ID# 4106T7.

Shortly after Barton opened his first blacksmith shop in Rochester, Henry Disston established his saw-making business in Philadelphia (1840). To what extent Barton used imported English cast steel is unknown. We do know that Henry Disston used Sheffield crucible saw steel until the Civil War. By 1840, Philadelphia was already the saw making center of the United States.

The saws produced in Philadelphia by Disston and other manufacturers were shipped throughout a rapidly expanding national landscape of growing cities and towns. Aside from supplying the needs of an expanding westward population, it was the saws produced by Disston and other American manufacturers that made Bangor, ME, the largest mid-19th century lumber-producing center in America. The forests of New England would soon be depleted, but there was another frontier for American toolmakers to supply.

New England, New York, and Pennsylvania toolmakers supplied the tools for the

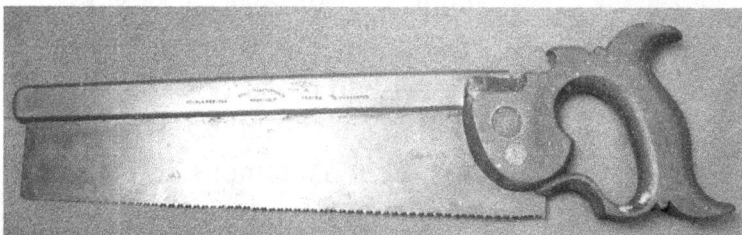

Figure 35. Disston back saw, spring steel, brass, and wood, 15 ¾" long, 12" blade, Davistown Museum MIV Collection ID# 10700T1.

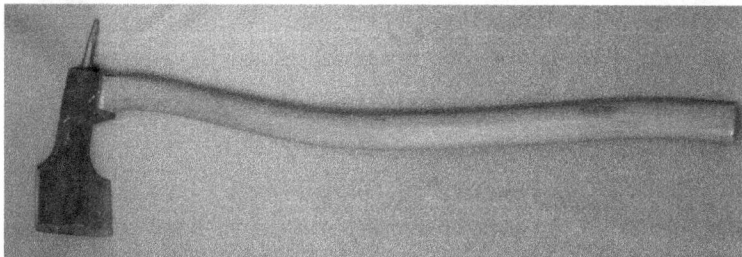

Figure 36. Collins & Co. shipwrights' adz, cast steel and wood, 10 ¾" long, 2 5/8" peen, 5" wide cutting edge, 31" handle, Davistown Museum IR Collection ID# 62406T4.

coopers, blacksmiths, and shipwrights of New England. They also supplied the hand tools and some of the machinery for the explosion of agricultural equipment production in the Midwest. Located midway between eastern tool factories and the midwestern agricultural equipment manufacturers was the most important of all components of the American factory system of toolmaking: the Pittsburgh iron and steel furnaces and forges.

Intelligent design must have once reigned in the geological evolution of the Midwest. To the east, the Allegheny and Monongahela rivers meet to form the Ohio River at Pittsburgh, PA, where some of America's richest coal, oil, and gas deposits were located. Pittsburgh and the Adirondacks region supplied high quality iron to the toolmakers of New England and, after the Civil War, became America's preeminent center of crucible steel production. The Ohio River, originating at Pittsburgh, was a geological fluke that allowed the efficient production and transportation of huge quantities of iron, and later steel, to the implement hungry hordes of westward moving Americans in the decades after 1820. It is in this geographical context that the American Industrial Revolution began moving west. Without westward expansion and the nearly unlimited agricultural opportunities of the Midwest, the Collins Tool Company of Canton, CT, (est. 1826) or the Douglas Axe Company of East Douglas, MA, (est. 1836) would have had a much more restricted market for the edge tools that they initially made out of imported English crucible steel. But was English cementation and crucible steel the only source of weld steel for the cutting edges of axes, adzes, and chisels as well as for shovels, plows, cultivators, mowing machines, and threshers? After 1850, it was not.

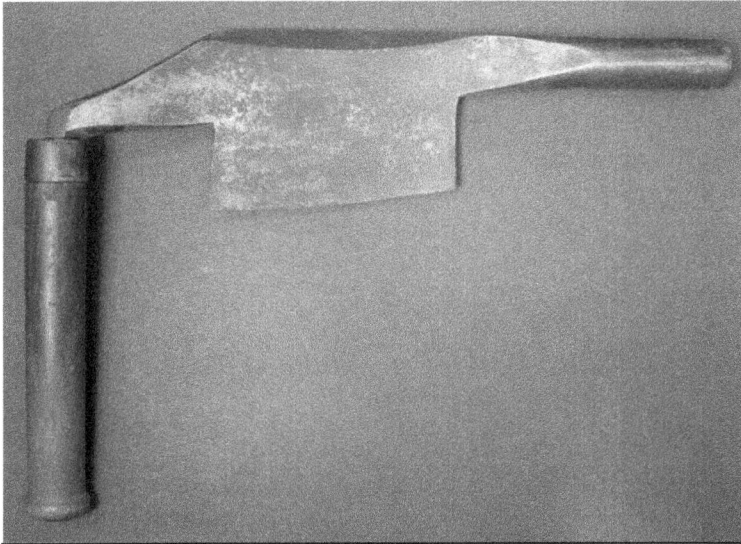

Figure 37. L. & I. J. White chamfering knife, cast steel and wood, 15" long, 5 ½" long cutting blade, 8" long handle, Davistown Museum MIV Collection ID# 41203T6.

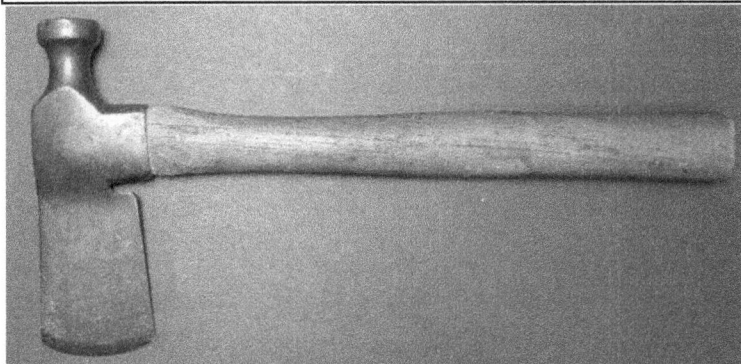

Figure 38. C. A. Williams & Co. lathing hatchet, cast steel and wood, 13" long, 2 ¼" wide blade, 1" diameter poll, Davistown Museum IR Collection ID# 43006T9.

In 1828, Richard Hoe, working in Pennsylvania and using imported English cast steel and a steam-powered punching machine, began mass production of the first circular saw blades. John Lane, an Illinois blacksmith, began making plows with steel blades that were more efficient in cutting the Midwest sod than the cast iron blades of older English designed plows. Cyrus H. McCormick invented the first wheat reaper in 1831 and produced the first model in 1833. Also using imported Sheffield steel for the fabrication of his blades but not his machinery, John Deere constructed the first mowing machines in 1837. In the same year, John and Hiram Pitts designed the first mechanical thresher. Other American inventions created in response to the challenges of the

American frontier included corn cultivators, hay and grain rakes, grain drills, corn shellers, hay bailers, and cultivators. Specialized steel for the cutting edges of many of these tools was imported from Sheffield and Birmingham before the Civil War, but almost all other components of these machines were made in American factories, often owned by the inventors of the tools being produced. Many were made of cast iron, but, after 1837, the development of a wide variety of malleable cast irons greatly assisted tool and machinery makers by providing a more durable and machinable steely cast iron that was easy to cut and shape. At this time the first cast iron plane appeared in America, i.e. the Knowles pattern joiner's plane, and it was soon followed by the much more durable malleable cast iron planes pioneered by Leonard Bailey and the many small patented planemakers described by Roger K. Smith (1981, 1992) in his comprehensive *Patented Transitional & Metallic Planes in America* volumes 1 and 2. By 1850, a vigorous, increasingly mechanized, American toolmaking industry was well established in New England and gradually spread to the Midwestern states during the last half of the 19th century.

Figure 39. Sun plane, blade signed "White 1837", cast steel and wood, 14" long, 3" wide, 2" wide blade, Davistown Museum MIII Collection ID# 100400T6.

Steelmaking Reconsidered

The fourth decade of the 19th century was a revolutionary period of industrial activity in America. Worldwide demand for high quality iron and steel tools was rapidly expanding in a peacetime economy that stimulated growing demands for hand tools, agricultural equipment, and industrial machinery in Europe and the Americas. There was no location

Figure 40. Holly's patent plane c. 1852, 9 ¼" long, 2 1/8" wide blade, Davistown Museum MIV Collection ID# 111106T1.

where Samuel Collins might not send his axes or the Underhill clan ship their edge tools. The unresolved question remains as to whether Sheffield crucible steel supplied the rapidly increasing demands for high quality special purpose steel. While Samuel Collins and Henry Disston continued to use imported Sheffield crucible steel for their axes and saws well into the mid-century (Tweedale 1987), numerous other New England

toolmakers began making malleable cast iron tools after 1840. Could the English factories at Sheffield, the most important center of English steel production in the early 19th century, supply the burgeoning demand for high quality steel in America? And what about the sudden rise in demand for malleable cast iron for hand planes and other tools

Figure 41. Mayo's plow plane, identical to the original patent drawing that was submitted on Sept. 14, 1875, including the decorated iron fence. Photo courtesy of Rick Floyd.

after Hazard Knowles made the first cast iron plane in 1827?

There soon followed a spectacular florescence of American malleable cast iron planemakers, the most famous of which was Leonard Bailey, who began working in Boston in 1855, making his split frame jack planes, which are now so sought after by collectors. He was preceded by such planemakers as Birdsill Holly (1852), Thomas Worraro (1854), and William Foster (1843) who were then followed by a multiplicity of other hand plane and toolmakers (Figure 40 and Figure 41). Smith (1981, 1992) provides an excellent survey of this uniquely American enterprise in his monumental two volume publication,

Patented Transitional & Metallic Planes in America. Volume 2 includes an extensive list of patents for planes, shaves, and inventions of every kind, beginning with Woodward's 1812 patent for shaving leather and continuing with every important innovation in hand plane design into the early decades of the 20th century. The plane blades of many of these tools were made and imported from Sheffield, but the remaining components were designed and made in America's burgeoning tool factories of the mid-19th century. A brief review of the roots and intricacies of crucible steel production helps set the context for the rapid evolution of America's robust 19th century toolmaking industries.

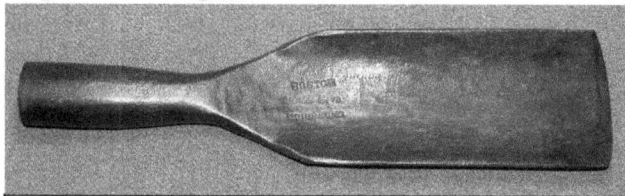

Figure 43. Gouge marked "UNDERHILL" and "BOSTON", cast steel, 13 7/8" long, 3 1/8" wide, Davistown Museum MIV Collection ID# 112303T2.

By the second quarter of the 19th century, England's steel industry was concentrated in one location, Sheffield. When Benjamin Huntsman rediscovered the ancient process of producing cast steel in 1742, he was actually manufacturing the highest quality steel by a chemical process. Cast steel had the most uniform microstructure of any form of steel, characterized by homogenous carbon distribution and an absolute minimum of contaminants. The purity and microstructural uniformity of crucible cast steel gave it a plasticity

24   *Sheffield steel and America*

Table 1.2 *Sheffield and Pittsburgh crucible steel prices 1863*

	Cents per lb		
	Sanderson Bros New York	Singer, Nimick Sheffield Works	Jones, Boyd Pittsburgh Works
Best cast steel	22	21	20–21
Extra cast steel	23		
Round Machinery	14	13–15	13–16
Swage cast steel	25		
Best double shear steel	22		
Best single shear steel	19		
Blister first quality	17½		
Blister second quality	15½	} 8–12	
Blister third quality	12½		
German steel best	15½		
German steel Eagle	12½	} 9–12	} 9–11
German steel third quality	11½		
Sheet cast steel 1st quality	22		
Sheet cast steel 2nd quality	18	} 15–21	} 15–23
Sheet cast steel 3rd quality	16		
Shovel steel best	14½		
Shovel steel common	13½		
Sheet cast steel for hoes	14½	11½	11½
Mill saw steel	15½	14	14
Billet web steel	17½		
Cross-cut saw steel	17½	18	18
Best cast steel for circulars to 46 in.	25	23	23
Toe corking best	10	9¾	9¾
Spring steel best	11		
Spring steel 2nd quality	10	} 9–10¾	} 9–10¾
Spring steel 3rd quality	8¼		

Source: SCL Marsh Bros. 249/24, 28–9.

Figure 42. Sanderson Price List (Tweedale 1987, 24).

that resulted in ease of fabrication as hot rolled bars and sheets of steel in Henry Cort's newly designed rolling mills (1784) or for domestic tool production by individual English smiths working in small tool factories using time tested forging, quenching, and tempering techniques. However, Americans beat the British at their own game, designing and implementing drop-forging machinery (1840f.) that would shape exquisite cast steel tools for the next ninety years. Before 1860, cast steel imported from England dominated the market. After 1860, the mammoth net of America's crucible steel companies provided a domestic alternative to expensive imported cast steel. German steel imports continued, but they played no significant role in edge tool production after the Civil War.

Until the secret of its chemistry was unraveled in the mid-19th century, crucible steel was made from blister steel. The tedious production of blister steel in converting furnaces was labor- and energy-intensive and, after blister steel bar stock was broken up into pieces and put into small crucibles for cast steel production, the result was a very expensive product. The wide variety of steel types produced before the Civil War to meet the growing 19th century demand for steel of all types is illustrated by the Sanderson Brothers price list of 1863 (Figure 42). Sanderson Brothers was one of Sheffield's largest steel producers and a pioneer in developing alloy steels. The Sanderson price list illustrates the diversity and variety of steel being utilized by American toolmakers and machinists just prior to the classic third period of the mature Industrial Revolution, signaled by the advent of the Bessemer pneumatic and Siemens-Martin, and then Siemens open hearth bulk process steel production. It also illustrates the infancy of the Pittsburg steel industry, which was just being constructed in the wilderness of western Pennsylvania. While the shipsmiths of coastal Massachusetts, New Hampshire, and Maine were still hand forging the iron fittings for wooden sailing ships, the world of ferrous metallurgy was changing as fast the industrial landscape was expanding.

One of the items listed on the Sanderson price list, German steel, had been produced in European fineries since late medieval times by decarburizing cast iron. Henry Cort's puddling furnace provided an even more convivial environment for producing steel from decarburized cast iron. Barraclough (1994a) provides a description of this process:

> It slowly became clear that by altering the conditions within the puddling
> furnace, particularly during the boil, it might well be possible to remove most
> of the impurities from cast iron and still retain sufficient carbon for the
> product to have some of the properties of steel, rather than those of wrought
> iron. It also became evident that considerable niceties of judgment were
> involved, and the development of a suitable and reproducible technique for the
> production of steel in this way took many years, with many valiant attempts
> ending in failure. (Barraclough 1984a, 93)

Barraclough does not note that it was easier to adapt the puddling furnace to steel production in southern Germany and Austria for two reasons: the long German and Celtic tradition of the direct-process manufacturing of natural steel, and then, the later production of steel from Spiegeleisen (manganese-laced cast iron) using siderite ore from the Styrian Erzberg in Austria (ore mountain), the manganese content of which facilitated the decarburization process.

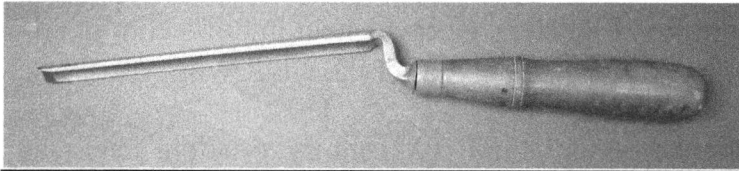

Figure 44. Buck Brothers crane-head-necked gouge, cast steel, brass ferrule, 12" long including 5 1/4" wooden handle, 1/12" wide, Davistown Museum IR Collection ID# 42904T4B.

The Sanderson Brothers price list thus illustrates the three major types of steel available to toolmakers in the United States in the years before the Civil War: crucible cast steel, shear and blister steel, and German steel. The highest quality and most expensive steel was crucible cast steel; shear steel was highly refined blister steel, piled, bundled, reheated, and reforged. So called because blister steel bar stock was "sheared" during the repiling and reforging process, shear steel was considered the best quality steel available before the evolution of the Sheffield cast steel industry, at least in England. Made from reforged blister steel, it still retained vestiges of alternating bands of high and low carbon iron and was the ideal material for Sheffield's famed cutlery manufacturers.

Figure 45. Back saw, marked "C. H. BILL & SON WALTHAM MASS", spring steel, brass, and apple wood handle, 15" long, 10" blade, Davistown Museum IR Collection ID# 040103T6.

The German steel on Sanderson's list was produced by decarburizing cast iron in either puddling furnaces or in large high shaft fineries. Before the widespread production of English cementation, and then crucible steel, German steel had dominated the world's steel markets since late medieval times. As with blister steel, it could be refined again into a variety of special purpose steels. By the time of the compilation of the Sanderson price list, English-made cast steel was supplanting both shear and German steel. A prime example of this is saw steel. Henry Disston used crucible steel and many of his best saws are labeled "London Spring Steel," which was probably the highest grade of carefully refined, forged, and re-rolled shear steel but also may have been made from rolled sheet cast steel. Nonetheless, back saws and hand saws occasionally appear stamped with the term "German steel," recalling the long tradition of German steel production, even though when they were stamped in

English it usually indicates that the saw was made from shear steel in England, not Germany.

John Russell may have imported both crucible steel and shear steel to make his famed skinning knives in Greenfield, MA, in the 1830s. He certainly had a wide variety of steel types to choose from. On the Sanderson price list, shear steel is among the most expensive. Not produced in crucibles but from reheated, forge welded bars of blister steel, the manufacturing of shear steel may have required more finesse than crucible steel production, but its unique microstructure could not be produced by any other methods. Barraclough (1984a) attributed a German origin for shear steel. Its production in Sheffield was perfected by a shipwrecked German steelmaker from Remscheid, Wilhelm Bertram. He produced five types of shear steel in 1693 from blister steel at Newcastle, an important English steelmaking center, which flourished between 1675 and 1750, prior to the rise of Sheffield. Shear steel production was later introduced into Sheffield in 1767 (Barraclough 1984a, 66). The key to its successful production was sorting blister steel bars by fracture prior to reforging. For almost a century, before English crucible steel became available in large quantities, shear steel was the top grade of English edge tool steel. Individual ironmongers in America would also have produced it, especially for the flourishing swordsmithing trade, which received a boost from the French and Indian War, and then from the American Revolution and the War of 1812.

Temper	Name	Approximate mean carbon content	Fracture appearance
1	Spring heat	0·60–0·70%	80% sap
2	Cutlery heat	0·75–0·85%	60% sap
3	Shear heat	0·90–1·00%	40% sap
4	Double shear heat	1·05–1·15%	20% sap
5	Steel-through heat	1·20–1·30%	Uniformly fine
6	Melting heat	1·40–1·60%	Uniformly coarse
7	Glazed heat	1·70–2·00%	Very coarse and faceted

Figure 47. Varieties of blister steel. Reprinted with permission from K. C. Barraclough, 1984, Vol. 2, *Steelmaking before Bessemer: Crucible Steel, the Growth of Technology.* The Metals Society, London. pg. 45.

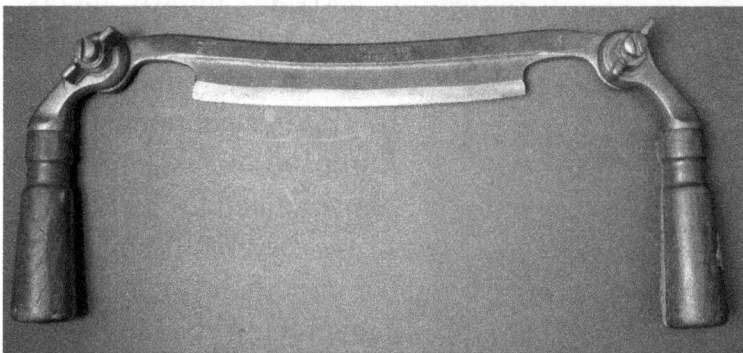

Figure 46. Folding drawknife marked "A. J. WILKINSON & CO. MAKERS - BOSTON MASS.", cast steel, iron, and wood, 10 ½" long, 6" blade, Davistown Museum MIV Collection ID# 52403T4.

The Sanderson price list (Figure 42) illustrates the continuing importance of shear steel in the era of cast steel. American toolmakers were utilizing not only crucible steel for their tools but also shear steel, blister steel, and German steel. The cheaper spring steel listed by the Sanderson Brothers was yet another variation of blister steel. Barraclough (1984b) provides a concise summary of the varieties of blister steel (Figure 47), which remained an important component of English and American steel production well into the late 19th century.

For the very best cast steel on the Sanderson list, crucible steel manufacturers would have selected only the very best blister steel. Other types of blister steel could be chosen to produce the wide variety of special purpose cast steels, such as saw steel, also listed on the Sanderson list. The list is detailed enough to separate shovel steel from sheet cast steel for hoes, illustrating the sophistication of tool production techniques before the modern era of bulk process steel. But American toolmakers were utilizing English steel to make high quality American tools in factories more efficient in their operation than those of hide-bound English toolmakers. The English retained a monopoly on cast steel production until the rise of the Pittsburgh steelmakers. Competition from Jones Boyd, the American manufacturer in Pittsburgh, is indicative of the birth of an American cast steel industry that would gradually, but not completely, supersede the Sheffield cast steel industry until the demise of both in the third decade of the 20th century.

Expensive English cast steel couldn't possibly supply the need for steel in the rapidly growing market economy of America in the first four decades of the 19th century. American ironmongers didn't produce high quality cast steel until the Civil War, due to the lack of temperature resistant clay crucibles, but in America, a land of immigrants, European iron- and steel-producing technologies were well known, and probably widespread but small in scale and also poorly documented. Hartley (1957) notes the state-of-the-art status of the Saugus Iron Works in the 1640s, illustrating how quickly any improvements in iron and steel production techniques originating in England or continental Europe would be implemented in the American colonies. Given the importance of cementation steel in the early 18th century and the availability of imported English crucible steel after 1750, important questions

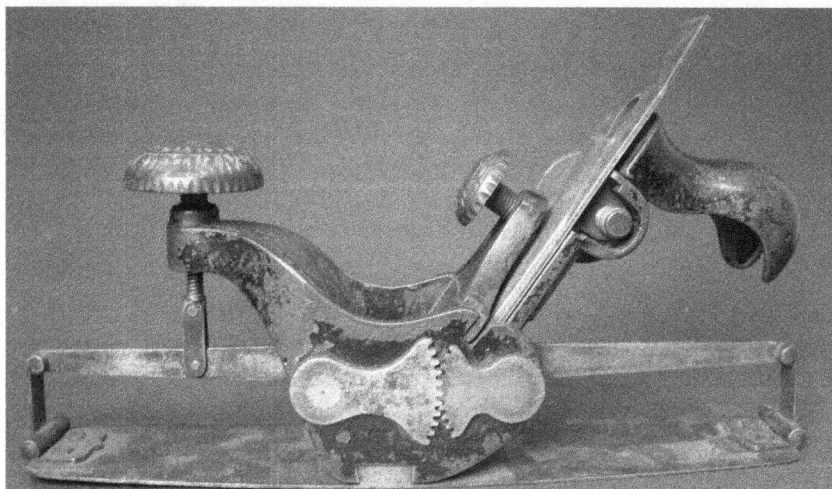

Figure 48. Stanley model 113 circular plane, marked "STANLEY RULE & LEVEL CO. | PATENTED SEPT. 25, 1877", cast iron, Japanned, nickel-plated trim, 10 ½" long, 1 ¾" wide blade, Davistown Museum IR Collection ID # 31808PC3.

nonetheless linger about the sources of steel used in the early Republic. Where did the Ames Shovel Company of Easton, MA, (est. ± 1790) obtain their shovel steel? This same question can be asked of the Underhill clan, D. R. Barton, the Buck Brothers, Timothy Witherby, Bailey and Chaney, Henry Disston, the Stanley Tool Company, and the many other American toolmaking enterprises that were established between 1827 and 1860. The Ames Company was one of the longer established American toolmakers and an investigation into the sources of its shovel steel in its early years provides a significant insight into steel production in the United States 1790 – 1840 (Figure 29). Before the spread of railroads between Massachusetts and Pennsylvania in the 1850s it is unlikely that the Ames Company imported significant quantities of steel from the Sanderson clan when it was so close to the robust bog iron industry of southern New England, which had the capacity to make puddled steel of suitable quality for Ames shovels, especially after the proliferation of small reverbatory furnaces and blister steel cementation furnaces in New England in the early 19th century.

At some future date, advances in the study of the isotopic components of archaeometallurgical samples of 18th and 19th century tools may reveal more information about the sources of their ferrous constituents and the technologies used in their smelting and forging. As the onset of the age of biocatastrophe begins, numerous chapters in the history of ferrous metallurgy that precede the classic period of American toolmaking remain lost in the ancient fog-bound labyrinths of the historic past in our era of growing ahistorical functional illiteracy.

The End of the Wooden Age

Henry Clifton Sorby wrote his treatise on the microstructure of metals in 1864 and presented it at the British Association Meeting in Bath, England, but he did not publish it until 1887 (Smith 1960, 74). In between these dates, working in the obscurity of the forests of Dean on the north side of the River Devon, Robert F. Mushet helped Henry Bessemer solve the problem of the efficient production of durable low carbon steel by the pneumatic process. A chemist and metallurgist, Mushet knew what Bessemer did not and added a manganese-laced cast iron called "spiegeleisen" to the molten cast iron to avoid over oxygenation. This was the first in a number of major innovations that allowed the bulk production of massive quantities of low carbon steel. R. F. Mushet was also responsible for another creative innovation in ferrous metallurgy, the development of the first important and widely used alloy steel, Self-Hard or R.M.S. (8% tungsten, 2% carbon, 1% manganese). William Siemens soon followed with a third innovation, his improved design of the Siemens-Martin open hearth furnace. After a bit of delay, Sidney Gilcrest Thomas, an obscure postal clerk, solved the puzzle of phosphoric iron smelting by improving furnace design to include refractory linings as well as lime additions to the Bessemer converting process (Brack 2008b, 98), now called the basic refractory furnace, which facilitated the smelting of the vast majority of the world's deposits of phosphoric iron deposits. Henry Cort's reverbatory furnace and rolling mills and their efficient production of malleable iron, puddled steel, and rolled crucible cast steel had facilitated the wide application of that now old-fashioned prime mover, Watt's steam engine. This steam engine ran both the larger blast furnaces and the iron puddling furnaces that produced malleable and wrought iron in such large quantities. Initially, the larger blast and puddling furnaces filled the demand for more iron and steel in the early years of the 19th century but could not satiate the growing need for steel from 1840 onward when ships made of iron, and then, steel, especially warships, began making their appearance on oceanic horizons.

It took a half a century for machine-made ships to put an industrial ending to the age of the wooden ship. Among the most important uses of both the Bessemer pneumatic furnace and the Siemens open hearth furnace was making the steel plates for the ships of the English and American navies. Wrought iron, then steel, rails finally transported the coffin of the wooden shipbuilding era to its gradual, obscure oblivion. Why do work with your hands when boiling water will do the work for you? So much for beating out beams with a broad ax, the dubbing of the adz, the slice of the mortising gouge and slick, the sawdust and inconvenience of the pitsaw. The narration of the history of the wooden age ends with the appearance of steam-powered prime movers, including the railroad engine that simultaneously facilitated westward expansion and the demise of the wooden sailing vessel. Where is the statue memorializing the cast iron tilting band saw needed to remind us of the fate that befell the wooden shipwright?

Technological Overlap: Hand-forged or Drop-forged?

When Henry Bessemer was inventing his unique technique for quickly decarburizing liquid cast iron into malleable low-carbon iron bar stock, (not quite wrought iron, because his process burned out the silicon,) the end of the greatest period of American shipbuilding, marked by the panic of 1856, had just occurred. This year is symbolic in that it is a point in time midway from the beginning of the American factory system of mass production in the early 19th century and the final demise of the tradition of handmade tools and the wooden ships they built in the late 19th century. No one specific year can mark what was almost a century of technological overlap.

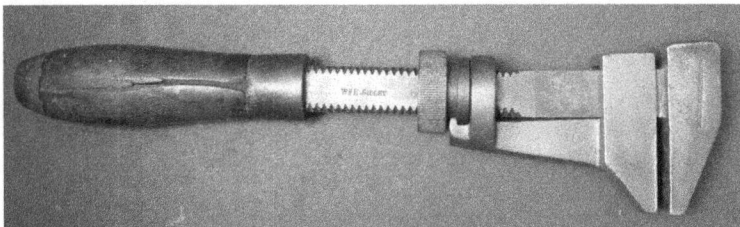

Figure 49. Wagon wrench, marked "S. MERRICK'S PATENT" plus owner signature "Wm E. SIBLEY", drop-forged steel and iron with a wood handle, 10" long, Davistown Museum MIII collection ID# 62406T6.

Only a few more clipper ships were to be built in East Boston and in Maine, yet the era of the Penobscot Bay bulk cargo carriers, the Downeasters, was two decades in the future. Bessemer had not even dreamed of "steel production." The idea that low-carbon malleable iron with a ±0.3% carbon content would soon be called "mild steel" had not yet been imagined by one of the pioneers of the bulk steel era. But whatever you called it, "malleable iron" or "mild steel," there was already a huge new market for this product. The all-consuming railroad, iron and steel ships, bridges, steam engines, and huge factories and their machinery were quickly altering landscapes with their mercury sulfide-emitting smoke stacks. If you had soft, mild steel, you could efficiently drop-forge useful hand tools of every kind in some of the many factories now dotting a growing industrial landscape. The exception to the mass production of drop-forged tools were edge tools needed for woodworking and the steel hand files so necessary to shape the machinery of the Industrial Revolution, including those drop-forged hand tools and interchangeable parts made before the era of the grinding machine (1875). Why forge weld a coach wrench when one could drop-forge a monkey wrench, made, for example, by the Loring Coes and Aury Gates wrench factory in Worcester, MA (1841), for the soon to proliferate railroad mechanics. For almost 40 years of technological overlap, hand-forged, and then, drop-forged files tediously finished the final forms of the products of the American factory system. Ironically, the last toolmakers to forge weld or cut and shape the tools they were making by hand were the file-makers of Sheffield, Birmingham, and Lancashire, England. Their files were still hand-wrought a quarter of a century after innovative American toolmakers had already figured out how to use machines to make files. The essence of the florescence of American toolmaking was the innovative use of machinery designed by the English industrial revolutionaries to mass-produce tools, machinery, and especially, firearms with interchangeable parts.

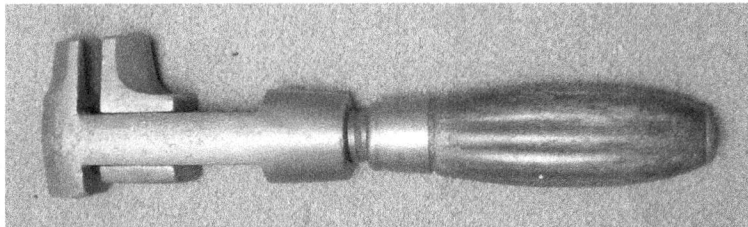

Figure 50. American wrench, marked "PAT SEPT 8.1885 AUGUSTA, ME", steel. A typical drop forged hand tool. Photo courtesy of Liberty Tool co.

The continued presence of isolated edge toolmakers in many New England communities is obscured by the rapid evolution of sophisticated factories like those founded by Thomas Witherby (1848) and the Buck Brothers (1854), where expert edge toolmakers began hand tool manufacturing on an industrial scale. The increased use of machinery obscures their reliance on the ancient skills of the ironmonger: quenching, tempering, hammering, and annealing. The shipsmiths and edge toolmakers of Maine and New England continued their stubborn production of the ferrous necessities of late 19th century shipbuilding for almost a half century after Thomas Witherby and the Buck Brothers began their historic contribution to the classic period of American toolmaking. Technological overlap was an essential ingredient in this florescence.

The Search for American Steelmakers

In England, use of any steel other than high quality crucible cast steel for late 18th century and early 19th century chisels, plane blades, and other edge tool production was inconceivable. In contrast, out of necessity, American edge toolmakers in the same period utilized blister and shear steel from a wide variety of sources to supplant imported English cast steel.

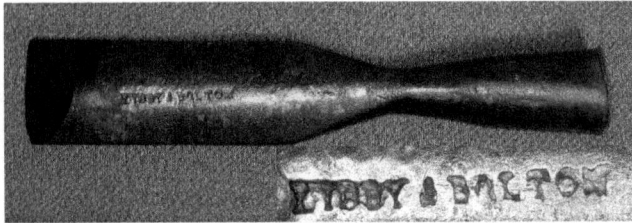

Figure 51. Libby & Bolton iron chisel made in Portland, ME, in the late 1850s, 8 ¼" long, 1 ½" wide, Davistown Museum Maritime IV Collection ID# 42604T2.

Samuel Collins clearly stated his reliance on imported English steel (Muir 2000, Gordon 1996). Henry Disston pretended he was using English cast steel for his saws long after he began producing and using American cast steel due to the continuing high reputation of English steel. The date he began producing his own cast steel is uncertain. Disston felt that nobody would believe that American cast steel was equal to the English cast steel, but, after 1860 it was. The Stanley Tool Company (est. 1857) allegedly utilized English cast steel in their plane blades well into the 1930s (Tweedale 1987). The majority of small carving tools found in New England tool chests bear the imprints of Sheffield manufacturers. (See Figure 52 showing five English tools in the tool kits of the Willard family of American clockmakers.) But the conventional paradigm doesn't work for many other edge tools and steel implements. The larger timber harvesting and timber framing tools of New England lumbermen and shipwrights are seldom stamped with English touchmarks.

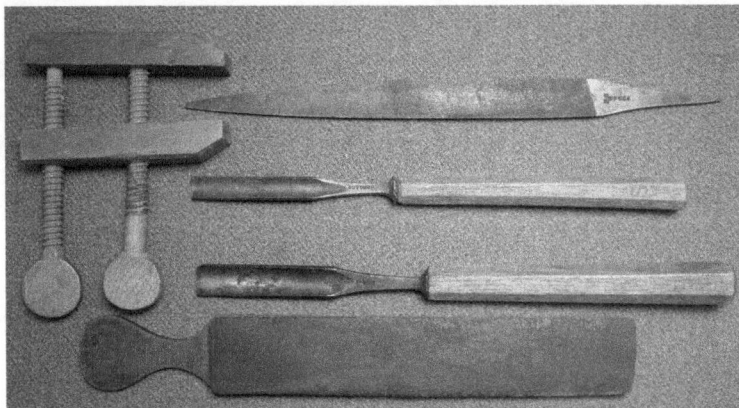

Figure 52. Group of tools found in the Simon Willard toolbox (Figure 3), including a Stubs file (Lancashire) and two signed Sheffield carving tools, Davistown Museum MIII Collection, ID#s 51201T9 - 51201T13.

Pennsylvania was the most important source of iron ore in the 19th century; Pennsylvania iron and coal deposits were the key reason why the huge steel industry at Pittsburg evolved after 1840.

Table 33 in Figure 53 from Paskoff's (1983) *Industrial Revolution* provides a snapshot of the iron production facilities so essential to the growth of New England's tool producing industries. In 1828, Nielson had invented the hot blast, which recycled heat from the furnace to create a heated air intake, which greatly increased furnace efficiency and productivity. Nonetheless, the ancient belief that cold blast, that is cold air, produced a higher quality iron than hot air, still lingered in iron producing regions of America and England. The types of production facilities listed in Paskoff, derived from an earlier 1859 publication, illustrate America's iron and steel industry on the cusp of change. The American iron industry, especially influenced by a rapid rise in demand for iron for railroad locomotives, rails, and other equipment, had proliferated throughout Pennsylvania by the mid-19th century. Ancient cold blast charcoal furnaces now operated next to the newer hot blast charcoal furnaces. The newest form of blast furnace used anthracite fuel, which was mined primarily in the eastern regions of Pennsylvania. Use of bituminous coal from western Pennsylvania and West Virginia in blast furnaces had yet to make an appearance. Charcoal forges were the primary method of refining of pig iron from the blast furnaces; numerous rolling mills were in operation, some of which manufactured wrought iron rails for the growing network of railroads. The hegemony of Bessemer steel rails was still three decades in the future. The large quantities of iron provided by the Pennsylvania iron and steel industry supplied an increasingly large network of American toolmakers, which had now spread from New England to New York, Pennsylvania, and Ohio. The classic period of American toolmaking is clearly documented by the extensive listings of the

TABLE 33

Regional Distribution of Production Facilities and Forms of Business Organization, 1849 and 1859 (in Percent)

Type of Facility	Eastern Region			Western Region			Whole State		
	I	P	Co.	I	P	Co.	I	P	Co.
Cold-blast charcoal furnace									
1849	60%	19%	21%	39%	34%	27%	48%	28%	24%
1859	33	30	37	44	25	31	37	28	35
Hot-blast charcoal furnace									
1849	44	25	31	28	33	29	40	27	33
1859	42	29	29	20	55	25	32	41	27
Anthracite blast furnace									
1849[a]	25	30	45	—	—	—	25	30	45
1859[a]	23	27	50	—	—	—	23	27	50
All furnaces									
1849	44	24	32	37	34	29	42	27	31
1859	28	28	44	31	42	28	30	31	39
Charcoal forge									
1849	62	27	11	33	33	33	61	27	12
1859[a]	51	28	21	—	—	—	51	28	21
Rolling mill									
1849	34	41	25	4	22	74	24	35	41
1859	33	38	28	17	28	55	28	35	37
All facilities									
1849	48%	28%	24%	30%	33%	37%	43%	29%	28%
1859	37%	30%	33%	25%	35%	40%	34%	31%	35%

Sources: Figures for 1849 are from table 28; figures for 1859 are from J. P. Lesley, *The Iron Manufacturer's Guide to the Furnaces, Forges and Rolling Mills of the United States* (New York: John Wiley, 1859), passim.
Note: I = individual; P = partnership; and Co. = company.
[a] The source lists none for the western region.

Figure 53. (Paskoff 1983, 107).

Early American Industry Association's *Directory of American Toolmakers* (Nelson 1999), and by the numerous high quality tools that can still be found in New England workshops, tool chests, and collections throughout the country.

Scattered among the surviving tools from these workshop tool hordes are forge welded timber framing tools of the highest quality that are neither marked "cast steel" nor have the typical stamps of English makers. The survival of so many of the larger-sized edge tools used by shipwrights and timber framers, such as slicks, mortising gouges, chisels, mast shaves, broad axes, and adzes, often stamped with the names and sometimes the locations of their American makers, is the primary evidence for the existence of a robust, indigenous edge toolmaking industry. It was this indigenous edge toolmaking industry that originated in New England and is documented in the previous volume of the *Hand Tools in History* series, *The Art of the Edge Tool*, which contains illustrations of numerous American-made edge tools, including some made in Maine.

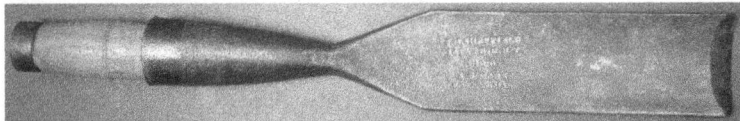

Figure 54. Gouge marked "J.M. SHEFFIELD", blister, cast, or German steel, wood, and iron, 14 7/8" long, Davistown Museum MIV Collection ID# 81101T9.

Figure 55. Corner chisel marked "J.Cray", cast steel and wood, 5" long, 1 1/16" cutting edges, Davistown Museum MIII Collection ID# 111001T2.

Figure 56. Chisel marked "F. DICKINSON", cast steel, 11 ½" long, Davistown Museum MIV Collection ID# 913108T42.

Figure 57. Chisel marked "G. H. TUCKER", iron and steel blade, brass ferrule, wood handle, 15 ¾" long, Davistown Museum MIV Collection ID# 913108T46A.

The Significance of Alloy Tool Steels

While much of the focus of the *Hand Tools in History* publication series is on the evolution of the edge tool manufacturing capabilities of American toolmakers, edge tools are only one component of the classic period of American toolmaking. The evolution of the American factory system of mass production is based on the invention, development, and practical application of machinery to make tools, guns, and metal consumer goods and the machinery that would gradually replace hand tools in the manufacture of the second and third generations of the machinery of the Industrial Revolution. Edge tools made none of this machinery, but linger on in importance well into the 20th century as the patternmakers' tools necessary for production of the wood patterns used for the sand molds of cast machinery and machinery components. Patternmakers' edge tools made by the Buck Brothers and others in large quantities played a hidden but equally important role in complementing the increasing variety and sophistication of machinist measuring tools used to make the machinery of the American factory system. Drop-forged wrenches repaired these machines, including that most important component of the spread of the American factory system, the steam-powered cast iron locomotive.

The spread of the railroads across America's landscape represented a second stage in the evolution of the American factory system, precursor to the evolution of an American consumer society, which was, in itself, the progenitor of biosphere-devouring global consumer society. The third stage in the evolution of the American factory system and the global consumer society to which it gave birth was the development of the internal combustion engine, the natural successor to the steam engine as consumer society prime mover. The evolution of the electric power grid, petrochemicals as prime mover, and a nuclear-powered global arms race follow these earlier stages in the globalization of industrial activities. All the developments in metallurgy that led to what is now a global consumer culture in crisis had roots in the evolution of sophisticated alloy steelmaking strategies. The critical ingredients in the growth of innovation and effective steelmaking strategies were the ever expanding knowledge of the use of heat, the hammer, and time to make a wide variety of variations in simple carbon-iron alloy mixtures. The periodic addition of other alloy elements to this brew is the story of a scientific revolution that unlocked the chemical and metallurgical secrets of steel- and toolmaking. There would be no classic period of American toolmaking without our Industrial Revolution, its factory system of mass production, and the evolving understanding of the use of alloys to alter and strengthen the microstructure of steel.

Machinists and mechanics tools were the dominant tool forms of the classic period of American toolmaking; they quickly replaced now irrelevant, and almost forgotten, edge tools and were the essential instruments of manual operation that gave birth to the factory system of mass production (machinists tools) and the global spread of the internal combustion engine (mechanics tools) that followed. The essential ingredient in the toolmaking activities of these successive Industrial Revolutions – a phenomenology of tools soon to end in the phenomena of biocatastrophe – is the evolution of alloy steels. An underlying theme of the entire Hand Tools in History series is the important role alloys played in the evolution of ferrous metallurgy, its successive Industrial Revolutions, and the ultimate devolution of industrial society they engendered. As noted in Volume 6, *Steel- and Toolmaking Strategies before 1870*, the microconstituents manganese and silicon played a critical role in the empirical evolution of ferrous metallurgy technologies despite the total ignorance of both the chemical and microstructural components of this industry before the mid-19th century.

Figure 58. Hub trimmer made by G. N. Stearns of Syracuse, NY, patent number 224,308, Feb. 10, 1880. Photo courtesy of Rick Floyd.

The rise of the American factory system and its growing use of machinists and mechanics tools correlate with the growing understanding of the role of alloys in hand tool and machinery production. Once toolmakers, with the help of R. F. Mushet, knew they could manipulate and improve the quality of their hand tools by the use of alloy steels, the classic period of American toolmaking was off and running. The rapid succession of the stages of industrialization and its transportation, energy production, and military systems and organizations was based on the growing awareness of the usefulness, in addition to manganese and silicon, of tungsten, cobalt, vanadium, chromium, and molybdenum. These alloys facilitated the drop-forged production of a rapidly increasing variety of hand tools, which were used to build the prototypes and components, as well as to repair, the increasingly complex manufacturing and transportation systems of global industrial society. The age of information technology and the fiber optic cables that are its prime mover are the penultimate development in the evolution of metallurgical technology.

The appearance of machines to manufacture other machines, motor vehicles for example, began rendering the now quaint but inventive designs of Victorian toolmakers obsolete. By 1930, with the help of industrialists such as Henry Ford, the classic period of American toolmaking was coming to an end. The appearance of computers, which provided the opportunity for the evolution of a global communications network, digital systems for fabricating machinery, and the technology necessary for the efficient waging of global warfare have now made hand tools of tertiary significance in the age of global consumer society. Nonetheless, a robust underground economy based on the use of hand tools in sustainable trades and horticultural activities show signs of stubborn persistence as a world financial system that suddenly ran out of money in September of 2008 crashes around us.

The Classic Period of American Machinists' Tools

The classic period of American machinists' tool production began its first half century of growth (1850f) just as the classic period of American woodworking tools began its last half century of robust productivity. There is, in fact, an intricate relationship between these two facets of toolmaking in America. Some of the highest quality edge tools made by the Buck Brothers in Millbury, MA, were for the patternmakers building the machinery for the Darling, Brown, & Sharp Company in Providence, RI as well as for other machinery and tool manufacturers. Laroy Starrett would soon make his appearance in Athol, MA; the classic period of American machinists' tools and the machines they built would quickly sink the wooden age. Exploration, conquest, and settlement of the new-found-lands would become the protohistory of an Industrial Revolution, which would impose itself on a biosphere-as-bank-account, culminating in the profit-driven consumer society of the 20th century.

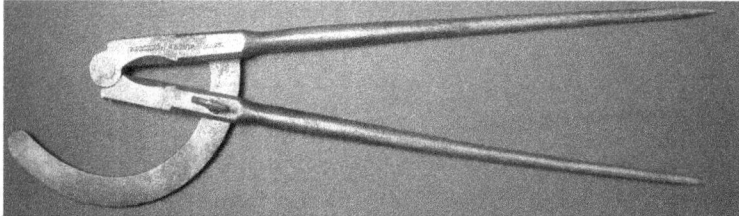

Figure 59. Dividers, marked P. PETERS. NATICK MASS., cast steel, 11 7/8" long, Davistown Museum MIII Collection ID# 81602T16.

Figure 60. T. F. Welch, Boston, MA, tap drill index, steel, 5 ¼" long, 1 5/8" wide, Davistown Museum IR Collection ID# 041505T30.

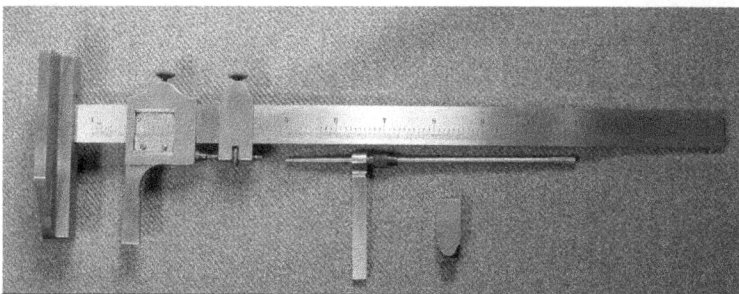

Figure 61. Height gauge, marked "THE L.S. STARRETT CO." "ATHOL,MASS.U.S.A.", and "No. 454", 14" long, 3 ¾" width, 1½" deep at base, photo courtesy of the Liberty Tool Co.

Many of the prototypes for machinists' and mechanics' tools originated in European communities, especially during the German Renaissance. Gunsmithing, watch and clock making, the manufacture of optical, navigation, and surveying instruments, and the production of early machines, such as printing presses, were made possible by numerous prototypes of 19th century machinists' tools not discussed or illustrated in these volumes. In Volume 7,

Art of the Edge Tool, brief notation is made of the beginnings of American machinists' tool production in Bangor, Maine. As with the inventive designs of the English industrial revolutionaries, earlier European tool prototypes were quickly adapted and often cleverly redesigned by American toolmakers. When Samuel Darling moved from Bangor, Maine, a most inaccessible location, to more accessible Providence, RI, in the late 1850s, he played a key role in the rapid growth of machine tool and machinery production. The huge brick megalithic factories that soon dotted the mill towns of New England quickly spread across the industrial landscape of North America.

Figure 62. J. R. Brown & Sharpe wire gauge, steel, 3 ½" diameter, Davistown Museum IR Collection ID# 032203T7.

While the Darlings, Sharpes, and Browns are among the most well known of the mid-19th century machinist tool manufacturers, numerous other smaller individual firms were now making tools throughout New England's river valleys, from Bangor, ME, on the Penobscot River, where the Crogan tape measure (Figure 63) was manufactured, to the many water privileges of southern New England. One of the most important and prolific manufacturers of machinist tools was the L. L. Davis Company (1867), soon to be called the Davis Level and Tool Co. (1875). Their finely made ornate cast iron levels and inclinometers are among the most sought after collectable tools produced during the second half of the 19th century. More well known is the

Figure 63. Crogan 25 foot tape measure, steel, 2 ¾" long, 2 ¾" wide, Davistown Museum IR Collection ID# 102503T5.

remarkable career of Laroy Starrett who was already at work designing his famous meat chopping machine, the hasher, patented May 23, 1865, which was produced by the Athol Machine Company, which he bought out in 1880, forming the L. S. S. Starrett Company.

Figure 65. L. L. Davis line level, cast iron and brass, 3 ½" long, Davistown Museum IR Collection ID# 41203T1.

Figure 64. Davis Level and Tool Co. level, marked "PAT SEP 17, 1867", wrought iron and brass, 24" long, 2 ¾" wide, Davistown Museum IR Collection ID# 120907T2.

The story of the evolution of the tool forms of the classic period of American toolmaking, unlike the earlier period of colonial and early American tool production, is easily retrievable as a result of the availability of a wide variety of written sources as well as a ubiquitous presence of tools from this era. Numerous 19th century tools have survived in New England workshops and reside in thousands of tool collections amassed in the 20th century by now aging, if not deceased, tool collectors. A vast literature on the design, production, and metallurgy of these machinists' tools is easily accessible to the determined researcher. Only a few key references and texts are cited in our bibliographies on this subject.

The Final Years

The edge tools in the collection of the Davistown Museum and those in many private collections and other museums clearly document the longevity of the indigenous toolmaking industry that culminated the classic period of American toolmaking. By the time of the growth and florescence of the Pittsburgh steel industry, which included the simultaneous production of high quality crucible steel at or near Pittsburgh furnaces, Thomas Witherby (1849), the Buck Brothers (Worcester, then Millbury, MA, 1853), the long-established Underhill clan, and a half dozen other relatively large manufacturers (Douglas, Leighton, etc.) were producing the majority of larger edge tools being used by New England shipwrights. American-made crucible steel was then being produced from domestically smelted charcoal bar iron, blister steel, or refined pig iron. By 1850, due to innovations in the smelting process, English cast steel was being produced by mixing high quality Swedish charcoal-fired pig iron with scrap steel and/or traditional Swedish charcoal-smelted malleable iron. These advances in steelmaking strategies and technologies also signaled the coming decline of the age of wooden shipbuilding. Iron, then steel, coal-powered steamships were already replacing the larger transoceanic wooden ships.

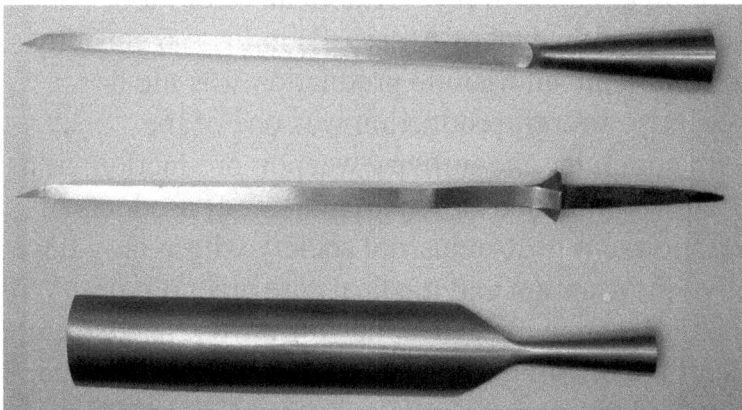

Figure 66. Three James Swan Company chisels marked "JAS. SWAN CO.U.S.A.", Davistown Museum IR Collection ID#s 32808DTM4, 32808DTM6, and 32808DTM25.

The series of photographs in Figure 51 to Figure 57 illustrate the progression of American edge tool users from their dependence on imported English tools and their use of the natural and German welded steel creations of 18th century New England shipsmiths and edge toolmakers to the widespread use of New England edge toolmakers' production of high quality tools made from either imported or domestically produced cast steel or reforged shear steel. This intermediate stage in American edge tool production was an important link between the earlier colonial edge toolmaking traditions and the later productivity of the still famed and highly sought after products of the Buck Brothers/Witherby/Underhill clan of edge toolmakers. As the classic period of American edge toolmaking drew to a close before World War II, the last famed maker of cast steel edge tools was the James Swan Company of Seymour, CT (Figure 66 and Figure 67).

Companies, such as Greenlee and Stanley, continued production of still highly regarded edge tools after World War II, not for shipbuilding but for general woodworking. Among the most sought after edge tools of the mid-20th century are the Stanley "Everlasting" chisel sets. These tools symbolize the end of the classic period of American edge

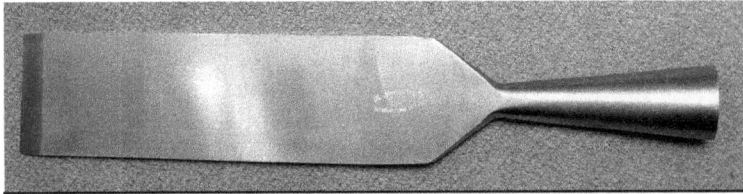

Figure 67. Chisel marked "THE JAMES SWAN Co. SEYMOUR,CONN. U.S.A." with a swan trademark, Davistown Museum IR Collection ID# 40408DTM5.

tool manufacturing, one last flourish by artisans working in the final years of three centuries of edge toolmaking by colonial and American toolmakers. A scattering of other toolmakers continued to make high quality hand tools for other trades, such as machinists (Brown & Sharpe, L.S.S. Starrett,), mechanics (Williams, Snap-On), plumbers (Rigid), sheet metal workers (Pexto), and other hand tools (Plumb, Vaughan, Simmons [Keen Kutter], Osborne), but, for our survey of edge toolmaking, the end of crucible steel production in the 1930s coincided with the end of the classic period of American toolmaking. Another half century of high quality toolmaking continued, but only a few of the modern toolmakers noted above were still in business at the end of the millennium. 1985 marked the beginning of the near total collapse of most American steel-, tool-, and machine-making industries, as well as much of the rest of America's industrial capacity. At the time of this book's publication, American automobile production was the latest industry to show decline. Until very recently, aircraft production was one of the industries that could be cited, along with computer and military weapon production, as a still viable American industry. The PEDE (brominated fire retardant-laced) products of Silicon Valley may be the last great creation of a post-industrial society whose only hope for survival is sustainable nondestructive information and medical care technologies. Will its bioengineers, medical magicians, and electronic elite save the world from pandemics; food and water stress; underemployment; the ongoing financial crisis, which is the legacy of a predatory shadow banking network; and the prospect of a collapse of global consumer society? Or will new nanotechnology and then picotechnology ecotoxins make the classic period of American toolmaking seem like a giant holiday before the age of chemical fallout became the age of biocatastrophe?

All major machinists' tool manufacturers and most other tool manufacturing companies documented the tools they made, often with illustrations and photographs in the form of tool catalogs. A brief survey of the most important of these companies and a selection of their catalogs, many reproduced by the indefatigable members of the Early American Industries Association, are contained in *Appendix C. 18th and 19th Century American Toolmaker Company Files.* It is hoped that some of the information in the *Hand Tools in History* publication series and in the individual company files in this volume will be of some interest to younger and future generations and artisans, who, by necessity, are faced with the daunting task of assuring the economic and physical survival of their families in the coming age of biocatastrophe. There is no other alternative than the creation of sustainable economies; knowledge of the history of hand tools and their use and manufacture will be a most important component of survival in the post apocalypse.

The listing of the toolmakers of what we call the classic period of tool manufacturing follows these introductory essays. The focus of these information files is two-fold. First, we have a natural inclination for documenting American toolmakers in view of the frequency of their appearance at the Liberty Tool Co. as a result of almost four decades of "tool picking" in New England workshops, cellars, and boat shops. Secondly, the special focus of the Davistown Museum collections is the edge tools utilized in America's wooden age. Many of these edge toolmakers are, with the exception of their listing in the Early American Industry Association's *Directory of American Toolmakers* (Nelson 1999), nearly forgotten. Our listings may not solve the mystery of where the steel for these essential implements of the wooden age was smelted, but at least a permanent public exhibition of their legacy can be maintained, even if in an obscure coastal New England hill country town (Liberty, Maine). This brief exploration of the industrial milieu and metallurgical history of their origins is intended to help students of American history understand the evolution of the wooden age and the brief hegemony of America's toolmakers as a pinnacle of the achievement of a western industrial society now facing the crisis of the age of biocatastrophe.

Appendix A. Time Line

The following time line and definitions help clarify the transition from ancient steel- and toolmaking strategies and techniques to those used by New England shipsmiths and shipwrights during the florescence of colonial and early American shipbuilding along the New England coast. It was the evolution of these strategies and techniques into the modern bulk steel manufacturing processes that characterizes modern industrial civilization.

Time line

Date	Event
1900 BC	First appearance of Chalybean steel smelted from self-fluxing iron sands of the Black Sea, until recently, often mistaken for meteorite-derived steel
1200 BC	Beginning of the Iron Age in the eastern Mediterranean; era of direct processed natural steel and malleable iron production
800 BC	First evidence of carburizing and quenching in the Near East (Barraclough 1984a)
750 BC	Halstadt: first Iron Age culture in Europe
700 BC	First appearance of malleableized cast iron in China
300 BC	Earliest documented crucible steel production (Wootz steel) in Muslim communities (Sherby 1995a)
200 BC	Iron and steel production by Celtic metallurgists in Noricum (Austria) begins supplying weapons to the Roman Republic
50 BC	Ancient Noricum is the main center of Roman Empire ironworks
125 AD	"Steel is made in China by co-fusion" (Barraclough 1984a)
150-250 AD	Roman armorers working in Britain possibly using steel derived from bloomery-produced cast iron to make armor (Brack 2008b)
700	Era of Merovingian swordsmiths utilizing currency bars transported from Austria to the Danube River by the Iron Road
1000	First documented Viking forge at L'Anse aux Meadows, Newfoundland
1350	First appearance of blast furnaces in central and northern Europe; probable beginning of bulk steel production by decarburizing cast iron
±1465	First appearance of blast furnaces in the Forest of Dean, England
1509	Natural steel made in Weald (Sussex, England) by fining cast iron (Barraclough 1984a)
1601	First record of the cementation process in Nuremburg, Germany (Barraclough 1984a)

1607	First shipsmith's forge in the American colonies used at Fort St. George, Maine (Brain 2007)
1613-1617	Cementation process is patented in England (Barraclough 1984a); blister steel produced by carburizing wrought iron in cementation furnaces
1629-1642	The great migration of Puritans from England brings hundreds of trained ironworkers to New England
1646	First integrated ironworks established at Quincy and Saugus, Massachusetts
1652	James Leonard establishes the first of many southeastern Massachusetts bog iron forges on Two Mile River at Taunton, MA
1686	First documented use of the cementation process in England
1709	Abraham Darby substitutes coke for coal to fuel blast furnaces
1720	First of the Carver, MA, blast furnaces established at Pope's Point
±1720	William Bertram invents manufacture of "shear steel" at Tyneside, England (Barraclough 1984a)
1722	René de Réaumur provides first European account of malleableizing cast iron
1742	Benjamin Huntsman adapts the ancient process of crucible steel production for his watch spring business in Sheffield, England
1763-1769	James Watt improves the Newcomb atmospheric engine to produce the first steam engine
1774	John Wilkinson invents a boring machine used to hollow out the cylindrical cavities of Watt's steam engine pressure vessels
1775	Bolton and Watt begin mass production of steam engines
±1783	Josiah Underhill begins making edge tools in Chester, NH
1784	Henry Cort introduces his redesigned reverbatory puddling furnace as well as groove rolling mills for producing bar iron stock from wrought and malleable iron
1798	Eli Whitney receives order for 10,000 guns and begins using interchangeable parts in his manufacturing process
1802-1807	Henry Maudslay invents and produces 45 different types of machines for mass production of ship's blocks for the British Navy
1815-1835	The factory system of using interchangeable parts for clock and gun production begins making its appearance in the United States
1818	Thomas Blanchard designs a lathe for turning irregular gunstocks
1827	Collins Axe Co. established in Collinsville, CT
1828	Adoption of the hot air blast improves blast furnace efficiency
1831	First production of malleable cast iron in US by Seth Boyden (NJ)
1832	D. A. Barton begins making axes and edge tools in Rochester, NY

1835	Mass production of steel by the puddling process in Germany
1835	First railroad is established between Boston and Worchester, MA
1837	Collins Axe Co. begins the first production of drop-forged axes
1837	Steam power rotary blowing engine introduced in England
1839	William Vickers of Sheffield, England, invents the direct conversion method of making crucible steel without using a converting furnace
1849	Thomas Witherby begins making edge tools in Millbury, MA
1850	Joseph Dixon invents the graphite crucible used in steel production
1853	The Buck Brothers organized the Buck Brothers Company in Rochester, NY, then moved to Worcester, MA (1856) and Millbury, MA (1864)
1856	Gasoline is first distilled at Watertown, MA
1856	Bessemer announces his invention of a new bulk process steel production technique at Cheltenham, England
1863	First successful work on the Siemens open hearth process of bulk steel production
1865	Bulk production of cast steel now ongoing at Pittsburg, PA, furnaces
1868	R. F. Mushet invents "Self Hard", the first commercial alloy steel
1874	Use of the tilting band saw revolutionizes shipbuilding at Essex, MA
1879	Sidney Gilchrist Thomas invents basic steelmaking
1906	First electric arc furnace is installed in Sheffield, England
1913	Harry Brearley invents stainless steel
1920-1930	Gradual end to crucible steel production in England and America
1939	Bain's (1939) first comprehensive elucidation of the microstructural complexities that result from the heat treatment of austenized steel

Appendix B. Definitions: Types of Iron and Steel

The definitions reprinted below are an excerpt from *Handbook for Ironmongers: A Glossary of Ferrous Metallurgy Terms* (Brack 2008c), Volume 11 in the *Hand Tools in History* series. The beginning of the classic period of American toolmaking is characterized by the overlap of 16th, 17th, and 18th century steel- and toolmaking strategies and techniques with those that characterized the advent of the factory system of manufacturing and the appearance of bulk steelmaking technologies after 1865.

Carbon content of ferrous metals: Sources vary widely in defining the *minimum* carbon content of steel, which ranges from 0.1 to 0.5% carbon. Please note the caveats that follow the definitions.*

Wrought iron: 0.01 – 0.08% carbon content (cc); soft, malleable, ductile, corrosion-resistant, and containing significant amounts of siliceous slag in bloomery produced wrought iron, with less slag in blast-furnace-derived, puddled wrought iron. Wrought iron is often noted as having $\leq 0.03\%$ carbon content.

Malleable iron 1): 0.08 – 0.2% carbon content (cc); malleable and ductile, but harder and more durable than wrought iron; also containing significant amounts of siliceous slag in bloomery produced malleable iron, with less slag in blast furnace derived, puddled malleable iron.

Malleable iron 2): > 0.2 – 0.5% carbon content (cc). Prior to the advent of bulk-processed low carbon steel (1870), iron containing the same amount of carbon as today's "low carbon steel" (see below) was called "malleable iron." Its siliceous slag content gave it toughness and ductility, qualities not present in modern low carbon steel, hence its name. Before 1870, a wide variety of common hand and garden tools and hardware were made from malleable iron with a significantly higher carbon content than wrought iron.

Natural steel: 0.2% carbon content or greater. Natural steel containing less than 0.5% cc is synonymous with the term malleable iron. Natural steel is produced only by direct process bloomery smelting and was the primary form of steel produced in Europe from the early Iron Age to the appearance of the blast furnace (1350). Small quantities of natural steel continued to be produced by bloomsmiths, especially in the bog iron furnaces of colonial New England and Appalachia until the late 19th century.

German steel: 0.2% carbon content or greater. Steel made from the decarburizing of cast iron in finery furnaces, as, for example, at the Saugus Ironworks after 1646. The strategy of making German steel dominated European steel production between 1400 and the advent of bulk process steel technologies, hence the term "continental method" as an alternative name for this type of steel production.

Wrought steel: 0.2 – 0.5% carbon content (cc); another name for malleable iron. Wrought steel was made from iron bar stock and was deliberately carburized during the fining process to make steel tools that are still commonplace today, such as the ubiquitous blacksmith's leg vise.

Low carbon steel: 0.2 – 0.5% carbon content (cc). Less malleable and ductile than wrought and malleable iron due to its lack of ferrosilicate, low carbon steel is harder and more durable than either and can be only slightly hardened by quenching. Some recent authors (Sherby 1995a) define low carbon steel as having 0.1% cc. Produced after 1870 as bulk process steel (e.g. by the Bessemer process), low carbon steel has all its siliceous slag content removed by oxidation. Before the advent of bulk process steel production, there was no such term as "low carbon steel." All iron that could not be hardened by quenching (< 0.5% cc) was known as "malleable" iron, more recently often referred to as "wrought" iron.

Tool steel: 0.5 – 2.0% carbon content (cc). Tool steel has the unique characteristic that it can be hardened by quenching, which then requires tempering to alleviate its brittleness. Increasing carbon content decreases the malleability of steel. If containing >1.5% carbon content, steel is not malleable, and, thus, not forgeable at any temperature. Such steel is now called ultra high carbon steel (UHCS). Palmer, in *Tool Steel Simplified*, provides this generic description of tool steel: "Any steel that is used for the working parts of tools" (Palmer 1937, 10).

Ultra high carbon steel (UHCS): 1.5 – 2.5% carbon content (cc); a modern form of hardened steel characterized by superplasticity at high temperatures and used in industrial applications, such as jet engine turbine manufacturing, where extreme strength, durability, and exact alloy content are necessary. Powdered metallurgy technology is frequently used to make UHCS.

Cast iron: 2.0 – 4.5% carbon content (cc); hard and brittle; not machinable unless annealed to produce malleable cast iron.

*Caveats to carbon content of ferrous metals

- Both modern and antiquarian sources vary widely in their definitions of wrought iron, malleable iron, and steel. Modern sources variously define steel and/or low carbon steel as iron having a carbon content greater than 0.08%, 0.1%, 0.2%, and 0.3%.

- Before the advent of bulk process steel industries (1870), which produced huge quantities of low carbon steel that could have a carbon content in the range of 0.08 – 0.5%, iron having a carbon content of < 0.5% cc was called malleable iron. Other generic terms for iron that could not be hardened by quenching (> 0.5% cc) were bar iron, wrought iron, and merchant bar.

- The 1911 edition of the *Encyclopedia Britannica* defines wrought iron as containing less than 0.3% carbon, cast iron as having 2.2% or more carbon content and steel as having an intermediate carbon content > 0.3% and < 2.2%.

- Gordon (1996) defines steel as having a carbon content > 0.2%. This cutoff point is probably the most appropriate to use in defining steel, but also poses a problem since most sources define wrought iron as having < 0.08% cc; therefore, leading to the confusion of iron with a carbon content > 0.08% but < 0.2% as being either wrought iron, low carbon steel or an orphan form of undefined iron.

In view of the long tradition of the use of the term malleable iron, this glossary resurrects the use of that term to cover this gray area of the carbon content of ferrous metals.

Appendix C. 18th and 19th Century American Toolmaker Company Files

Table of Contents

Introduction

A number of observations need to be made about the following toolmaker company files, the most important of which is the fact that our information files focus on the most important and productive 19th century and early 20th century toolmakers. Surviving examples of tools made by these companies have made, and continue to make, frequent appearances at the tool stores of the Jonesport Wood Co., Inc., including the Liberty Tool Company, and are highly sought after by the many customers still seeking these treasures from America's industrial past. Some of the most commonly found marks in New England workshops are noted as frequently found by the Liberty Tool Company (FFLTC) of Liberty, Maine, and its affiliated stores in Jonesport (1970-83), Hulls Cove Tool Barn in Hulls Cove (1983f.), and Captain Tinkham's Emporium in Searsport (1996f.)

The second note of importance is these files are not a representative selection of all American toolmakers but, rather, a manifestation of our physical location in the Norumbega backcountry of coastal Maine and our focus on the recovery of the edge tools made in New England for its vigorous shipbuilding industry. Numerous other information sources provide much more data on America's numerous toolmakers than room allows in this brief appendix. In particular, the *Directory of American Toolmakers* (Nelson 1999) is the most essential and comprehensive source of data. Many of our file descriptions are direct quotations from this indispensible text. Our observations or those of others follow this citation. A digital edition of the *Directory of American Toolmakers* was issued in 2007. The files in this appendix contain only a small percentage of the total citations in that text. As noted, our emphasis is on New England toolmakers whose tools have been used in New England shipyards and/or are frequently recycled by the Liberty Tool Co. for the many trades people and artisans who are our customers. Emil Pollak's (2001) *A Guide to the Makers of American Wooden Planes* (fourth edition revised by Thomas L. Elliott) is the definitive source of information on American wooden planemakers, only a few of which have been listed in these files. The same observation may be made about Roger Smith's (1981, 1992) two volume *Patented Transitional & Metallic Planes in America 1827 – 1927* (PTAMPIAI and II). Our focus instead is on the most important edge toolmakers working in New England, particularly those whose tools are in the Davistown Museum collection or have been frequently recycled to New England woodworkers during the last four decades. Also excluded from these information files are most of the Maine and Canadian toolmakers listed in our *Hand Tools in History* series, Volume 10, 5th edition of the *Registry of Maine Toolmakers* (Brack 2008d). Despite these omissions we hope the reader will find these files a useful source of information about toolmakers working in New England and elsewhere in the 19th and early 20th centuries.

Many of the catalogs republished in the 1970s by members of the Early American Industry Association or donated by Davistown Museum benefactors, especially those in the Elliott Sayward collection, are listed and, in some cases, illustrated in this Appendix under the company file names that published them. It is important for the reader to note that many other catalogs and all the important tool reference books used by the Davistown Museum and the Liberty Tool Company are listed in the bibliographies which follow. Most catalogs cited in these company files are present in the museum collection and available for perusal by visiting collectors, scholars, students, or readers. The museum's Center for the Study of Early Tools libraries, which contain these catalogs, are open to the public by appointment or chance. Please contact the museum (curator@davistownmuseum.org) to make an appointment if you wish to peruse our extensive collection of books on the history of metallurgy and the florescence of American toolmakers.

It should also be noted much more information about the individual companies listed in these files is available in the bibliographic citations, which follow this appendix, both as a component of the discussion of the evolution of the American factory system, and also as specific companies documented by such writers as Kenneth Roberts, Roger K. Smith, and the many other scholars and collectors who collectively have made this publication a viable endeavor. Our toolmaker company information files as well as our *Hand Tools in History* publication series are always being updated and may be accessed online at www.davistownmuseum.org/TDMtoolInfo18_19century.html.

The company files appendix also contains occasional photographs of representative tools, most of which are in the Davistown Museum collection, and are identified by their name, historic context, and museum identification number, which allows interested readers to look up more information about the particular tool in the museum's tool collection catalog (Volume 9: An Archaeology of Tools, http://www.davistownmuseum.org/publications/volume9.html). Interested readers should also note our peculiar delineation of the historic periods of the state of Maine, which we use to catalog incoming tools. The unique history of Maine was significantly influenced by the French and Indian Wars, which prevented settlement of large areas of Maine from the beginning of the King Philip's War in 1676 until the Treaty of Paris in 1763 finally established Maine's current borders. This note on the manner in which we organize our tool collections may help visiting readers from southern New England and other locations understand the abbreviations we use throughout the publication series, which are as follows.

Tools in the Davistown Museum collections are assigned to the following time periods, which are abbreviated as:

- MI: Historic Maritime I (1607-1676): The First Colonial Dominion
- MII: Historic Maritime II (1720-1800): The Second Colonial Dominion & the Early Republic
- MIII: Historic Maritime III (1800-1840): Boomtown Years & the Dawn of the Industrial Revolution
- MIV: Historic Maritime IV (1840-1865): The Early Industrial Revolution
- IR: The Industrial Revolution (1865f.)

The innovative creations of the classic period of American toolmaking are expressed by the tools themselves. Their iconography symbolizes the vigorous toolmaking, city building, forest depleting, smoke belching, carbon dioxide emitting, pyrotechnic society that designed some of the niftiest hand tools ever made. The following companies made hand tools in America; in so doing they helped build America. They made convivial tools, not the tools of warfare. These are not, in fact, the tools that created the age of chemical fallout. The age of polychlorinated biphenyls and other chemical fallout ecotoxins followed the classic period of America's toolmaking; both are now in the past.

Adams, Ezekiel
W. Boscawen and Hopkinton, New Hampshire, 1849-1867
Tool Types: Axes and Edge Tools
Remarks: This is thought to be the Adams of ADAMS & ROWELL (1846-1847). His locations were determined by censuses and directories from the period (Nelson 1999).

Aiken, Herrick
Dracut and Brighton, Massachusetts and Franklin, New Hampshire, 1823 – 1864
Tool Types: Awls, Cutlery, Handles, Knives, Leather Tools, Other, Saw Tools, and Tool Boxes
Identifying Marks: H. AIKEN'S/PATENT
Remarks: His primary products were saw sets and awls but he also had patents on a number of other items including leatherworking tools, a handle and a toolbox. His business was taken over by his son, Francis Herrick Aiken (Nelson 1999).

Allen, A. B. & Co.
New York City, New York, 1848-1853?
Tool Types: Agricultural Implements
Identifying Marks: A.B.ALLEN & CO/ &c
Remarks: While the DATM lists A.B. Allen as existing only in 1848 (Nelson 1999), the catalog in the Davistown Museum library is from 1849.
Links: http://meiszen.net/family/tree/manly/_FAM/fam_letters/1d.htm - Meissner, Loren P. Friedrich A. Meissner Letters. Meissner Family Website.
http://www.davistownmuseum.org/bioAllenAB.html

Almond Mfg. Co.
Fitchburg and Ashburnham, Massachusetts, 1915-?
Tool Types: Bevels, Levels, Machinist Tools, Rules, and Screwdrivers
Identifying Marks: SAWYER TOOL MFG. CO./ASHBURNHAM MASS (sometimes without the city/state)
Remarks: Almond Mfg. Co. was Sawyer Tool Co. prior to 1915. The "Mfg." in the title was not always used. See Sawyer Tool Co (Nelson 1999).
Links: http://www.americanartifacts.com/smma/advert/ay223.htm -- A device containing a piece manufactured by Almond Tool Co.

Almond & White
Newark, New Jersey, 1855-1856-?
Tool Types: Edge Tools
Identifying Marks: SAWYER TOOL MFG. CO./ASHBURNHAM MASS (sometimes without the city/state)
Remarks: The tools attributed to these makers could be either John White's or his son William White's (Nelson 1999).

Ames Shovel and Tool Co.
Ames, John
West Bridgewater, Massachusetts, -1777-1805

Ames, Oliver
Easton, Massachusetts, 1803-1844
Ames & Sons, Oliver
Easton, Massachusetts, 1844-1901
Tool Types: Shovels
Identifying Marks: O. Ames
Remarks: John Ames of W. Bridgewater, MA (d. 1805) made shovels from 1777 - 1805. Oliver Ames was John Ames youngest son and inherited his forge and land. Oliver was born in W. Bridgewater, April 13, 1779. He moved to Easton, MA in 1803. He ran two businesses after his father died. By 1807 he was in a hoe making partnership with Asa Waters (Ames, Waters & Co.). In 1807 he moved to Plymouth, MA and managed the shovel making plant of Plymouth Iron Works. In 1814 he moved back to Easton and expanded his operation there to South Braintree, MA. In 1844 he turned the management of the business over to his sons Oakes and Oliver and the name changed to Oliver Ames & Sons. Oliver Ames & Sons operated in Easton from 1844 - 1901. In 1901 they were reorganized as Ames Shovel and Tool Co. and merged with H. M. Myers Shovel Co., T. Rowland's Sons, Wright Shovel Co. and Elwood Steel Plant. In 1952 the name was changed to O. Ames Co. Brands used before 1900 include T.M. Porter and James Adams. The Oliver Ames Plow Co. was an affiliated company after 1860.
Links: http://www.stonehill.edu/sihc/Pictures/shovels.htm
http://www.stonehill.edu/sihc/Tofias/
http://www.nps.gov/gosp/research/ames.html

Appleton, Thomas L.
Boston and Chelsea, Massachusetts, 1878-1892
Tool Types: Wood Planes
Identifying Marks: THOsL APPLETON/BOSTON (2) THOsL APPLETON/CHELSEA (name curved)
Remarks: Thomas Appleton was one of New England's most prolific manufacturers of hand planes, especially tongue and groove planes (FFLTC).

Armstrong Mfg. Co., F.
Bridgeport, Connecticut, 1875-1922
Tool Types: Dies, Taps, Wrenches, and Others
Identifying Marks: F. Armstrong/Bridgeport, Ct.
Remarks: Patents include April 6, 1875 for taps and dies and November 19, 1885 for a pipe cutter. His name and the company name are used interchangeably in the catalogs (Nelson 1999).

Arrowmammett Works
Middletown, Connecticut, 1841-1860-
Tool Types: Wood Planes
Identifying Marks: ARROWMAMMETT/WORKS/MIDDLETOWN (top line curved);
ARROWMAMMETT WORKS/MIDDLETOWN (top line curved)
Remarks: This is the plant of the Baldwin Tool Co., which used the name as a trade name (Nelson 1999).

Atha Tool Co.
Newark, New Jersey, 1884-1913
Tool Types: Blacksmithing Tools, Hammers, Railroad Tools, Stone Working Tools, Farrier Tools, and Mining Tools
Identifying Marks: ATHA TOOL CO.; ATHA TOOL CO. (in a horseshoe figure with an A in the center), A.T.CO.
Remarks: There is some contradiction concerning the ontogeny of this particular company--contention exists concerning whether or not it was formerly Newark Steel Works. Benjamin Atha, who was part of the Newark Steel Works, incorporated the company in 1891. It may have been established as early as 1875. They bought out Emmett Hammer Co. in 1884, Hartford Hammer Co. in 1892, Eyeless Tool Co. in 1897, Clark Edge Tool Works in 1897, Buffalo Hammer Co. in 1898, and Yerkes Tool Co. in 1898. Finally, in 1913, Stanley Rule & Level Co. in 1913 bought them out (Nelson 1999). Stanley Rule & Level Co. retained the classic "horseshoe" logo after incorporating Atha Tool Co. on some of their products.

Athol Machine Co.
Athol Depot, Massachusetts, 1868-1920
Tool Types: Calipers, Dividers, Levels, Vises, Wrenches, and Other Household Tools
Identifying Marks: Variations of the name, city and state (variations in capitalization and abbreviation), sometimes with a patent date; A.M.CO.
Remarks: L.S. Starrett founded this company with a number of other partners in 1868, left in 1878, and bought it up in 1905. It produced a number of his patented items including food choppers, presses, washing machines and other items from 1865 to 1873. They also acquired (or possibly founded) the Standard Tool Co. (Nelson 1999). See Laroy S. Starrett Co.

Atkins & Co., Elias C.
Indianapolis, Indiana, and Hamilton, Ontario, 1857-1952
Tool Types: Axes, Files, Hammers, Knives, Saws and Saw Accessories, and Shaves
Identifying Marks: Variations of E.C. ATKINS & CO./INDIANAPOLIS, IND., sometimes with patent marks or different cities.
Remarks: Atkins learned to make saws from an uncle in Connecticut at 23. His business grew and branched out with sales divisions in a number of cities, including Chicago. Brands used included SILVER STEEL, PEERLESS, VARIETY, STERLING, EUREKA, PERFECTION, AAA (Atkins Always Ahead), REX, HEMLOCK KING, VICTOR, COMMON, TYEE, HOWATSON, REDWOOD KING, CEDAR KING, DEXTER, DIAMOND, TUTTLE, AMERICAN, THE KING, FLIPPEROR, CHIEF, T.R. ROBERTS, DAMASKENED, FOUR HUNDRED, SPEED KING, JUNIOR MECHANIC, STANDARD, SUPERIOR, RELIABLE, A1, CHIEF BUCKEYE, THE WINNER and LONE STAR. They also distributed tools from (and possibly bought out) Moore Bros., Ajax Mfg. Co., Cross & Speirs, and E.B. Rich & Son and owned numerous patents. The Indianapolis plant was called Sheffield Saw Works and the Hamilton plant was the Hamilton Saw Works. They were eventually bought out by Borg-Warner Corp in 1952, then by Nicholson File Co. in 1966, and by Cooper Industries in 1972 (Nelson 1999).

Auburn Tool Co.
Auburn, New York, 1864-1893
Tool Types: Clamps, Handles, Plane Irons, Rules, and Wood Planes
Identifying Marks: Variations of AUBURN TOOL CO./AUBURN N.Y., including one with two straight lines or the top line curved with a thistle or star under it, THISTLE
Remarks: This company succeeded Casey, Clark & Co. with Casey as its president, then became the Ohio Tool Co. in 1893. Other names under which they produced included New York Tool Co., Owasco Tool Co., Genesee Tool Co., Ensenore and Star. Some of their products were produced using Auburn state prison labor. Of particular interest is a "Phelps Combination Plane," a plane with a level vial and rule built in, possibly patented to a Frank Phelps of Auburn on February 9, 1892. (Nelson 1999) Auburn Tool Co. was part of a collective effort between H. Chapin's Son, Greenfield Tool Co. and Sandusky Tool Co. called the Plane Maker's Association, organized circa 1858 to fix prices. Thus, the prices in the pictured catalog pages (from 1869) were agreed upon between the leading plane manufacturers at the time. In 1866, Auburn Tool Co. lost their prison labor contract, outbid by J.M. Easterly & Co. which later became A. Howland & Co. Upon the dissolution of A. Howland Co. in 1874, Auburn Tool Co. resumed the use of prison labor. On November 14, 1893, they merged with Ohio Tool Company of Columbus, OH. While plane production continued in Auburn, it was under the Ohio Tool Company's name.
References:
The Clamp Guy. 2006. History of Auburn Tool Company in NY. *Wooden Clamp Journal* May 26, 2006. http://www.theclampguy.info/hist_au.htm (accessed August 19, 2007).
Auburn Tool Company. 1869. *Price List of Planes, Plane Irons, Rules, Gauges, Hand Screws, &C., Manufactured and sold by Auburn Tool Company, (Successors to Casey, Clark & Co.)*. Fitzwilliam, NH: Ken Roberts Publishing Co.
Links: http://www.davistownmuseum.org/bioAuburn.html

The Bailey Tool Co.
Selden A. Bailey (also the Bailey Wringing Machine Co.)
Woonsocket, Rhode Island, and New York City, 1872-1880
Tool Types: Metal Planes, Shaves, Washing Machines, and Other Household Tools
Remarks: Selden A. Bailey, born September 12, 1821, made a variety of hand tools under a number of patents from 1871 to 1875. Two of these patents were co-owned by a Joseph R. Bailey of Woonsocket but Joseph's significance is unknown. The Woonsocket office also sold washing and wringing machines under the EXCELSIOR brand. Stanley Rule & Level Co. bought them out in 1880 and Selden died in 1903 (Nelson 1999). It's important to note that Selden Bailey was not related to Leonard Bailey (see below) despite the fact that both had a relationship with the Stanley Rule & Level Co. Roger K. Smith provides detailed information about the history of all the makers of patented transitional and metallic planes in his definitive two volume survey of American steel planemakers cited below. The Smith texts are replete with detailed black and white and color photographs of the most important planes made during the classic period of American toolmaking.
References:
Smith, Roger K. 1981. *Patented transitional & metallic planes in America 1827 - 1927*. Lancaster, MA: North Village Publishing Co., pg. 61-9.
Smith, Roger K. 1992. *Patented transitional & metallic planes in America – Vol II*. Lancaster, MA: North Village Publishing Co., pg. 31-3.

Bailey, Leonard
Hartford, Connecticut, 1855-1884
Tool Types: Bevels, Household Tools, Metal Planes, and Shaves
Identifying Marks: Tools marked *BAILEY* were made by Stanley (or one of his other companies) and the name indicates Bailey's enduring fame as the designer and patentee of many of the basic frog and cap irons of the modern hand plane.
Remarks: Bailey, a prolific toolmaker, was born May 8th, 1825 and died February 5th, 1905. He was granted numerous patents, especially for metal plane components, and worked for Stanley Rule & Level Co. from 1869 to 1875, a time during which they bought out two other companies, resulting in his mark being on numerous tools (Nelson 1999). Until May of 1869, Bailey was the owner and proprietor of his own company and factory, Bailey, Chaney & Co., which he sold to Stanley Rule & Co. See: Bailey, Chaney & Co. and Stanley Co. Smith provides a detailed survey of Leonard Bailey's tool designs as well as a detailed chronological listing of "events as related to Leonard Bailey's manufacturing activities after he left Boston." (Smith 1981, 52-4) and the Bailey Chaney Co. ceased operation.

References:
Bailey, Leonard & Co. [1876] n.d. *Illustrated catalogue and price list of patent adjustable iron bench planes, try squares, bevels, rules, levels, hammers, &c., &c.* Fitzwilliam, NH: Roberts Publishing Co.
Bailey, Leonard & Co. [1883] 1975. Catalog: *Leonard Bailey & Co.'s patent adjustable iron bench planes, try squares, bevels, spoke shaves, box scrapers, &c.* Fitzwilliam, NH: Ken Roberts Publishing Co.
Smith, Roger K. 1981. *Patented transitional & metallic planes in America 1827 - 1927.* Lancaster, MA: North Village Publishing Co., pg. 23-54.
Smith, Roger K. 1992. *Patented transitional & metallic planes in America – Vol II.* Lancaster, MA: North Village Publishing Co., pg. 207.
Links: http://www.supertool.com/StanleyBG/stan0a.html
http://web.mit.edu/invent/iow/bailey.html
http://www.pasttools.org/articles/Bailey_patent_model.htm
http://www.davistownmuseum.org/bioBaily.html

Bailey, Chaney & Co.
Boston, Massachusetts, 1867-1869
Tool Types: Metal Planes and Shaves
Remarks: Leonard Bailey produced planes, scrapers, and spoke shaves for a short time prior to being bought out by Stanley Rule & Level Co. in 1869 (Nelson 1999). Bailey's split frame patented iron jack and fore planes are among the most sought after of all specimens of planes made during the classic period of American tool manufacturing. Excellent color and black and white photographs of Leonard Bailey's first planes produced by Bailey, Chaney & Co. can be found in Smith (1981) on pages 40, 42, 44, 45, and 46.
References:
Smith, Roger K. 1981. *Patented transitional & metallic planes in America 1827 - 1927.* Lancaster, MA: North Village Publishing Co., pg. 41-51.

Baldwin Tool Co.
Middleton, CT, 1841-1857
Tool Types: Carpenters' Tools, Knives, Plane Irons, Wood Planes, Joiners, Tool Boxes, and Other
Identifying Marks: BALDWIN/TOOL CO. (curved top line), BALDWIN TOOL CO./MADE FROM/BUTCHERS/CAST STEEL/WARRANTED (top and bottom lines curve outward), ARROWMAMMETT WORKS/MIDDLETON
Remarks: Austin Baldwin founded this company as the Arrowmammett Works, which was a prolific manufacturer of planes, including a circa 1845 crown molding plane recently recovered by the Liberty Tool Co. They were a prolific manufacturer of wooden molding planes (FFLTC). They were bought out in 1857 by The Globe Mfg. Co. and began to produce only plane irons (Nelson 1999).

Baldwin, Samuel (& Co.)
Bennington, New Hampshire, 1826-1870
Tool Types: Axes, Cutlery, Knives, Leather Tools, Screwdrivers, and Shaves
Identifying Marks: S BALDWIN
Remarks: Samuel Baldwin was a specialized blacksmith who, in addition to working under his own name, made tools for Baldwin & Whittmore from 1853 to 1855 and was the principal of this company from 1856 to 1857 (Nelson 1999).

Bangs, Rufus W.
North Bennington, Vermont, 1831-1852
Tool Types: Squares
Identifying Marks: R.W. BAN___
Remarks: Rufus W. Bangs and Stebbins D. Walbridge patented a method of rolling steel which became a widely utilized method for producing squares. His shop was flooded in 1852 and his assets were taken over by Hawks, Loomis & Co. (Nelson 1999).

J. & E. R. Barbour
Portland, Maine, 1801-1891?
Tool Types: Wrenches, Mechanical Rubber Goods, Engineer's Specialties, Steamboat, Railroad and Mill Supplies, Steam Machinery and Appliances, Boots and Shoes, and Others
Identifying Marks: Patent dates
Remarks: The DATM lists their data as confusing and notes the mention of a wrench for sale in an E.H. Barbour catalog (Nelson 1999). While the DATM lists the company as existing in 1891, a catalog in the Davistown Museum collection gives a date of establishment as 1801, noting they have "90 years in business. J. & E. R. Barbour, Engineering and Rubber Goods, Lubricating Oils, &c., Nos. 8 and 10 Exchange Street."
References:
Barbour, J. & E. R. n.d. *J. & E. R. Barbour, dealers in mechanical rubber goods, engineers specialties, steamboat, railroad, and mill supplies, contractors for steam machinery & appliances. Nos. 8 and 10 Exchange Street, Portland, Maine.* Mechanic Falls, ME: Poole Bros. Printers and Electrotypers.
Links: http://www.davistownmuseum.org/bioBarbour.html

Bartlett & Young
Portsmouth, New Hampshire, circa 1812
Tool Types: Adzes, Axes, Chisels, Draw Knives, Hatchets, Plane Irons, and Stone-working Tools
Remarks: This was probably the collaborative name of blacksmiths James Bartlett and Asa Young (Nelson 1999).

Barton Tool Co., David R.
Rochester, New York, 1832-1874
Tool Types: Adzes, Axes, Carpenter Tools, Chisels, Cooper Tools, Draw Knives, Hammers, Hatchets, Picks, Tinsmith Tools, and Wood Planes
Identifying Marks: "D.R. BARTON", sometimes with "& CO." and "ROCHESTER", sometimes with "N.Y.", sometimes with "1832" and a star, sometimes with line, circle, or half-circle marks
Remarks: David Barton was a partner in many different toolmaking companies, all in Rochester, New York. He was born on July 4, 1805 and died April 26, 1875. The Barton Tool Company group was one of the most prolific manufacturers of hand tools outside of New England during the middle decades of the 19th century. Great quantities of Barton tools were used by settlers migrating to the western states; Barton tools make a frequent appearance in New England tool chests and collections. Barton tools are often easily identified by the ubiquitous mark "1832", which was stamped on their tools throughout the 19th century. Below is a chronological list of D. R. Barton's companies:

Barton & Guild: dates not known
Tool Types: Edge Tools
Remarks: This is probably one of the short lived partnerships that David R. Barton was in before forming D. R. Barton & Co. D. R. Barton supposedly took over Henry W. Stager and Charles Guild's Stager & Guild in 1832. It's possible that Charles Guild is the Guild of this partnership.

Barton & Babcock: 1832 - 1834
Tool Types: Axes and Edge Tools
Identifying Marks: D.R. BARTON // J.H. BABCOCK.
Remarks: D. R. Barton & Co. marks 1832 on some tools as if it was their starting date, which is based on Barton's involvement in this partnership. John H. Babcock later worked as a blacksmith for the D. R. Barton & Co.

Barton & Smith: 1842
Tool Types: Bits, Edge Tools, and Wooden Planes
Identifying Marks: BARTON & / SMITH / ROCHESTER (with both S's backwards)
BARTON & SMITH / ROCHESTER.
Remarks: This was probably a partnership with Albert H. Smith.

Barton & Belden: -1844-1848-
Tool Types: Axes, Cooper's Tools, Drawknives, and Edge Tools
Identifying Marks: D.R. BARTON / I. BELDEN / ROCHESTER
Remarks: This company consisted of David R. Barton and Ira Belden. Belden was also a Rochester hardware dealer operating as Ira Belden & Co. It is not clear if that business succeeded this partnership or was concurrent with it. An English toolmaker, Ash, was so impressed with Barton &

Belden tools that he copied them and marked the copies "Rochester Pattern."

David R. Barton & Co.: 1849 - 1874.
Tool Types: Adzes, Axes, Carpenter Tools, Chisels, Cooper's Tools, Drawknives, Hammers, Hatchets, Picks, Tinsmith Tools, and Wooden Planes
Identifying Marks: D. R. BARTON (with and without the CO.) and ROCHESTER (with and without the N.Y.). Marks include "1832" and on some a star figure. There was also a variety of shapes used with the lettering: straight line, oval, and half-circle.
Remarks: Two partners, William R. Mack and Royal L. Mack, took the company over in 1874 and renamed it Mack & Co. They continued to use the original name as a trademark until 1923. See the entry on Mack.

Barton & Milliner: 1863
Tool Types: Edge Tools
Remarks: This partnership was formed while Barton was with D. R. Barton & Co. The relationship between the two companies is not clear. Joel P. Milliner (or Millener) had earlier had his own edge tool business in Canada (Joel P. Millener & Co.).

David R. Barton Tool Co.: 1874 - 1880.
Tool Types: Augers, Axes, Bits, Cooper's Tools, Edge Tools, and Wooden Planes
Identifying Marks: D.R. BARTON / 1832 / ROCHESTER N.Y. (in an oval shape with top and bottom lines curved); D. R. BARTON TOOL CO.
Remarks: David R. Barton and his sons, Charles and Edward, formed this company after D. R. Barton & Co. was taken over by the Macks. It was also bought out in 1880 by Mack & Co. who again continued to use the marks. So the mark D.R. BARTON / 1832 / ROCHESTER N.Y. was used by D. R. Barton & Co., D. R. Barton Tool Co. and Mack & Co. (Nelson 1999).
References:
Barton, D. R. & Co. 1873. *Catalogue and revised standard list of mechanics' tools and machine knives, manufactured by D. R. Barton & Co.*, Rochester, NY: D.R. Barton & Co.
Kosmerl, Frank. 1995. Rochester, New York: A 19th-century edge tool center: Part 2. *The Chronicle.* 48:7-12, 24, 27.
Links: http://www.davistownmuseum.org/bioBarton.html -- A short history submitted by James Stewart

Barton & Whipple
Vermont, 1830
Tool Types: Squares
Identifying Marks: B & W
Remarks: Gardner Barton and Stephen Whipple made Hawes patent squares in the Bennington/Shaftsbury area of Vermont (Nelson 1999).

Basset, John
Norton and Taunton, Massachusetts, circa 1760
Tool Types: Wood Planes
Identifying Marks: IOHN-BASSET/OF-NORTON, IOHN-BASSET

Remarks: Tools with these marks were possibly made by John Basset, a joiner from Taunton and Norton, father of Elijah Bassett (with two Ts), another possible planemaker (Nelson 1999).

Bates Mfg. Co.
Orange, Massachusetts, circa 1870
Tool Types: Augers and Bits
Remarks: Unrelated to another Bates Mfg. Co. of Orange, New Jersey. One of the most prolific manufacturers of wood bits (Nelson 1999).

Baxter Wrench Co.
Newark, New Jersey, and New York City, 1868-1883-?
Tool Types: Wrenches
Remarks: This company was one of a number making wrenches patented by William Baxter. Patent dates specific to this company include those issued on February 12, 1856, December 1, 1858, February 9, 1859, February 12, 1870, and July 17, 1883. The New York City address was solely a sales office. Baxter himself lived from November 22, 1822 to October 17, 1884 (Nelson 1999).
Links: http://www.wrenchingnews.com/MVWC/wrench-of-the-month.html -- Contains some drawings of a Baxter patent wrench.
http://www.alloy-artifacts.com/other-makers-p3.html -- Photo of a Baxter wrench

Beatty & Son Co., William M.
Chester, Pennsylvania, 1839-1882
Tool Types: Axes, Chisels, Edge Tools, and Household Tools
Identifying Marks: Variations of "BEATTY & SON" with nothing, "W.," or "WM." before it and nothing, "CHESTER," "CHESTER, PA" or "MEDIA" after/below it. Figures of a cow, eagle, or the date "1806" occasionally appear.
Remarks: The "sons" referred to in "W.M. Beatty & Sons" is contradictory to the singular "& SON" of their maker's mark. This could refer to John C. Beatty who later joined William P. Beatty. While "& Sons" was added to the name in 1839, John doesn't appear to have joined until 1840. While directories show them in Philadelphia and Oakdale, PA in 1870, Media, PA has been found on tools from that period. Due to John C.'s frequent moves, it is difficult to pin down exactly when the company's transition from Springfield, PA to Chester, PA took place. John C. left in 1850, placing William P. Beatty and Samuel Ogden in charge until 1860 when John returned and bought out Ogden. John left again in 1867 to be succeeded by Thomas W. Woodward who was succeeded by another William H. Beatty (possibly a son or brother to John C.). In 1875, John C. was the proprietor of Chester Edge Tool Works and referred to as "the surviving partner of Beatty & Sons," regardless of whether this signifies that the two were the same (Nelson 1999). One of America's most prolific edge toolmakers, including broad axes used by shipwrights.
Links: http://www.yesteryearstools.com/Yesteryears Tools/Beatty Axe Markings.html
http://www.davistownmuseum.org/bioBeatyson.html

Belcher & Bros.
Camptown, New Jersey, and New York City, circa 1843-1852
Tool Types: Rules, Bevels, and Squares
Identifying Marks: BELCHER/BROS.&CO/NEW YORK, T & W BELCHER MAKERS (star at

beginning and end), BELCHER BROTHER MAKERS NEW YORK

Remarks: This name and variations thereof (including "Belcher Bros. & Co." and "T.&W. Belcher") were used by Thomas, William, and/or Charles Belcher, all brothers. Around 1843, this company moved its factory to Camptown but continued to sell their tools out of New York (Nelson 1999).

Belden Machine Co.
New Haven, Connecticut, circa 1885
Tool Types: Hammers, Other Stone-working Tools, and Wrenches
Identifying Marks: Belden Machine Co., "DEN"
Remarks: This company manufactured a nail puller, a slaters' hammer, a slaters' stake, and an adjustable wrench with an 1885 patent mark (Nelson 1999). Amongst slate roofers, Belden is famous for their slate hammers. The Davistown Museum has a Belden slate hammer in their collection.
Links: http://www.beldenmachine.com/ -- Belden Machine Company's website.
http://www.davistownmuseum.org/bioBelden.html

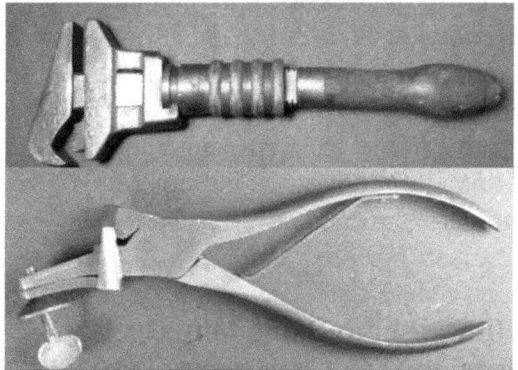

Bemis & Call Co.
Springfield, Massachusetts, 1844-1930
Tool Types: Calipers, Dividers, Knives, Saw Tools, Scales, and Wrenches
Identifying Marks: Variations of the company name with the city, state and patent names and dates; B&W surrounded by a circle
Remarks: Stephen C. Bemis of S.C. Bemis & Co. and Amos Call formed this company and changed its name to

A typical Bemis & Call adjustable wrench and saw set from the Davistown Museum MIV Collection ID# 32708T47 (top) and 30202T2 (bottom).

Bemis & Call Hardware & Tool Co. in 1855. They made wrenches with patents belonging to A.D. Briggs on April 8, 1853, William C. Bemis (Stephen's son) on December 2, 1873, and calipers patented May 28, 1861 by James H. Call (Nelson 1999). One of New England's most prolific toolmakers.
Links: http://www.alloy-artifacts.com/other-makers.html#b+c

Bennett, N.
Middleboro, Massachusetts, circa 1775-1777
Tool Types: Plane Irons or Scientific Instruments
Identifying Marks: N. BENNETT/MIDDLEBORO, MA.; N. BENNETT
Remarks: There are contradictory reports of an N. Bennett producing tools in Middleboro: The first was a smith making plane irons for E. Clark and gives Nebediah as the first name; the second reports a Nehemiah producing survey instruments (Nelson 1999).

Billings & Spencer Co.
Hartford, Connecticut, 1873-1950
Tool Types: Chisels, Hammers, Machinist Tools, Pliers, Screwdrivers, Vises, and Wrenches
Identifying Marks: Various arrangements of the company name, city, state (sometimes as CT or

CONN), sometimes with U.S.A. or a B in a triangle

Remarks: Founded as Roper Sporting Arms Co. in 1869 by Charles E. Billings and Christopher M. Spencer, this company's name was changed in 1873 despite retaining its 1869 "est." date. Spencer left shortly thereafter. They produced numerous tools patented under C.E. Billings with other holders including Hayden and Lowe. This company purchased the Coes Wrench Co. from Bemis & Call in 1939 (Nelson 1999).

Links: http://home.comcast.net/~alloy-artifacts/billings-spencer-company.html
http://www.davistownmuseum.org/bioBillingspencer.html

Billings, Charles Ethan
Hartford, Connecticut, 1865-1915
Tool Types: Machinist Tools
Identifying Marks: C.E. Billings
Remarks: Charles Ethan Billings lived from December 5, 1835 to June 5, 1920. His career path started with apprenticing at Robbins & Lawrence Co. in Windsor Vermont, moving on to Colt Patent Fire Arms Co. of Hartford, Connecticut, in 1856, then to E. Remington & Sons of Utica, New York from 1862 to 1865, moving up to superintendent of the Weed Sewing Machine Co. in Hartford from 1865 until 1869 when he formed his own company, Billings & Spencer Co. Despite holding numerous patents, Billings has never worked alone, always producing under another company (Nelson 1999).

Birmingham Plane Mfg. Co.
Birmingham, Connecticut, 1855-1891
Tool Types: Metal and Wood Planes, Plane Irons, and Shaves
Remarks: This company is also known as Birmingham Plane Co. and Birmingham Conn. Plane Co. prior to the name changing to Derby Plane Mfg. Co. in 1891. They made a variety of planes, including those patented to Solon R. Rust and G.D. Mosher (Nelson 1999). Several photographs of Birmingham planes in the Davistown Museum collection are in the *Hand Tools in History* series volumes 7 and 8.

Links: http://www.supertool.com/etcetera/deadends/hayworth.htm

Bishop & Co., George H.
Lawrenceburg, Indiana, and Cincinnati, Ohio, 1886-1919
Tool Types: Saws
Identifying Marks: SPEED & EASE, OH, GREYHOUND, HIGH SPEED, BULLDOG, variations of "George" including GEO., cities and states, patent dates
Remarks: This company made a variety of saws with patents as early as 1886 (Nelson 1999).

Bliss & Co., John
New York City, 1854-1870
Tool Types: Scientific Instruments
Remarks: Products made by this company include watches, transits, and a combination parallel rule and protractor (Nelson 1999).

Bliss & Co., Rufus
Pawtucket, Rhode Island, 1845-1935
Tool Types: Awls, Clamps, Hammers, Handles, Tool Boxes, and Others
Identifying Marks: R. BLISS MF'G. CO./PAWTUCKET.RI
Remarks: Bliss worked independently prior to 1845 when he formed this company with A.N. Bullock, though he left around 1850 when E.R. Clark and A.C. Bullock joined. His name remained with the company for some time before being changed to R. Bliss Mfg. Co. (Nelson 1999).

Blodgett Edge Tool Mfg. Co.
Manchester, New Hampshire, 1853-1862
Tool Types: Adzes, Axes, Edge Tools, Hatchets, and Shaves
Identifying Marks: BLODGETT/EDGE TOOL MFG. CO./CAST STEEL/WARRANTED/H.C. REYNOLDS AGENT
Remarks: This company name sometimes appears without "MFG." Its name was changed to Amoskeag Ax Co. in 1862 but the original name persisted. George Reynolds, patent holder on an axe pole making machine, was superintendent from 1856-1860; Henry C. Reynolds took over as agent in 1863 (Nelson 1999).
References: Browne, George Waldo. (1915). *The Amoskeag Manufacturing Company of Manchester, New Hampshire.* Amoskeag Manufacturing Co., Manchester, NH.
Garvin, James L. and Garvin, Donna-Belle. (1985). *Instruments of change: New Hampshire hand tools and their makers 1800 - 1900.* New Hampshire Historical Society, Canaan, NH.
Klenman, Allen. (December 1998). Amoskeag Ax Company of Manchester, New Hampshire. *The Chronicle.* 51(4). pg. 129-30.
Mayer, John. (1994). The mills and machinery of the Amoskeag Manufacturing Company of Manchester, New Hampshire. *Journal of the Society for Industrial Archeology.* 20(1/2). pg. 69-79.

Boker, H. & Co.
Imports, Valley Forge, 1837-1969
Tool Types: Bits, Braces, Dividers, Knives, Pliers, Saws, and Wood Planes
Identifying Marks: Variations of a tree logo; H. BOKER & CO./MADE IN U.S.A. (text forms an oval)
Remarks: Boker imported most of its tools from Germany through U.S. affiliates. Their name appears with variations of a first initial, "H.," but it is unclear what this stood for (Nelson 1999).
Links: http://www.boker.de/us/ -- Boker's current website
http://www.northamericanknives.com/page/39062202 -- History of Boker Knives

Bonney Vise & Tool Works
Philadelphia and Allentown, Pennsylvania, -1886-1910-
Tool Types: Augers, Machinist Tools, Vises, and Wrenches
Identifying Marks: Variations on "BONNEY" including city, state, brand names, and patent dates; a "B" in a shield
Remarks: All dated marks on tools made by this company refer to Philadelphia. Brand names used include VIXEN, HERCULES, MASTERPIECE, and STILLSON on Stillson type wrenches (Nelson 1999). Bonney has continued to make high quality hand tools well into the 20th century (FFLTC).
Links: http://www.alloy-artifacts.com/bonney-forge-tool.html

Boston Metallic Plane Co.
Boston, Massachusetts, 1872-1873
Tool Types: Metal Planes
Remarks: Patents on planes made by this company included September 24, 1872 and September 23, 1873 by Cyrus H. Hardy; September 23, 1873 by Francis Smith; and November 25, 1873 by Joseph F. Baldwin, an officer in the company whose patent was also used by Meriden Malleable Iron Co. Baldwin and Hardy's patents were assigned to a John Sully, apparently a company official (Nelson 1999). For additional information see Smith's PTAMPIAI and II (1981, 1992).

Bowles, Thomas Salter
Portsmouth, New Hampshire, -1806-1825-
Tool Types: Calipers, Rules, Surveying Instruments, Sextants, Telescopes, Wantage Rods, and Board Rules
Identifying Marks: T.S.BOWLES/PORTSMOUTH, N.H.; Made by.Thomas S. Bowles.Porstmouth,NH
Remarks: Bowles moved to Portland, Maine, in 1825 and then to Bath. It is unknown whether he produced tools after the move (Nelson 1999).

Bradley, Gershom W.
Weston and Westport, Connecticut, 1834-1911
Tool Types: Adzes, Axes, Chisels, Draw Knives, Hammers, Hatchets, Hoes, knives, Picks, and Shaves
Identifying Marks: G.W. BRADLEY and variations, including patent dates, a B with an arrow through it, or "CAST STEEL" or "AXE CO." added; possibly "MILES BRADLEY'S"
Remarks: MILES BRADLEY'S might be a lower cost line of tools by Gershom. The W in his name may stand for Wakeman or Warren (Nelson 1999).

Brady & Son, William
Mt. Joy and Lancaster, Pennsylvania, 1868-1898-
Tool Types: Axes, Chisels, Draw Knives, Agricultural Tools, Hammers, Hatchets, Household Tools, Picks, and Stone-working Tools
Identifying Marks: WM. BRADY & SON/MT.JOY PA.; BRADY/CAST STEEL
Remarks: The Brady family was numerous and prolific, with three generations of toolmakers producing tools under this company and family name. David Brady had two sons David Jr. and William N., who worked with him in Mt. Joy. William had at least three sons: Henry Austen, W. Scott, and William N. David Jr., who also had an unknown number of sons, possibly including Louis P., Christian H., and Israel, all of whom were involved in the business in some way. In 1875, the company moved to Lancaster, where yet another son (probably William N. Jr.) joined up, and the name had "& Sons" possibly added (there are no tools with this mark). The original William Brady died in 1890, but his name was still used by the company under W. Scott Brady until 1898 (Nelson 1999).

Broad & Co., Elisha
Milltown, New Brunswick, -1871-1883
Broad & Sons, Elisha

116

St. Stephen and Milltown, New Brunswick, 1883-1895
Broad & Co., Hewlett
St. John, New Brunswick, -1881-1901
Tool Types: Adzes, Axes, Draw Knives, Hatchets, and Other Edge Tools
Identifying Marks: Elisha Broad & Co: E.BROAD/MILLTOWN; Elisha Broad & Sons: E.BROAD & SONS/MILLTOWN; E.BROAD & SONS/ST. STEPHEN, sometimes with S or N.B. added; Hewlett Broad & Co.: H.BROAD/ST.JOHN
Remarks: Elisha worked with his brother Hewlett prior to using his own name for his company, renamed E. Broad & Son in 1883 when he took a son into the business. In 1885 the name changed again from "Son" to "Sons." One of the sons was named Harry W.; the other son or sons, however, are unknown. The same year, Elisha moved to a Douglass Axe Mfg. Co. factory in St. Stephen, where the company remained until he died in 1895. The name was then changed to St. Stephen Edge Tool Co. (Nelson 1999).
References:
Brack, H. G. 2008d. *Registry of Maine Toolmakers*. Hulls Cove, ME: Pennywheel Press. (See Appendix G: Reference list of Maritime Toolmakers).

Brombacher & Co., A.F.
New York City, New York, -1760-1922
Tool Types: Cooper Tools, Gauging Tools, and Others
Identifying Marks: A.F.BROMBACHER&CO./29&31 FULTON ST.N.Y.
Remarks: This company dealt in tools made by D.R. Barton, L.&I.J. White, and others. Their est. date is 1760, when they may have succeeded an unknown 18th century toolmaker or vendor. An 1870 directory lists them as "Swan and Brombacker" at 33 Fulton in New York. One or more Jacob Brombachers were likely involved in this company around 1835 to 1900 (Nelson 1999).

Brown & Sharpe Co.
Providence, RI, 1833-1853
Brown, J. R. & Sharpe 1853-1860
Darling, Brown & Sharpe 1866-1892
Brown & Sharpe Mfg. Co. 1868-2004-
Tool Types: Micrometers, Machinist Tools, Levels, and Others
Identifying Marks: Brown & Sharp Mfg. Co./Providence, RI; J.R.BROWN & SHARP.PI/PROVIDENCE R.I.

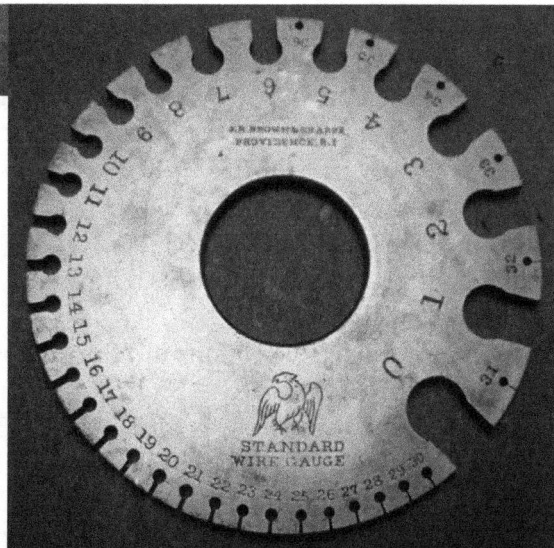

Davistown Museum MIV Collection ID# 50402T7

Remarks: The dates of operation of the various Brown, Sharpe, and Darling businesses overlap and are somewhat confusing. The business now conducted by the Brown & Sharpe Mfg. Co. was founded in 1833 by David Brown and his son

Joseph R. Brown. David Brown retired in 1841 and the business was continued by Joseph R. Brown until 1853, when Lucian Sharpe became his partner, and the firm of J. R. Brown & Sharpe was formed. The Brown & Sharpe Mfg. Co. was incorporated in 1868. Joseph R. Brown began manufacturing steel rules and other tools of precision in 1850. In 1852, a similar line of work was begun by Samuel Darling and, in 1866, the partnership of Darling, Brown & Sharpe was formed. The business carried on under that name until the partnership was dissolved by the purchase of Mr. Darling's interest in 1892 (Brown & Sharpe 1941, 4). The name J.R. Brown & Sharpe, the original company name, was still used as late as 1899 (Nelson 1999). Brown & Sharpe is now a brand of the Hexagon Metrology Group. Also see more biographical information about Samuel Darling in the Davistown Museum online essay on Precision Toolmaking.

References:

Brown & Sharpe. 1902. *Brown & Sharpe Mfg. Co.: Machinery and tools*. Providence, RI: Brown & Sharpe.

Brown & Sharpe. 1941. *Brown & Sharpe small tools: Catalog no. 34*. Providence, RI: Brown & Sharpe.

Brown & Sharpe. n.d. *The mircrometer's story: 1867 - 1902*. Providence, RI: Brown & Sharpe.

Cope, Kenneth L. *A Brown & Sharpe Catalogue Collection, 1868, 1887, 1899*.

Brack, H. G. 2008d. *Registry of Maine Toolmakers*. Hulls Cove, ME: Pennywheel Press. See Samuel Darling, Darling and Bailey, and Darling & Schwartz for information on Darling before he moved to Connecticut.

Roe, Joseph Wickham. 1916. *English and American Tool Builders*, New Haven, CT: Yale University Press. Reprinted by McGraw-Hill, New York and London, and by Lindsay Publications, Inc., Bradley, IL.

Links: http://www.brownandsharpe.com/
http://www.netrorg/RIToolmakers/BROWN-SHARP/BROWNSHARPE.html
http://www.davistownmuseum.org/Inventoryofpictures/WebInfoBrownSharpe.html
http://www.roseantiquetools.com/id44.html
http://www.davistownmuseum.org/bioBrownSharpe.htm
http://www.ask.com/bar?q=when+did+brown+%26+sharpe+get+sold&page=1&qsrc=0&ab=4&u=h ttp%3A%2F%2Fwww.answers.com%2Ftopic%2Fbrown-sharpe-manufacturing-co

Buck Bros.
Millbury, Massachusetts, 1853-1972

Tool Types: Awls, Axes, Chisels, Draw Knives, Shaves, Hammers, and Iron Planes

Identifying Marks: Variations of "Buck Bros.," often with "CAST STEEL" or "WARRANTED."

Remarks: Buck Brothers was formed in 1853 by John, Charles and Richard T. Buck in Rochester, NY, after emigrating from England and working for D. R. Barton. In 1856, they moved to Worcester, Massachusetts, then in 1864 to Millbury, MA. John stayed in Worcester as part of Buck & Reeves, later rejoining Buck Brothers as only an employee. Charles Buck had a fight with his brother and started his own edge tool company in 1873. Old timers say

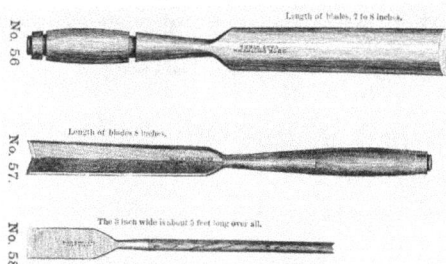

Buck Brothers. (1890). *Price list of chisels, plane irons, gouges, carving tools, nail sets, screw drivers, handles, &c. manufactured by Buck Brothers*. Riverlin Works, Millbury, MA. Reprinted in 1976 by Ken Roberts Publishing Co., Fitzwilliam, NH.

Charles' edge tools were very slightly superior to those of his brother. There are several different stories of their later activities: one states they both used the Buck Brothers name until an early 1880s lawsuit forced Charles to stop; another says Charles regained an interest in Buck Brothers in 1877 while still maintaining his own business. Richard Buck was born in 1832, died in 1893. After his death, his sons-in-law E. M. Ward and William L. Proctor ran the company. Proctor bought out Ward in 1913 and Charles Buck's separate business in 1915. The business continued until 1972. The company made edge tools (awls, axes, chisels, draw knives), hammers and plane blades. Marks consisted of several arrangements of their name and such touchmarks as a deer head, CAST STEEL and WARRANTED. A facility name RIVERLIN WORKS was also used. Charles Buck started his own business called the Charles Buck Edge Tool Co. and/or the Millbury Edge Tool Works in 1873. He used the mark CHARLES BUCK CAST STEEL. DATM states he made chisels, but the Museum also has an example of a gouge with this mark (Nelson 1999). Nooduck Chisel Co., also known as Nuduck Chisel Co., of North Grafton, MA, made chisels for Buck Brothers under a subcontract around 1890.

References:
Buck Bros. (1890). *Price list of chisels, plane irons, gouges, carving tools, nail sets, screw drivers, handles, &c. manufactured by Buck Brothers*. Buck Bros: Millbury, MA. Reprinted in 1976 by Ken Roberts Publishing Co., Fitzwilliam, NH.
Kebabian, John S. Early American factories: Buck Brothers, Millbury, Mass. *The Chronicle.* 25(1): 10-11.
Links: http://www.geocities.com/sawnutz/buck/index.htm -- Buck Brothers history
http://www.craftsmanstudio.com/html_p/BuckBrosChisels.htm -- Information on the modern Buck Brothers
http://www.davistownmuseum.org/bioBuckBrothers.html

Buff & Buff Manufacturing Company
Green St., Jamaica Plain, Massachusetts
Tool Types: Transits and Telescopes
Remarks: (Buff and Berger) (C L Berger & Sons Manufacturing Co.) The Buff & Buff Manufacturing Co. in Boston was an important member of a group of New England manufacturers of surveying transits and equipment that had its roots in the classic period of American machinist tools. The sophisticated products of America's newly discovered ability to build complex machines such as surveyor's transits using the machinery invented and constructed in the Industrial Revolution (rather than hand work and hand tools) resulted in products such as the Buff & Buff transit on display in The Davistown Museum IR collection. No modern computerized circa 2000 surveyor's transit equals the Buff & Buff specimen in beauty and quality of construction and materials, advances in efficiency and measuring capabilities notwithstanding. Also in the collection of The Davistown Museum is a six color lithograph of the museum's transit. This lithograph, one of perhaps thirty, was recovered from the Green St. factory by Liberty Tool Co. (c. 1984) after the Buff & Buff facility had closed and disposed of most of its equipment. The transit in the museum collection was purchased from a Marblehead, Massachusetts, estate in 1999. At the time of the Liberty Tool Co. salvage operation, all that remained in the Green St. factory were hundreds of transit and telescope level bubbles and parts, grinding stones and miscellaneous equipment. The lithographs were hidden away on a high shelf in one of the dusty abandoned offices.
References:

Buff & Buff Mfg. Co. 1938. *Surveying instruments: For civil and mining engineers.* Boston, MA: Buff & Buff Mfg. Co.
Links: http://www.jphs.org/20thcentury/buff-buff-manufacturing-company.html
http://www.davistownmuseum.org/bioBuff.html

Call, A.
Springfield, Massachusetts
Tool Types: Machinist Tools, Including Scaled Beam Points
Remarks: Possibly Amos Call of Bemis & Call (Nelson 1999).

Callender & Co., Benjamin
Boston, Massachusetts, 1862-1887
Tool Types: Screwdrivers and Wood Planes
Identifying Marks: B.CALLENDER & CO/CAST STEEL; B.CALLENDER & CO/BOSTON
Remarks: Callender was a Boston agent for the American File Co. of Rhode Island and a probable vendor who remarked tools made by others (Nelson 1999).

Calley, Ela D.
Franklin, New Hampshire, -1868-1881-
Tool Types: Shaves
Identifying Marks: E.CALLEY/FRANKLIN, N.H.
Remarks: Calley was a patternmaker and machinist who worked at "Aiken's Mill." His mark has been found on spoke shaves recovered by the Liberty Tool Co. (Nelson 1999).

Canney, Wesley J.
Tuftonboro, New Hampshire, -1870-1872-
Tool Types: Axes, Knives, and Other Edge Tools
Remarks: Canney worked in Melvin Village within Tuftonboro (Nelson 1999).

Cantelo, J. S.
Boston, Massachusetts, 1891-
Tool Types: Draw Knives
Identifying Marks: J.S.CANTELO/BOSTON (sometimes with PAT.1891 or "WARRANTED"
Remarks: (Nelson 1999). Cantelo's mark appears on a number of folding draw knife designs, including some with spring hinges (FFLTC).

Card Mfg. Co., S. W.
Mansfield, Massachusetts, 1874-1908-
Tool Types: Dies and Taps
Identifying Marks: S.W.CARD, MANSFIELD, MASS; S.W. Card Mfg. Co./Mansfield Mass./USA (Nelson, 1999).
Remarks: One of the most prolific makers of diestocks, often in wooden box sets.
References:
S. W. Card Manufacturing Co. (1928). *S. W. CARD MFG. CO. DIVISION UNION TWIST DRILL*

COMPANY, CATALOG 32. Mansfield, MA.

Copeland, Jennie F. (1936). *Every day but Sunday: The romantic age of New England Industry.*

Carey, George
Sunderland, Vermont, -1885-1890
Tool Types: Chisels and Shaves
Identifying Marks: GEO. CAREY Mass./USA (Nelson 1999)

Carpenter Tap & Die Co., J. M.
Providence and Pawtucket, Rhode Island, 1870-1902-
Tool Types: Dies and Taps
Identifying Marks: J M CARPENTER/TAP & DIE CO/PAWTUCKET, R.I. USA; Small C inside a large V
Remarks: This company was acquired by Union Twist Drill Co. at an unknown date after 1900 (Nelson 1999).

Carr, James
Goffstown, New Hampshire, -1768-1771-
Tool Types: Axes and Bits

Carr & Co., WM. H.
Philadelphia, Pennsylvania, 1838-
Tool Types: Augers, Bits, and Forks
Links: http://www.davistownmuseum.org/bioCarr.html

Casey & Co., George
Auburn, New York, 1857
Tool Types: Plane Irons and Wood Planes
Identifying Marks: CASEY & Co./AUBURN/N.Y. (top two lines curved); CASEY & Co/AUBURN N-Y/EXTRA STEEL
Remarks: Casey was a part of Casey, Kitchel & Co. (1847-1856); Casey, Clark & Co. (1858-1864); and the Auburn Tool Co. (1864 – 1880), all of which used Auburn prison labor (Nelson 1999).

Chandler & Farquhar
Boston, Massachusetts, 1882-present
Tool Types: Machinist Tools and Others
Remarks: Frank Chandler and Charles S. Farquhar dealt in tools for machinists, blacksmiths, factories, mills, and other various metalworkers. They added "Co." to their name in 1904. The company was operating as recently as 2009 out of Randolph, Massachusetts (Nelson 1999).
References:
Chandler & Farquhar Co. 1924. *Chandler & Farquhar Company Machinists' Tools And Supplies, Mill Supplies, General Hardware.* Chicago, IL: R. R. Donnelley & Sons Co.
Links: http://cf.dreamscape.com/ -- Chandler and Farquhar's current website
http://distribution.activant.com/press/classic-revival.html -- An article on the recent buyout of

Chandler & Farquhar
http://www.davistownmuseum.org/bioChandlerFarqhuar.html

Chandler, C. E.
Boston, Massachusetts, 1883-1895
Tool Types: Machinist Tools
Identifying Marks: MF'D BY/C.E.CHANDLER/BOSTON,MASS,USA/PATENTED/JULY 31, 1883
Remarks: This company made a Charles H. Fowler patent speed indicator and a micrometer stand as well as numerous other hand tools (Nelson 1999).

Chandler, William
Henniker, New Hampshire, -1878-1886-
Tool Types: Axes and Edge Tools (Nelson 1999)

Chapin & Co., Nathaniel
New Hartford, Connecticut, and Westfield, Massachusetts, -1840-1870-
Tool Types: Wood Planes
Identifying Marks: N CHAPIN & Co.; EAGLE FACTORY/WARRANTED/N CHAPIN & Co. (top two lines curved)
Remarks: Nathaniel was Hermon's older brother and worked with him prior to forming this company. His factory moved from New Hartford to Westfield in 1847 (Nelson 1999). The Chapin clan was among southern New England's most prolific planemakers (FFLTC).

Chapin, David B.
Newport, New Hampshire, 1827-1870-
Tool Types: Edge Tools
Remarks: Chapin added "& CO." to his name in 1830 but did not keep it. It is possible David B. was the Chapin of Chapin & Kelsey (Nelson 1999).

Chapin, E. M.
Pine Meadow, Connecticut, 1868-1876-
Tool Types: Levels and Planes
Identifying Marks: E.M.CHAPIN/PINE MEADOW, CONN, sometimes with PATd JUNE 6, 1876
Remarks: This Chapin received a plane patent with Solon Rust on March 31, 1868. It is possible this was Edward M. Chapin, Hermon Chapin's son (Nelson 1999).

Chapin, Hermon
New Hartford and Pine Meadow, Connecticut, 1828-1860
Tool Types: Wood Planes
Identifying Marks: H.CHAPIN, UNION FACTORY/WARRANTED/H.CHAPIN (top two lines curved)
Remarks: Hermon Chapin was associated with a number of other Chapins in the toolmaking industry of his time. He began as a part of Copeland-Chapin from 1826-1828, was the brother of Nathaniel Chapin and father to Edward M., George W., and Philip E. Chapman who succeeded him

as H Chapin & Sons in 1860. Out of this line sprung H. Chapin's Sons, H. Chapin's Son, and H. Chapin's Son & Co. Planes marked "Baltimore" were sold to his brother Philip (in Baltimore) but were probably not manufactured there. Pine Meadow and New Hartford were likely the same town with different, interchangeable names, not two separate locations (Nelson 1999; FFLTC).

References:

Chapin, Hermon. [1853] 1976. *Cataloque and invoice prices of rules, planes, gauges, &c. manufactured by Hermon Chapin. Union Factory, Pine Meadow, Conn.* Fitzwilliam, NH: Ken Roberts Publishing Co.

Links: http://www.davistownmuseum.org/bioHermonChapin.html

Chapin-Stephens Co. Union Factory
Riverton and Pine Meadow, Connecticut, 1901-1929

Tool Types: Planes, Rules, Calipers, and Combination Tools
Identifying Marks: THE CHAPIN - STEPHENS CO PINE MEADOW CONN. U.S.A. (sometimes split between lines with or without the top line curved upward), THE C-S CO.
Remarks: H. Chapin's Son Co. (est. 1897) and D.H. Stephens & Co. (est. 1861) merged to form this company in 1901. A number of mergers and buyouts including the Chapin name began with Hermon Chapin, a maker of wood planes. He was the father of Edward M. Chapin, George W. Chapin and Philip E. Chapin, who succeeded him as H Chapin & Sons in 1860, then H. Chapin's Sons, H. Chpain's Son, and H. Chapin's Son & Co. See Hermon Chapin.

References:

The Chapin-Stephens Co. ca. [1914] n.d. *Catalog No. 114: The Chapin-Stephens Co. Union Factory: Rules planes gauges plumbs and levels hand screws handles spoke shaves box scrapers, etc.* Fitzwilliam, NH: Ken Roberts Publishing Co.

Roberts, Kenneth D. 1978. *Wooden planes in 19th century America, volume II: Planemaking by the Chapins at Union Factory, 1826 - 1929.* Fitzwilliam, NH: Ken Roberts Publishing Co.

Links: http://www.davistownmuseum.org/bioChapin.html

Chase, Amos
-1850-1864-
Chase, David G.
1856-1887
Chase, John Winslow
1846-1877
N. Weare, New Hampshire

Tool Types: Augers, Edge Tools, Handles, and Leather Tools
Identifying Marks: A.CHASE,WEARE,NH; J.W.CHASE/WEARE.N.H.
Remarks: These brothers all worked out of New Hampshire at roughly the same time. Amos worked as a machinist starting in 1836 and made a variety of tanner and currier tools. He holds an invalid patent on a currier's arm board from April 6, 1864. David G. (middle initial sometimes recorded as J. or S.) specialized in knives and tool handles. John Winslow made currier and cobbler tools as well as hollow augers and punches. He holds patents on a skiving machine in 1859 with J.A.

Safford and 1864 with his brother Charles F. Amos. Charles was not known to produce tools in New Hampshire, but a Charles F. Amos in Dixfield, Maine, who produced agricultural implements, could be the same man (Nelson 1999).

Chase, James

Gilmanton (changed to Gilford in 1812), New Hampshire, -1797-1812
Tool Types: Coopers' Tools, Hammers, Handles, Rules, Squares, and Wood Planes
Identifying Marks: J.CHASE
Remarks: James was a carpenter who specialized in cabinets and joining. He made his own mallets and yardsticks (Nelson 1999).

Chase, Parker & Co.

Boston, Massachusetts, 1873-1939-
Tool Types: Hammers, Farrier Tools, and Others
Remarks: This company bought out Dodge, Haley & Co. circa 1928 (Nelson 1999).

Chatillon & Sons, John

New York City, New York, -1867-1894-
Tool Types: Scales
Identifying Marks: JOHN CHATILLON & SONS NEW YORK Y.S.A.;
CHATILLON'S/IMPROVED CIRCULAR/SPRING BALANCE//WARRANTED/NEW YORK
Remarks: John began producing scales under his name in 1835 and continued at least until 1858. The "& SONS" may not have been added until after his death. Patent dates include May 10, 1867, December 10, 1867, October 1872, January 4, 1876, May 1878, January 6, 1891, and January 26, 1892 (Nelson 1999).

Chelor, Cesar

Wrentham, Massachusetts, 1753-1764-
Tool Types: Wood Planes
Identifying Marks: CE.CHELOR/LIVING*IN/WRENTHAM (sometimes without "Living In");
CESARCHELOR/LINING*IN*WRENTHAM
Remarks: Cesar began as a slave producing planes under Francis Nicholson so he probably produced some planes bearing Nicholson's mark. He continued producing when Nicholson's will freed him in 1753 and worked with Jethro Jones from around 1765 to 1769 and Sambo Freeman from around 1758 to 1761 (Nelson 1999).
References:
Avila, Richard T. (1999). *Cesar Chelor and the world he lived in.* Smithsonian Institution, Anacostia Museum, Washington, DC.
Links: http://anacostia.si.edu/Online_Academy/Academy/artifacts/objects/object_3_frame.htm
http://findarticles.com/p/articles/mi_qa3983/is_200103/ai_n8942589/
http://www.memorialhall.mass.edu/turns/view.jsp?itemid=6256&subthemeid=15

Cheney Hammer Co., Henry
Little Falls, NY, 1856-1878 (possibly until 1949)
Tool Types: Hammers and Axes

Identifying Marks: CHENEY, H. CHENEY/Ha___

Remarks: Henry Cheney was born in Ostego, NY on January 12th, 1821, where he possibly made hammers before moving to Little Falls in 1856. He holds a 4 July 1871 patent on a hammer with a nail holding and starting feature. In 1874, he bought the S. H. Farnam factory and also made axes (though a 1904 catalog in our collection shows hammers for sale). A Cheney Hammer Co. or Henry Cheney Hammer Corp. was in business in 1949 with a founding date of 1836 (Nelson 1999).

References:

Cheney, Henry Hammer Co. [1904] n.d. *Illustrated Catalogue of the Henry Cheney Hammer Co.* The Special Publications Committee M-WTCA.

Links: http://www.rootsweb.com/~nyherkim/littlefalls/waterpower2.html -- A website concerning a dispute over use of the river in Little Falls involving Henry Cheney Hammer Co.

http://www.threerivershms.com/lf2.htm -- History of Little Falls

http://www.davistownmuseum.org/bioCheney.html

Child, John Edwin
Providence, Rhode Island, 1852-1875

Tool Types: Edge Tools and Wood Planes

Identifying Marks: J.E. CHILD; J.EDWIN CHILD

Remarks: Child produced plane bodies and contracted for Greenfield Tool Co. and probably worked with or under Isaac Battey in 1850 (Nelson 1999).

Clark
Chatam, New Jersey, -1860-

Tool Types: Axes and Edge Tools (Nelson 1999)

Clark & Co., Alex
Quincy, Massachusetts, 1890-1908

Tool Types: Bevels, Blacksmiths' Tools, Chisels, Drills, Hammers, Handles, Jacks, Pliers, Rules, Squares, and Stone-working Tools

Remarks: Clark, a blacksmith from Scotland, started by making tools for local granite quarries. By 1900, he was also making blacksmithing tools. In 1908, the name was changed to Vulcan Tool Mfg. Co. and it has been run by Clark descendants ever since (Nelson 1999). The current name is Vulcan.

Links:

http://www.vulcantools.com/html/about.html

Clark Edge Tool Works
-1897

Tool Types: Edge Tools

Remarks: They have also been reported as Clark Edge Tool Co. Atha Tool Co. bought out Clark Edge Tool Works in 1897 (Nelson 1999).

Clark, William A.
Connecticut, 1858-1920-

Tool Types: Augers and Bits

Identifying Marks: CLARK/EXPANSIVE BIT

Remarks: It is unclear whether the patents issued to variations of William Clark in Connecticut were all the same person or multiple individuals. These patents include an expansive bit from May 11, 1858; hollow bit augers from July 12, 1859 and June 12, 1860 out of Bethany; countersinks on February 2, 1869 and December 12, 1871 out of Woodbridge; and an ice auger from June 10, 1873 out of New Haven (Nelson 1999). The 1858 patent was the first of the modern expansion bits still used today.

Cobb, William
-1816-1819
Rochester and Rome, New York
Cobb & Thayer
1820-1821-
Rochester, New York
Tool Types: Adzes, Axes, Chisels, Draw Knives, Plows, Scythes, and Other Edge Tools
Identifying Marks: COBB & THAYER/CAST STEEL
Remarks: Cobb sometimes used "& Co." The move from Rome to Rochester occurred in 1816; in 1820, the company name changed to Cobb & Thayer. William Cobb and Lawson Thayer were succeeded by Cobb & White (Nelson 1999).

Coburn (& Son), Franklin Watson
New Durham, New Hampshire, 1856-1910
Tool Types: Axes, Cutlery, Hammers, Knives, and Shaves
Identifying Marks: F.W.C.
Remarks: Franklin Watson Coburn used "& Son" on the end of his company title from 1887 to 1890. This may refer to Franklin Watson Jr. or his other son, Alonzo G., is unclear. In 1911, it became F.W. Coburn & Co. under Franklin Watson Coburn Jr. (1856-1918) (Nelson 1999).

Coes, L. & A. G.
1841-1869
Coes & Co., Loring
1869-1881-
Coes & Co., A. G.
1869-1881
Coes Wrench Co.
1885-1901

Davistown Museum MIV Collection ID#11301t12.

Worcester, Massachusetts
Remarks: http://www.davistownmuseum.org/BioPics/CoesPriceList.jpg
L. & A. G. Coes: In 1836, Loring and Aury Gates Coes bought the wool machine business of Kimball & Fuller and continued making these machines until 1839 (apparently using some other name.) In 1841 they formed the L. & A.G Coes partnership and started making wrenches under a 16 April 1841 patent. The mark used was L. & A.G. COES | WORCESTER, MASS. In 1853, they bought the shear-blade and knife business of Moses Clement. In the early 1860s they bought the Taft & Gleason wrench business. In 1869 they separated forming the two businesses described below. Herb Page adds: "During the period of 1848 to 1852 the firm of L & A.G. Coes contracted with the firm of Ruggles, Nourse & Mason on a 5 year term to market the entire production of wrenches produced by this fledgling firm. R.N.& M. had branches in both Worcester and Boston

and the wrenches produced during this time period were stamped with 1) 'L. Coes Patent', 2) 'Ruggles, Nourse & Mason' if space permitted, depending on size of wrench and 3) 'Boston & Worcester' indicating the sales outlets of the marketing firm. These wrenches were manufactured in Worcester at the firm of L & A. G. Coes."

Loring Coes & Co.: Loring Coes was born in 1812 and died in 1907. His company's working dates are from 1869 to 1900 in Worcester, Massachusetts. He was formerly part of L. & A. G. Coes. L. Coes & Co. began with the shear blade and knife part of L. & A. G. Coes, but later resumed making wrenches. Coes Wrench Co. merged back into L. Coes & Co. in 1888 and the company continued using both names. The company mark is L. COES & CO. | WORCESTER, MASS. Loring Coes had wrench patents dated: 10 Nov. 1863, 23 Feb. 1864, 23 March 1869, 1 June 1869, 10 Aug. 1869, 26 Oct. 1869, 9 Jan. 1877, 6 July 1880, 8 July 1884, 12 July 1887, 15 Dec. 1891, 29 Dec. 1891 and 14 Aug. 1894. The Davistown Museum has received communications from an owner of a wrench with the L. Coes mark and a pat'd date of Apr. 30, 1895. A second owner has a bar wrench (crescent wrench) with this patent date.

Aury Gates Coes & Co.: Aury Gates Coes was born in 1817 and died in 1875. His company made wrenches from 1869 to 1881 in Worcester, Massachusetts. He was formerly part of L. & A. G. Coes. His sons continued the business after his death until changing the name to Coes Wrench Co. in 1881 or 1885 (sources differ). The company mark is A.G.COES & Co. | WORCESTER | MASS with the name line curved. His wrenches were also commonly marked with just his name and the patent date 6 March 1866 or 26 Dec. 1871.

Coes Wrench Co.: This company's working dates are from 1885 or 1881 to 1928 in Worcester, Massachusetts. It was originally A.G. Coes & Co. and made both knives and wrenches. The company merged back into the L. Coes & Co. in 1888, but both names continued to be used as marks. At some time, the company was acquired by Billings & Spencer or Bemis & Call (sources differ.) The mark was different configurations of the maker name, city and state (Nelson 1999). Coes Reservoir is a 100-acre property at the Worcester headwaters in the Tatnuck Brook Watershed. The historic Coes Knife Company formerly occupied the site. For more information on Coes wrenches see the discussion in the Davistown Museum online essay on Boston wrenches (http://www.davistownmuseum.org/bioBostonWrench.htm).

References:

Cope, K. 1992. The Coes Wrench Company. *The Gristmill*. 69: 16.

Page, Herb (Mr. Oldwrench). 2001. Reach for the wrench: Coes key model. *The Fine Tool Journal* 51(2): 6-8.

Page, Herb (Mr. Oldwrench). 2001. "No name" wrenches. *The Fine Tool Journal* 50(4): 23-24.

Page, Herb (Mr. Oldwrench). 2002. Reach for the wrench: The evolution of baby Coes wrenches. *The Fine Tool Journal* 52(1): 15-17.

Page, Herb (Mr. Oldwrench). 2002. Reach for the wrench: The song of the monkey-wrench. *The Fine Tool Journal* 52(2): 17-18.

Page, Herb. 2004. *The brothers Coes and their legacy of wrenches.* Davenport, IA: Sunset Mercantile Enterprises.

Herb Page has produced numerous other articles and research on Coes and many other wrench makers.

Links: http://www.hhpl.on.ca/GreatLakes/scripts/Page.asp?PageID=3825 – A page from the Blue Book of American Shipping: Marine and Naval Directory of the United States

http://www.chicago-scots.org/clubs/History/Names-McD-Mu.htm - The Chicago Scotts club adds

"he was granted patent #38316 for improvements in screw wrenches. He was offered $500 for his patent."
http://www-personal.umich.edu/~pfs/tool/re10.html
http://www.davistownmuseum.org/bioCoes.htm

Coffin, John T.
Center Harbor, New Hampshire, -1881-1892-
Tool Types: Knives, Shaves, and Other Edge Tools
Remarks: Child produced plane bodies and contracted for Greenfield Tool Co. and probably worked with or under Isaac Battey in 1850. Coffin worked under "John T. Coffin & Son" from 1884 to 1886 (Nelson 1999).

Collins & Co.
Collinsville and Canton, Connecticut, 1826-1957
Tool Types: Adzes, Axes, Hammers, Wrenches, Machetes, and Swords
Identifying Marks: The maker's name in various incarnations, city, state and "Made in U.S.A.."; Sometimes a crown figure with an arm and hammer nestled in it and the word "Legitimus" curving upward under it.
Remarks: According to their 1921 catalog, the mark "Collins & Co., Hartford" is used on their most premium products whereas "R. King," "Bx Swift," "Bv Wise" and "Charter Oak" are used on their budget lines of products. At this time, they manufactured over 1,100 products. It also purports that shoddy imitations of their products have been manufactured in Germany with their exact mark and the marks "B. Collins," "D. Collins" or "H. Collins" were made by other American manufacturers. The "Legitimus" crown and arm logo arose in response to these imitators.
References:
Collins & Co. [1921] n.d. *Illustrated catalogue of axes, hatchets, adzes, picks, sledges, hoes, wrenches, bush hooks, etc., etc. manufactured by Collins & Co. established in 1826.* Long Island, NY: The Early Trades & Crafts Society.
Collins & Co. [1935] n.d. *A Brief Account of the Development of The Collins Company in the Manufacture of Axes, machetes and Edge Tools.* Fitzwilliam, NH: Ken Roberts Publishing Co.
Links: http://www.visitcollinsville.com/visitcollinsville_002.htm
http://www.cantonmuseum.org/
http://www.davistownmuseum.org/bioCollins.html

Collins, David
Hartford, Connecticut, 1809-1825
Tool Types: Joiner Tools and Carpenter Tools
Remarks: David Collins was somehow related to Robert J. Collins Jr. In 1825, he began working with a Samuel Collins, possibly another relative, and together they founded Collins & Co. (Nelson 1999).

Collins, Robert Johnson, Jr.
Collinsville and Canton, Connecticut, 1805-1835
Tool Types: Wood Planes
Identifying Marks: COLLINS/HARTFORD; COLLINS/UTICA; R.J.COLLINS/ROCHESTER

Remarks: Robert Johnson Collins Jr. worked closely with his son, Robert Johnson Collins III, and their work is often indistinguishable. Robert Jr. worked with Leonard Kennedy as Kennedy & Collins from 1803 to 1805 prior to working in Hartford. It is unclear how long he lived and worked in Utica and Rochester and, while he died in Ravenna, it's unknown whether he worked there (Nelson 1999).

Colt Co., Samuel
Hartford, Connecticut
Tool Types: Axes
Remarks: Presumptively an earlier incarnation of Colts Patent Fire-Arms Mfg. Co., though it is unknown whether they manufactured guns (Nelson 1999). The dates of operation of Samuel Colt's ax factory are currently unknown.

Colvin & Bro., E.
Pawlet, Vermont, -1870-1874-
Tool Types: Axes (Nelson 1999)

Cook, Martin
Kingston, Massachusetts, -1849-
Tool Types: Knives, Leather Tools, and Shaves
Remarks: The knives he produced were specifically for cobbling but the type of shave is unclear (Nelson 1999).

Cooley, William
Boston, Massachusetts, -1832-1849-
Tool Types: Wood Planes and Other Edge Tools
Identifying Marks: W.COOLEY/LINCOLN ST.//BOSTON; W.COOLEY BLACKSTONE ST.BOSTON
Remarks: Cooley worked as Cooley & Montgomery in 1844. The edge tools may not have been produced after 1834 when he began making planes (Nelson 1999).

Copeland & Chamberlain
Worcester, Massachusetts, 1872-1901
Tool Types: Calipers, Dividers, and Machinist Tools
Identifying Marks: S COPELAND/PAT MAY 24 1887
Remarks: This company, formed by Samuel Copeland (1815-1891) and Charles W. Chamberlain, produced an Albert A. Cook's extension divider/caliper, patented December 12, 1871, and an Copeland's extension divider, patented May 24, 1887. The name changed to Copeland Hardware Mfg. Co. in 1901 (Nelson 1999).

Copeland & Chapin
Pine Meadow, Connecticut (now Hartford), 1826-1828

Tool Types: Wood Planes
Identifying Marks: S COPELAND/PAT MAY 24 1887
Remarks: This company consisted of Daniel Copeland and Hermon Chapin (Nelson 1999).

Copeland, Daniel
Hartford, Connecticut and Huntington, Massachusetts, -1827-1842-
Identifying Marks: D.COPELAND/HARTFORD (sometimes without HARTFORD)
Tool Types: Carpentry Tools and Wood Planes, Possibly Other Joiner Tools
Remarks: Copeland worked with D. & M. Copeland with his brother Melvin from 1822 to 1825 and again sometime after 1842 out of Huntington, MA, as well as with Copeland & Chapin (with Herman Chapin) from 1826 to 1828; he also produced tools under his own name (Nelson 1999). Numerous Copeland planes have been recovered by the Liberty Tool Co.

Couch, John
Salisbury, New Hampshire, -1862
Tool Types: Chisels and Other Edge Tools
Remarks: John Couch was a blacksmith known for his edge tools who possibly worked with his brother Samuel, as evidenced by a gouge marked J.&S. Couch (Nelson 1999).

Craddock, Thomas
Lockport, New York, -1826-
Tool Types: Edge Tools (Nelson 1999)

Crescent Tool Co.
Bridgeport, Connecticut, -1883-1902
Tool Types: Machinist and Carpentry Tools, Screwdrivers, Nail Pullers, and Others
Identifying Marks: Crescent/Bridgeport
Remarks: This company also used the names GIANT, Kennelly & Cain, and Crescent Tool Works. They made a Patrick Kennelly protractor, patented October 2, 1883 (Nelson 1999). This important tool company stayed in business until well into the 20th century and are famous for their high quality adjustable wrenches (FFLTC).

Crossman (& Son), Amory W.
W. Warren, Massachusetts, 1850-1883-
Tool Types: Chisels, Draw Knives, Metal Planes, Scythes, and Others
Identifying Marks: A.W.CROSSMAN/CAST STEEL, sometimes with WARRANTED
Remarks: Amory W. Crossman and his son Amory Jr. made a draw knife attachment patented by Amory W. Crossman, October 16, 1883, and a plane patented by Benjamin A. Blandin of Charlestown, Massachusetts, May 7, 1867. Amory Sr. produced under his own name 1850 to 1866 (Nelson 1999). A prolific maker of fine draw knives.

Cumings (Cummings), Allen
Boston, Massachusetts, -1848-1854-
Tool Types: Wood Planes
Identifying Marks: A.CUMINGS/BOSTON (top line curved)

Remarks: Cumings sold plane stocks to Greenfield Tool Co., was part of M. Read & Co. from 1844 to 1845, and part of Read & Cumings from 1846 to 1847. He was born in New York and may have worked there prior to moving to Massachusetts (Nelson 1999).

Currier, Moses F.
N. Weare, New Hampshire, -1850-1870-
Tool Types: Augers, Bits, Dies, and Other Edge Tools
Remarks: Currier worked with his brother, Daniel G. Moses, around 1853 and may have been part of Glover & Currier (Nelson 1999).

Dalrymple, James
Newark, New Jersey, -1861-1880-
Tool Types: Edge Tools
Remarks: Dalrymple worked as a blacksmith from 1849 to 1860, probably working as a part of Dalrymple & White. His son James W. was a part of Forgie & Dalrymple in 1879 (Nelson 1999).

Danforth, Jacob
Jaffrey, New Hampshire, -1792-1811
Tool Types: Axes
Remarks: Danforth lived in Amherst, New Hampshire prior to living in Jaffrey but whether he made axes there is unknown (Nelson 1999).

Daniels, George Washington
Boston, Massachusetts, 1850-1886
Tool Types: Calipers, Dividers, Vises, Watchmaking Tools, Small Bench Vises, and Eyelet Tools
Identifying Marks: G.W.DANIELS/WALTHAM,MASS.
Remarks: Daniels was born December 22, 1830 and died May 9th, 1886 (Nelson 1999).

Darling, Brown & Sharpe
Providence, Rhode Island, 1866-1892-
Tool Types: Bevels, Calipers, Machinist Tools, Rules, and Squares
Identifying Marks: D.B.&S.; DARLING,BROWN&SHARPE PROVIDENCE, R.I.
Remarks: This company consisted of Samuel Darling, Joseph R. Brown, and Lucian Sharpe and was a merger of J.R. Brown & Sharpe and Darling & Schwartz. The Brown & Sharpe Mfg. Co. formed without Darling in 1868 and the two operated together until 1892 when Darling's share (and the original company) were bought out. The name persisted until at least 1879. Aside from patents owned by the founders, they were known to produce tools with a July 6th, 1852 patent by Nathan Ames, an August 2, 1887 patent by Alton J. Shaw, and a September 24 patent by C. E. W. Dow (Nelson 1999). See Brown & Sharpe.

Davis & Co., George W.
Nashua, New Hampshire, 1863-1897
Tool Types: Machinist Tools
Remarks: This company succeeded George A. Rollins & Co. Rollins remained a partner until 1879 (Nelson 1999).

Davis & Furber Machine Co.
North Andover, Massachusetts, 1832-1974
Tool Types: Textile Machines, Wrenches, and Other Tools
Remarks: The wrenches and other tools made by this company were for servicing their textile machines. This company may have been founded under another name (Nelson 1999).

Davis, Leonard L.
Springfield, Massachusetts, -1867-1875
Tool Types: Levels, Metal Planes, and Tools
Davis Level & Tool Co.
Springfield, Massachusetts, 1875-1892

Tool Types: Levels, Awls, Bits, Braces, Calipers, Dividers, Machinist Tools, Metal Planes, Saw, Screwdrivers, Squares, and Vises

Davistown Museum IR Collection ID# 102501T1.

Remarks: The ornate levels and measuring tools of L. L. Davis and the Davis Level & Tool Company of Springfield, MA, are among the most sought after examples of the American toolmaker in the second half of the 19th century. There is no direct connection between the Davis family of the Davistown Plantation and Leonard Davis. The Leonard L. Davis Co., started in 1867, "...became the Davis Level & Tool Co. in 1875. Leonard L. Davis was born 21 Feb. 1838 and died 13 Aug. 1907. He had 17 March 1868 and 17 Sept. 1867 patents for inclinometer levels made by this company (but marked with his name only) and the successor. He also had a 31 Aug. 1875 metal plane patent, and a 21 Nov. 1871 level patent; the planes were made, but it is not certain if the level was. (Note: Different Davis levels are marked with 17 March 1867 and 17 March 1868 patent dates; documentary sources indicate that only the 1868 date is valid.)"

The Davis Level & Tool Co., (1875 - 1892) "...made inclinometer levels with 29 May 1877 (F.T. Ward & T. Bedworth), 17 Sept. 1867, 17 March 1868, and 22 Sept. 1868 patents; metal planes patented by Charles E. Torrance 2 Jan. 1872; a universal square patented by Joseph C. Marshall 2 Jan. 1877; braces with patent dates of 17 April 1883, 14 Oct. 1884, and 1886 (John Bulen); and a 4 Dec. 1866 R. Hathaway**Error! Bookmark not defined.** patent combination gauge (also made by J. Stevens & Co.). Davis sold his patent rights to the M.W. Robinson Co. of NY." (Nelson, 1999). Numerous Davis levels and inclinometers are in the collection of the Davistown Museum and illustrated in the *Hand Tools in History* publication series.

References:
Clark W. Bryan & Co. 1874. *Springfield City Directory and Business Advertiser for 1874 - 75. For the Year Commencing July 1, 1874.* Clark W. Bryan & Co.

Inland Massachusetts Illustrated. *A Concise Résumé of the Natural Features and Past History of the Counties of Hampden, Hampshire, Franklin, and Berkshire, their Towns, Villages, and Cities, Together with a Condensed Summary of their Industrial Advantages and Development, and a Comprehensive Series of Sketches Descriptive of Representative Business Houses.* 1890. The Elstner Publishing Company, Springfield, MA.

National Publishing Co. *Commerce, Manufactures & Resources of Springfield, Mass.: A Historical, Statistical & Descriptive Review.* 1883. National Publishing Co.

Springfield City Directory and Business Advertiser for 1874 - 75. For the Year Commencing July 1, 1874. 1874. Published by Clark W. Bryan & Co. pg. 412.
Links: http://www.antiquetools.com/levels/davislevel.html

http://www.sydnassloot.com/Brace/Davhtm#LDavis
http://www.melmillerantiquetools.com/davis_level_toolpage.htm
http://www.davistownmuseum.org/bioDavis.htm

Dean, Henry N.
New Bedford, Massachusetts, -1870-1871-
Tool Types: Coopers' Borers
Identifying Marks: H.N.DEAN (Nelson 1999)

Dean, S.
Dedham, Massachusetts, 1775-1820
Tool Types: Wood Planes
Identifying Marks: S*DEAN/DEDHAM; S.DEAN
Remarks: The planes marked S*DEAN/DEDHAM are significantly earlier than those marked S.DEAN; the former are attributed to Samuel Dean (1700-1775), a joiner in Dedham circa 1740, while the latter belong to his descendent and fellow Dedham resident, Samuel H. Dean (1767-1825) (Nelson 1999).

Dearborn, Warren
Sandwich, New Hampshire, 1831-1862
Tool Types: Churns, Washboards, Other Household Tools, Handles, Rules, Saw Frames (but not complete saws), Squares, and Wood Planes
Remarks: Dearborn, a carpenter, worked with Dearborn & Skinner from 1828 to this company's founding (Nelson 1999).

Demeritt, John
Montpelier, Vermont, 1859-1896
Tool Types: Cutlery and Other Edge Tools (Nelson 1999)

Denison, John & Lester
Saybrook, Connecticut, 1832-1840-
Identifying Marks: J.&L.DENISON/SAYBROOK
Denison, John
Saybrook, Connecticut, -1845-1860-
Identifying Marks: JOHN DENISON/SAYBROOK; JOHN DENISON; J. DENISON
Denison & Co., Gilbert Wright
Winthrop, Connecticut, 1868-1890
Tool Types: Augers, Bits, Carpenter Tools, and Wood Planes
Identifying Marks: G.W.DENISON&Co./WINTHROP/CONN. (all lines curve upward); G.W.DENISON&Co/WINTHROP.CONN.; G.W.DENISON/WINTHROP.CONN
Remarks: It should be noted that Deep River and Winthrop are villages in the East and West ends of Saybrook. John & Lester Denison worked with Jeremiah Gladding from 1836 onward. When the company dissolved, Lester became a turner and John continued his career, working under his own name. Gilbert Wright's relationship to the other Denisons is unclear as he married into the family through Sarah Denison in 1865 and it is unknown how she was related to them (possibly a niece or

daughter of John or Lester) (Nelson 1999). Denison planes are frequently found by the Liberty Tool Co.

Derby Plane Co.
Derby, Connecticut, 1891-1900
Tool Types: Metal Planes and Shaves
Remarks: Formerly the Birmingham Plane Mfg. Co., this company was owned by George D. Mosher. Patents produced by and assigned to the company included a spoke shave patented by Mosher on September 19, 1876, a "B. Plane" he patented on October 22, 1889, a plane patented by Oliver R. Hayworth of Tarkio, Missouri, on November 7, 1893, and two July 14, 1891 plane patents by Charles F. Young of Birmingham, Connecticut (Nelson 1999).

Dewey, A. G.
Woodstock, Vermont, 1855-1873-
Tool Types: Axes, Straw Knives, Scythes, and Other Farm and Edge Tools
Remarks: Dewey succeeded D. Taft & Sons and was listed in Taftsville, a subdivision of Woodstock (Nelson 1999).

Dickinson, Porter
Amherst, Massachusetts, -1838-1849-
Tool Types: Axes, Chisels, Hammers, Hatchets, and Knives
Remarks: Dickinson's tools were sold by Kennedy & Way (Nelson 1999).

Dimond, Ephraim
Goffstown and Antrim, New Hampshire, -1825-1857
Tool Types: Scythes and Other Edge Tools
Remarks: Ephraim Dimond (sometimes spelled "Diamond") was probably a nephew of Israel Dimond and possibly worked with him in Goffstown (Nelson 1999).

Disston & Sons, Inc., Henry
Philadelphia, Pennsylvania, 1840-1955
Tool Types: Hammers, Knives, Levels, Marking Gauges, Pliers, Saw Tools, Saws, and Squares
Identifying Marks: Variations of the company name with city, state, patent dates, brands, and figures including scales, keystones, and "KEYSTONE TOOL (or saw) WORKS"
Remarks: The Disston company was established in 1840 by Henry Disston as the Disston Saw Works and initially only made saws. In 1855 he cast the first crucible saw steel ever made in America. He started also making files in 1865. See DATM pg. 227 - 229 for a complete listing of the Disstons and their tool manufacturing operations.
References:
Baker, Phil. (Fall 2008). Phil's saws: Henry Disston and the ergonomic backsaw. *Fine Tool Journal*. 58(2). pg. 10-2. IS.
Henry Disston & Sons. (1876). *Price list*. Reprinted in 1994 by Astragal Press.
Henry Disston & Sons, Inc. [1902] 1994. *Handbook for lumbermen with a treatise on the construction of saws and how to keep them in order*. Mendham, NJ: The Astragal Press.
Disston, Henry & Sons. 1920. *The Disston history: History of the works, Vol. 2*. Philadelphia, PA:

Disston Saw Co.

Disston, Henry & Sons. [1916] 1978. *The saw in history*. Philadelphia, PA: Midwest Tool Collector's Association and the Early American Industries Association.

Disston, Henry & Sons. [1922]. *Saw, tool and file book*. Philadelphia, PA: Henry Disston & Sons.

Disston, Henry & Sons, Inc. n.d. *Disston saw, tool and file manual: How to choose and use tools*. Philadelphia, PA: Henry Disston & Sons, Inc.

Disston, Henry & Sons, Inc. 1942. *Disston saw, tool and file manual*. Philadelphia, PA: Henry Disston & Sons, Inc.

Kebabian, John S. 1970. Early American factories: The Disston factory, Philadelphia, PA. *The Chronicle* 23(3): 39-40.

Morgan, Paul W. 1985. The Henry Disston family enterprise: Part one of a three-part article. *The Chronicle* 38(2): 17-20.

Taran, Pete. 1998. Disston, the good, the better, the best. *The Fine Tool Journal* 48(1): 13-15.

Taran, Pete. 2001. Disston type study: The medallions: Part one. *Fine Tool Journal* 50(3): 10-13.

The Tool Chest. 1999. Mr. Disston's view of saw nibs. *The Tool Chest*. 5: 32.

Links: http://www.disstonprecision.com/ - Official website

http://www.disstoninstitute.com/index.html - Online Reference of Disston Saws.

http://www.roseantiquetools.com/index.html - Rose Antique biography on Disston

http://sawshq.com/disstonsaws/ - Everything Saws, from Air Saws to Zero Clearance contains many links to sites both about Disston or selling their saws.

http://www.vintagesaws.com/cgi-bin/frameset.cgi?left=main&right=/library/library.html -Vintage Saws Library contains reprints of many references written by Pete Taran in *The Fine Tool Journal* on Disston saws and other saw-related information.

http://www.davistownmuseum.org/bioDisston.htm

Dixon Crucible Co., Joseph
Jersey City, New Jersey, 1827-1931-
Tool Types: Pencils and Crucibles
Remarks: This company's name was changed to Dixon Industrial Markers sometime in the mid-1900's (Nelson 1999). The role of Joseph Dixon in the evolution of American production of cast steel is noted in the *Hand Tools in History* series volumes 7 and 8. Dixon's manufacture of heat-resistant clay crucibles was the key factor in the evolution of America's ability to smelt cast steel of equal quality to the crucible cast steel being manufactured in Sheffield, England.

Doggett, Simon
Middleboro, Massachusetts, -1762-1775-
Tool Types: Wood Planes and Other Tools
Remarks: Doggett Simon (January 4, 1738 – May 6, 1823) was a joiner whose Tory leanings during the American Revolution negatively impacted his business, forcing him into farming after 1775 (Nelson 1999).

Douglas Axe Manufacturing Co.
E. Douglas, Massachusetts, 1836-1897
Tool Types: Adzes, Axes, Hoes, Knives, and Picks
Identifying Marks: Variations of the maker's name, "W.HUNT" or "WHUNT&CO.," one of the

cities and/or "CAST STEEL WARRANTED"

Remarks: Information surrounding this company is cloudy and contested at best owing to the "Hunt" references on the tool markings. Theories include that Warren Hunt was a founder and major stockholder or that Hunt was an axe maker who was bought out by Douglas Co. A Canadian subsidiary running from 1866 to 1885 had the same name until sold to E. Broad & Sons. Their business address was in Boston by 1870, but had other factories stamping HOWE, NEWSHOPS, LOVET WORKS, UPPER WORKS, GILBOA and EAST PLANT on tools. Other brand names used include E. MOORE, L. STONE, D. SHARP and L. QUIN, all company employees (Nelson 1999).
Links: http://findarticles.com/p/articles/mi_qa3983/is_200009/ai_n8925288/
http://www.davistownmuseum.org/biodouglas.html

Douglass Mfg. Co.
Seymour, Connecticut, and Arlington, Vermont, 1856-1894-
Tool Types: Augers, Bits, Boring Machines, Chisels, Draw Knives, Handles, Screwdrivers, and Taps
Identifying Marks: DOUGLAS MFG. Co.; D.M.CO.
Remarks: This company was owned by F.L. Ames (1856-1873), Thomas Douglass and Richard Bruff (1873), Russell & Erwin (1874-1877), and finally James Swan. It is possible this company had a branch in Bridgeport due to a knife-like tool with the maker's mark, "D.M.CO./BPT,CONN." (Nelson 1999). A number of Douglass tools were restamped and sold by James Swan to at least one retailer in Boston, Massachusetts.
References:
Bishop, Leander J. (2006). *History of American Manufactures 1608 to 1860*. Kessinger Publishing, Whitefish, MT.
Links: http://www.flickr.com/photos/22280677@N07/3310139375/

Dover Stamping Co.
Boston, Massachusetts, 1833-1891-
Tool Types: Hammers, Household Tools, Ice Tools, Knives, Picks, Saws, Shovels, and Tinsmith Tools
Remarks: This company lists an 1833 date of establishment but it may have had an earlier name. Products produced include waffle irons, coffee mills, fly traps, a sausage stuffer patented by A.W. Hale on March 15, 1859 which was later made by Peck, Stow & Wilcox, and egg beaters patented under MAMMOTH, FAMILY, and TUMBLER on May 31, 1870; May 6, 1873; August 26, 1876 (invalid); April 3, 1888; and November 24, 1891 (Nelson 1999).

Drew & Co., C.
Kingston, Massachusetts, 1837-1937-
Tool Types: Augers, Bits, Chisels, Hammers, and Others
Identifying Marks: Variations of the maker's name, city, state, MADE IN USA, and CAST STEEL
Remarks: Christopher Prince Drew was a famous maker of caulking tools, caulking mallets, shingle rips, and cat's

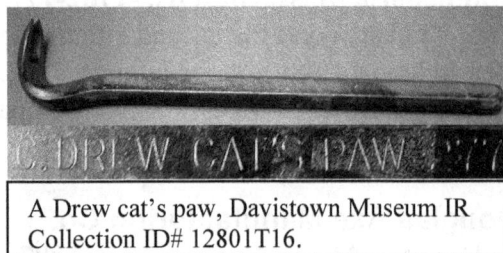

A Drew cat's paw, Davistown Museum IR Collection ID# 12801T16.

paws in Kingston, MA. Christopher Drew is also known for the high quality of his caulking mallets, the best of which were made from live oak. He also used black mesquite, but caulkers noted that it was particularly slippery to handle and preferred the live oak. Drew caulking mallets were either made with malleable iron or cast steel ferrules; the more durable cast steel caulking irons were marked with a triple O after the company mark, hence the name "triple ought" for the best quality caulking irons (Ed Shaw, personal communication). His company used the mark "C DREW & CO" and "C DREW & CO KINGSTON MASS". In 1970, the company was sold to the Kingston Tool Company owned by Robert W. MacWilliams of Ashburnham. The factory burned to the ground shortly after the sale.

References:

Drew, Emily Fuller. (1937). *A century of C. Drew & Company*. Transcript of speech given before a regular meeting of the Jones River Village Club, Inc., on the evening of December the eleventh, 1937. Original located at the Kingston Public Library, Kingston, MA.

C. Drew & Company. (1972). *Catalog No. 34: C. Drew & Co. established 1837: factory at Kingston, Mass. Plymouth county.* The Marine Historical Association, Inc. and Antique Trades and Tools of Connecticut.

Larsen, Ray. (1984). The art of the draw forger. *The Chronicle* 37(4): 57-60.

Larsen, Ray. (1985). The art of the draw forger - part II. *The Chronicle* 38(1): 5-9.

Larsen, Ray. (2001). "A satisfied customer is the best advertisement." C. Drew & Co.'s caulking irons. *The Chronicle* 54(3): 102-109.

Silver Lake News. (1970). "Famed 'Cat's Paw' tool firm reorganizes: C. Drew sold for $99,000." *The Helen D. Foster Papers,* January 29, Clippings, Kingston Public Library, Kingston, MA.

Welcker, Peter, Welcker, Merrill L. and Welcker, Anne P. 2000. Generations of augermakers in Kingston, Massachusetts. *The Chronicle* 53(3): 120-122.

Links: http://www.numismalink.com/drew1.html

http://pages.sbcglobal.net/djf3rd/page12.html - C. Drew caulking mallet: A description of its use.

http://www.davistownmuseum.org/bioDrew.htm

Dunlap, Samuel
Henniker, Bedford, and Salisbury, New Hampshire, -1779-1830
Tool Types: Braces, Cheese Presses, Rakes, Rolling Pins, Saws, Textile Tools, Washboards, Wood Planes, and Others
Remarks: Samuel worked for his brother, John, before working on his own. He worked in Bedford from 1773 to 1779, Hennicker from 1779 to 1797, and in Salisbury from 1797 to 1780, but is only known to have made tools on his own in the latter two (Nelson 1999).

Dutcher, Elihu
Pownal, Vermont, -1844-
Tool Types: Metal Planes
Identifying Marks: ELIHU DUTCHER/POWNAL VT//PATENT
Remarks: DATM (Nelson 1999) lists his birth date as 1802 and death date as 1854. Two views of his plow plane are on the frontispiece.

Dwight & Co., Timothy
Seymour, Connecticut, -1836-
Identifying Marks: T.DWIGHT&CO., sometimes with a Jb or Jr at the end.

Dwights & Foster
Seymour, Connecticut
Identifying Marks: DWIGHTS & FOSTER/SEYMOUR, CONN., sometimes with "CAST STEEL" instead of SEYMOUR,CONN. or no second line

Dwights, French & Co.
Seymour and Humphreysville, Connecticut, -1849-1900-
Identifying Marks: DWIGHTS & FRENCH; DWIGHTS FRENCH & CO
Tool Types: Augers, Bits, and Plane Irons
Remarks: Timothy Dwight was somehow associated with Upson Mfg. Co. Either he or John Dwight were involved in Dwights, French & Co. and possibly Dwights & Foster. The "Dwights" in Dwights, French & Co. probably refer to Timothy and John Dwight (Nelson 1999).

Eagle Ratchet Co.
Holliston, Massachusetts, 1858-1868-
Tool Types: Drills and Wrenches
Remarks: This company made a ratchet patented on June 29, 1858 by H.H. Packer, the company's agent and supervisor (Nelson 1999).

Eagle Square Manufacturing Co./ Millington and George
Shaftsbury, VT, 1859-1874, 1874-1881-
Tool Types: Boring Machines and Squares
Remarks: The Directory of American Toolmakers has two listings for the Eagle Square Co. The first is Eagle Square Co., South Shaftsbury, VT, 1859 - 1874. "This is a consolidation of the former steel square making activities of Dennis George, Jeremiah Essex, Heman Whipple, Lewis Beach, and the Hawks, Loomis & Co. Other square makers Stephen A. Whipple, Milo Pierce, and Norman A. Douglass were also incorporators of the company, but were not listed as contributors to its initial inventory. It also used equipment and machinery formerly used by R.W. Bangs. In 1874 its name was changed to Eagle Square Mfg. Co." The second listing, Eagle Square Mfg. Co. continued in business at least until 1881.

Eagle Squares can be traced back to the 1817 patent of Silas Hawes who was the first of many square makers working in the Shaftsbury and South Shaftsbury area; DATM indicates Silas Hawes made squares in Shaftsbury, VT, 1814 - 1828, but that several other local makers also marked their squares "HAWES PAT." These were predecessors to the famous Eagle Square Co. organized in 1859. Along with the square makers noted above, DATM also notes May and Blackmer as a square maker working before the formation of Eagle Square. Jeremiah Essex also made squares in Bennington, Vermont, 1830 – 1859, before merging with the Eagle Square Co. in 1859. Of particular note is that the square makers who followed Silas Hawes often only marked their squares with either "HAWES PAT" or "S. HAWS PATENT WARRANTED STEEL" (Nelson 1999). An example of the latter is in the Davistown Museum Collection ID# 121906T1. Also in the Davistown Museum Collection is a second framing square ID# 040103T9 marked "HAWES patent 1825" and ID# 63001T3, a "J. Essex CAST STEEL WARRANTED No. 1." A curious aspect of the square manufacturing activities of makers preceding the organization of the Eagle Square Co. is the relatively common appearance of S. HAWS as the makers mark, rather than the also common HAWES PATENT. Why many of the early Vermont makers changed their mark is unknown.

Also of interest is the change in the metallurgical composition of the squares; both examples of the museum's early Hawes squares are made of malleable iron; the later Essex square is clearly stamped CAST STEEL. An ongoing project for the Davistown Museum will be to examine incoming examples marked Eagle Square (many have been sold by Liberty Tool Co. in the last 30 years) to see if they also are marked cast steel. More information about any Eagle Squares with the mark cast steel would be appreciated. The Shaftsbury, VT, square-making community is also important for the foundations laid for future measuring tool manufacturing activities. Major changes were occurring in the way tools were manufactured during the 1830s, 1840s, and 1850s. This Millington & George, 1853, patent model for a dividing machine is currently owned by Rick Floyd. Its dimensions are roughly 8 ½" X 5" X 6 ½" tall. This patent model represents a landmark in the evolution of the Industrial Revolution as its use in the marking of framing squares meant that all the marks on a square

Photo courtesy of Rick Floyd.

previously hand stamped could be done by this machine in one step. This patent model dates from the same period of time -- the 1850s -- as the invention of the micrometer, the milling machine, the Robbins and Lawrence Armory's first production of interchangeable rifle components, and signals the advent of factory production of framing squares. For a more detailed and very interesting description of Silas Hawes tedious hand stamping of squares in the earlier years of the 19th century, use the University of Vermont link below and peruse their historical notes. The wooden model illustrated here was reproduced in a much larger size, presumably in cast iron, to become one of the earliest dividing machines to facilitate rapid, accurate production of measuring tools. Information on other dividing machines utilized by other toolmakers during the 19th century would be greatly welcomed by the Davistown Museum.

Links: http://www.rootsweb.com/~vermont/BenningtonShaftsbury.html - Shaftsbury Township Information.
http://cdi.uvm.edu/findingaids/collection/eaglesquare.ead.xml -- 1847-1962, Special Collections, University of Vermont Library.
http://www.shaftsbury.net/images/eagle_square.htm -- Photo of the plant
http://www.davistownmuseum.org/bioEagleSq.htm

Eaton, E. &. E.
Enfield, New Hampshire, 1849
Tool Types: Edge Tools
Eaton, Eben
Tool Types: Coopers' Adzes, Axes, and Knives
Eaton, Edward
-1850-1860-
Tool Types: Axes, Chisels, and Edge Tools
Eaton, Edward Jr.
-1850-
Remarks: E. & E. Eaton was possibly Edward Eaton Sr. and Jr. (Nelson 1999).

Eaton, Ephraim
Fisherville (later called Penacook) and Concord, New Hampshire, -1852-1853-
Tool Types: Anvils
Remarks: Fisherville was a village in Concord. Ephraim Eaton only steel plated, finished and hardened anvils in his Concord shop, having them cast elsewhere. He may be the leather crafting toolmaker from Concord named E. Eaton and/or the E. Eaton making anvils in Troy, New Hampshire (Nelson 1999).

Elliot, Henry
Taunton, Massachusetts, -1870-1871-
Tool Types: Edge Tools (Nelson 1999)

Ellsworth, G. F.
(South) Gardner, Massachusetts, -1868-1871-
Tool Types: Edge Tools (Nelson 1999)

Emerson Edge Tool Co.
Woodstock, Vermont, and East Lebanon, New Hampshire, -1874-1900-
Tool Types: Axes, Agricultural Tools, Hoes, Scythes, Shovels, Sickles, and Hay, Corn, and Straw Knives
Identifying Marks: EMERSON EDGE TOOL CO./TAFTSVILLE, VERMONT
Remarks: The Vermont plant, formerly used by D. Taft & Sons, closed in 1883. The "Taftsville" mark comes from the name of the part of Woodstock it occupied. It may or may not be related to Emerson & Cummings, Emerson & Kimball, or A.V.&M.W (Nelson 1999).

Essex, Jeremiah
Bennington, Vermont, -1830-1859
Tool Types: Squares
Remarks: He merged with Eagle Square Co. in 1859. In 1860 he was making cotton belting (Nelson 1999).

Eyeless Tool Co.
-1897
Tool Types: Picks and Railroad Tools
Remarks: This company became part of Atha Tool Co. in 1897 (Nelson 1999).

Fairbanks & Co., E. & T.
St. Johnsbury, Vermont, and New York City, New York, 1828-1916
Tool Types: Hoes, Plows, and Scales
Remarks: Erastus and Thaddeus Fairbanks are chiefly known as scale makers, starting with their 1828 patent, but the company actually consisted of a hardware store and produced hoes and plows. The "& Co." may have been removed at some point. In 1916 they became the Fairbanks, Morse Co. (Nelson 1999).

Faxon, Richard

Braintree, Massachusetts, -1795-

Tool Types: Adzes, Axes, and Other Edge Tools

Identifying Marks: FAXON

Remarks: Tools marked FAXON may have been made by a different Faxon who died in Boston in 1824 and was succeeded by Jesse J. Underhill (Nelson 1999). The Jonesport Co. recovered two Faxon broad axes from a Braintree, MA, shop lot, c. 1975, and has recently recovered two other Faxon signed tools from the B. F. Cutter farm in South Pelham, NH. Both of these two latter tools, a clearly marked coopers' adz and an offset hewing (vine?) ax are illustrated along with commentary about the possible significance of the Faxon family as one of the many clans of New England toolmaking families working in the period from 1650-1900.

Fay & Co., J. A.

Keene, New Hampshire, Norwich, Connecticut, and Cincinnati, Ohio, 1830-1899-

Tool Types: Carpentry Tools, Grinders, Saw Tools, and Saws

Remarks: George Page began producing in 1830 and founded this company with Jerub Amber Fay and Edward Joslin in 1834, adopting this name officially later, using Page's 1830 est. date. Their product line included powered, hand/foot powered and non-powered items. They were joined by C.B. Rogers circa 1848 until 1861 when his plant in Norwich became C.B. Rogers & Co. and the Keene plant moved to Cincinnati (Nelson 1999).

Fay, Charles P.

Springfield, Massachusetts, 1884-1887

Tool Types: Calipers and Dividers

Identifying Marks: Mf'd. by C.P.Fay/Sp'f'd.Mass.USA

Remarks: This company sometimes used "& Co." or the brand YANKEE. L.S. Starrett bought him out in 1887 and kept using names associated with his patents as late as 1898 despite his leaving Starrett to become a VP of J. Stevens Arms & Tool Co. in 1896. Patents include calipers on June 2, 1885 and January 1886, and a number of patents he assigned to Starrett and J. Stevens Arms & Tool Co. He also made calipers patented by Samuel B. Dover on November 7, 1882 and James H. Bullard on February 9, 1886 (Nelson 1999).

Fisher, John

Lowell, Massachusetts, -1832-1836-

Tool Types: Railroad and Other Adzes and Other Edge Tools

Identifying Marks: J.FISHER/CAST STEEL/LOWELL

Remarks: Railroad adzes with this mark may be from a later maker with the same name (Nelson 1999).

Fisher, Mark

Levant, Maine, and Trenton, New Jersey, 1843-1847

Tool Types: Anvils

Identifying Marks: FISHER MAKER/PATENT APRIL 24 1877

Remarks: Fisher is possibly the first commercial American anvil maker. He moved to New Jersey from Maine in 1847, became a part of Fisher & Norris under the name Eagle Anvil Works, and fathered Clark Fisher who later ran Eagle Anvil, Vise & Joint Works (Nelson 1999).

Farrington, I. B.
Brooklyn, New York, 1870-1879
Tool Types: Braces, Carpenter Tools, and Saws
Remarks: Farrington sold food-powered scroll saws and other tools. It's unclear whether he manufactured anything or just dealt in tools (Nelson 1999).
References:
Farrington, I. B. [1879] n.d. *Price List of I.B. Farrington's Ornamental Designs for Scroll Sawing and all kinds of Scroll Saw Machines*. Early Trades & Crafts Society.

Folding Sawing Machine Co.
Chicago, IL and Essex Center, Ontario, 1883-1942
Tool Types: Saws
Identifying Marks: F.S.M. CO., patent dates
Remarks: Marvin O. Smith was the founder of this company, holding patents on a one-man sawing machine from October 31, 1882; July 22, 1884; February 17, 1885; January 28, 1890 and a Canadian patent on November 25, 1885. Marvin's widow Mary was president of the company until 1904. The company underwent some sort of change that year and, despite remaining in business until 1942, did not appear in Chicago directories (Nelson 1999). The Chicago factory manufactured all orders except those from Canada; this allowed the sale of equipment duty-free within Canada. The tool catalog in the collection includes many written testimonials to the quality of these saws from both America and abroad.
References:
Folding Sawing Machine Co. [1897] n.d. *Folding Sawing Machine*. The Mid-West Tool Collectors Association.
Links: http://www.davistownmuseum.org/bioFolding.html

Fowler Co., Ltd., Josiah
Saint John, New Brunswick, Canada, 1860-1922
Tool Types: Axes
Identifying Marks: JOSIAH FOWLER CO. EXTRA AXE STEEL
Remarks: To quote an article in *The Chronicle*, "Just over the Maine border is the Canadian area of New Brunswick. In the nineteenth century it was supported by the huge ship building trade. Upwards of five hundred wooden ships, including some world renown sailing ships were made there. Among the workmen in the area were a number of top quality blacksmith families. These master iron workers and edge toolmakers came to Canada as United Empire Loyalists, American colonists who remained loyal to the British crown after the united States War of Independence. Josiah Fowler was a third generation U.E.L. He opened his first shop in 1860 and in a number of partnerships, was active in St. John as late as 1922. Good specimens of his axes are still located by sharp-eyed collectors. There was a large trade of edge tools between the United States and Canada over eight decades at least, up until World War I." (Klenman 1998, 25).
One of Josiah's descendents, Betty Dunfield, came across some of his records in her reconstruction of the family tree, including some letters he wrote while serving in the American Civil War as a

bugler.

When John Gardiner of the Mystic Seaport Museum visited the Jonesport Wood Co. store (now leveled) in W. Jonesport, Maine, in the late 1970s, he provided similar information about Josiah Fowler. He also indicated that J. Fowler was an important New England area ship's carpenter toolmaker whose adzes were, by oral tradition among Maine shipwrights, considered to be the finest ever made.

The DATM lists Josiah Fowler as working in St. John 1881 - 1920.

The text *St. John the Metropolis of NB* published in 1908 lists on pg. 56-7 that Fowler made tools as early as 1864.

References:

Klenman, Allen. 1998. Josiah Fowler of New Brunswick. *The Chronicle* 51(1): 25.

Brack, H. G. (2008a). *Art of the edge tool*. Pennywheel Press, Hulls Cove, ME.

Links:

http://news.google.com/newspapers?nid=37&dat=19000915&id=jW8DAAAAIBAJ&sjid=tykDAA
AAIBAJ&pg=4231,1180431

http://www.davistownmuseum.org/bioFowler.html

Fray, John S.
Bridgeport, Connecticut, -1859-

Tool Types: Braces and Combination Tools

Identifying Marks: JOHN S. FRAY/SPOFFORDS PAT; J.S.FRAY & CO. BRIDGEPORT CT. USA; J.S.&J. FRAY'S PAT. BRIDGEPORT.Ct.U.S.A.

Remarks: Fray also used "Co." and "& Co." He may have been a part of Fray & Pigg. Patents include the Spofford patent brace on November 1, 1859, a brace attachment from May 11, 1869, a bit brace from August 20, 1872, and a brace on January 8, 1889 (Nelson 1999). The Fray Co. and his patents were bought out by the Stanley Rule and Level Co. in the late 19th century.

Links: http://www.findagrave.com/cgi-bin/fg.cgi?page=gr&GRid=16891438 – Fray's Grave
http://www.sydnassloot.com/Brace/Fray.htm

French & Co., Raymond
Kinneytown (later Seymour and/or Derby), Connecticut, 1844-

Tool Types: Augers, Bits, Chisels, and Plane Irons

Remarks: This company consisted of Raymond French and John and Timothy Dwight. The name was soon changed to Dwights, French & Co. (Nelson 1999).

French, Swift & Co.
Derby, Humphreysville, and Seymour, Connecticut, 1847-1866

Tool Types: Augers

Remarks: This company consisted of Warren French, Charles Swift, Henry B. Beecher, and three other partners. In 1866, Beecher and/or his son F.H. Beecher took it over (Nelson 1999).

French, Walter
Mansfield and Seymour, Connecticut, 1812-1838-

Tool Types: Augers and Bits

Identifying Marks: WALTER FRENCH

Remarks: Walter, a pioneer manufacturer of screw augers and bits, was born January 5, 1781. He moved from Mansfield to Seymour around 1810 but whether he made tools prior to the move is unknown. He managed the Clark Wooster business in Humphreysville, Connecticut before becoming a part of French & Robbins in Westville, Connecticut (Nelson 1999).

Gage, John H.
Nashua, New Hampshire, 1838-1850
Tool Types: Edge Tools and Machinist Tools
Remarks: Gage used Nashua Machine Co.'s machine shop. He was part of Gage, Warner & Whitney from 1851 to 1862 and the first president of Underhill Edge Tool Col from 1852 to at least 1858 (Nelson 1999).

Gage Tool Co.
Vineland, New Jersey, 1883-1919
Tool Types: Planes

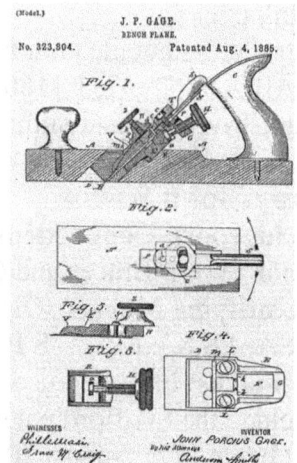

Remarks: John Porcius Gage formed this company and owned it until 1917. In 1919 it was sold to Stanley Rule & Level Co. who continued to use its name. J.P. Gage had plane patents on 4 August 1885, 13 April 1886 and 8 November 1892. The 30 January 1883 patent of David A. Ridges was also used. This company was known for its transitional planes with metal tops and wood bottoms (Nelson 1999). The transitional planes manufactured by this company began the incorporation of varying amounts of metal into the design of the planes themselves. See Carl Bopp's lecture on tracing down this movement.

References:
Aber, R. James. 1978. *Some notes on Gage planes*. CRAFTS of NJ meeting.
Bopp, Carl. 1978. *The Gage Tool Company and the Gage plane*. CRAFTS of NJ meeting transcription by R. James Aber.
United Brotherhood of Carpenters and Joiners of America. (1915). *Carpenter*. United Brotherhood of Carpenters and Joiners of America, Indianapolis, IN.
Links:
http://craftsofnj.org/toolshed/articles/Gage%20Block%20Plane%20by%20Welsh/Gage%20Black%20Plane%20by%20Welsh.htm
http://www.davistownmuseum.org/bioFowler.html

Germantown Tool Works
Philadelphia, Pennsylvania, 1884-1894
Tool Types: Axes, Hammers, Hatchets, Pliers, and Other Edge Tools
Identifying Marks: GERMANTOWN/TOOLWORKS/SOLID CAST STEEL (top and bottom lines form an oval); GERMANTOWN TOOL WORKS//PHILADELPHIA, PA; GERMANTOWN/MASTER BUILDER (in a keystone); G T W (in a keystone)
Remarks: This company's name was later changed to Griffith Tool Works. This company is possibly a brand name used by Yerkes & Plumb, who succeeded Jonathan Yerkes in 1857, Griffith's cited date of establishment (Nelson 1999).

Gilmore, Hiram & Leonard
Claremont, New Hampshire, 1826-1841-
Tool Types: Axes, Edge Tools, and Scythes
Remarks: These brothers were cited by an 1895 source as the Gilmore Edge Tool Works but they are not known to have used that name (Nelson 1999).

Gladwin & Appleton
Chelsea, Massachusetts, 1873-1877
Tool Types: Wood Planes
Identifying Marks: GLADWIN&APPLETON/BOSTON (first line curved)
Remarks: This company consisted of Porter A. Gladwin and Thomas L. Appleton who worked alone and with other partners (Nelson 1999) and was a prolific maker of molding planes, which are frequently encountered in new tool collections (FFLTC).

Goldblatt Tool Co.
Kansas City, Kansas
Tool Types: Axes, Hatchets, and Remodeling Tools
Identifying Marks: GOLDBLATT TOOL CO/KANSAS CITY KAN
Remarks: America's most prolific 20th century manufacturer of high quality masonry tools, the Goldblatt Co. may have started in Kansas City, date unknown, in the late 19th century as a maker of axes and hatchets (Nelson 1999).
Links: www.davistownmuseum.org/PDFsforInventory/WebMaritimeIV_PDF.pdf -- Includes an entry of a lot including Goldblatt tools.
http://goliath.ecnext.com/coms2/merc-compint-0000609903-Goldblatt-Tool-Co.html -- A company listing of a modern Goldblatt Tool Co. in Kansas City.
http://www.davistownmuseum.org/biogoldblatt.html

Goodell Co.
Antrim and Bennington, New Hampshire, 1875-1911
Tool Types: Agricultural Tools, Cutlery, Handles, Household Tools, Leather Tools and Knives, Wrenches, Apple Parers, Peach Parers, Cherry Pitters, Corers, Slicers, Can Openers, Seed Sowers, Pencil Sharpeners, and Others
Identifying Marks: Goodell/ANTRIM.N.H. (Goodell in cursive); GOODELL CO./ANTRIM.NH (sometimes on one line, sometimes with patent dates)
Remarks: This company consisted of David Harvey Goodell and the Woods Cutlery Co. The Bennington plant opened in 1879 but continued to mark tools ANTRIM. They used a number of brand names, including ACME, BONANZA, DANDY, EUREKA, FAMILY BAY STATE, FAMILY CHERRY STONER, NEW LIGHTNING, TURNTABLE, VICTOR, WHITE MOUNTAIN, and WINESAP. Patent dates include E.L. Pratt, Boston, October 6, 1863; E.L. Pratt, August 23, 1864, May 10, 1870; August 4, 1874; April 6, 1880, May 3, 1881, January 5, 1885 (invalid), April 27, 1886; November 6, 1886; November 16, 1886; March 13, 1888; May 24, 1898; and other unspecified patents in 1886 and 1893. They also made tools based on D.H. Whittemore patents on August 10, 1869 and January 11, 1871 (invalid) (Nelson 1999).
Links: http://www.industrialhistory.org/museumwebsite_002.htm -- Goodell Company Listing

Goodell Mfg. Co.
Greenfield, Massachusetts, 1899-1923-
Goodell Bros.
Shelburne Falls and Greenfield, Massachusetts, 1888-1899
Tool Types: Clamps, Drill Chucks, Drills, and Screwdrivers
Identifying Marks: Variations of the maker name, sometimes with Co., a city/cities, and/or patent dates

Goodell Tool Co.
Worcester, Shelburne Falls, and Greenfield, Massachusetts, 1888-1925
Tool Types: Braces, Glass Cutters, Wrenches, and Others
Identifying Marks: GOODELL TOOL CO//SLEBURNE/FALLS MASS.
Remarks: Goodell Bros. company was formed by Albert D. and Henry E. Goodell. Albert sold his interest to Dexter W. Goodell after the company moved from Shelburne Falls to Greenfield in 1892. In 1899, William Pratt bought them out and changed the name to Goodell Pratt Co. Other patents by Goodell Bros. included spiral screwdrivers patented July 22, 1890 and November 17, 1891. Sometime after being bought out in 1899, possibly as late as 1903 when it was incorporated, Henry E. Goodell formed Goodell Mfg. Co. Pratt eventually bought this company out as well in 1923 and merged it with the existing Goodell Pratt. Albert formed the Goodell Tool Co. with his son Frederick after his departure and moved from Worcester to Shelburne Falls in 1893. Albert had patents from July 14, 1868; June 10, 1873; July 29, 1873; and February 1884, some of which were assigned to Millers Falls Co. This company also made braces with 1892 and 1894 patents and a butt gauge with a December 18, 1894 patent. The company was slowly bought out by the Goodell Pratt Co., which owned 50% in 1907 and took it over entirely by 1925 (Nelson 1999).

Goodell Pratt Co.
Greenfield, Massachusetts, 1899-1931
Tool Types: Calipers, Drill Chucks, Drills, Saws, and Wrenches
Identifying Marks: Various arrangements and combinations of the maker name, city, state, and "TOOLSMITHS." Alternately, "G P/Co" in a shield outline.
Remarks: A successor to the Goodell Bros., this company acquired Stratton Bros., Coffin & Leighton Co. and Lavigne Micrometer Co. prior to being acquired themselves by Millers Falls in 1931. Stockpiles of marked tools continued to be retailed a few years after their acquisition. The owner, William Pratt (1867-1946), worked for H.H. Mayhew Co. and Wells Bros. Co. prior to founding Goodell Pratt Co. and was possibly related to Henry L. Pratt, a founder of Millers Falls Co. (Nelson 1999). Albert Goodell was involved in the patenting and manufacture of several drill braces while in collaboration with Goodell Pratt. He formed Goodell Bros. in 1888, one of the original companies that formed Goodell Pratt Co.

Links: http://www.owwm.com/mfgindex/detail.asp?ID=386 -- OWWM entry on Goodell Pratt
http://oldtoolheaven.com/related/goodell-pratt-history.htm -- Old Tool Heaven entry
http://www.sydnassloot.com/Brace/Goodell.htm -- Information on Albert Goodell, a contributor to

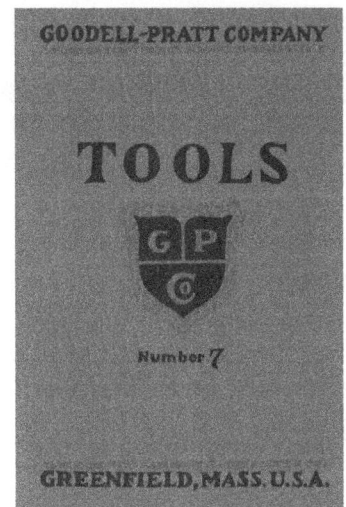

Goodell Pratt.

http://www.davistownmuseum.org/bioGoodelpratt.html
http://www.wkfinetools.com/hUS/boringTools/goodPratt/tools/handDrills/gpHD-5.5Series/gpHD-5.5B/gPrattHD_No5.5B.asp -- An article on some circa 1905 Goodell Pratt drills
http://preservationgreenfieldma.org/placesindustrial.html -- Greenfield Company listings

Goodnow & Wightman
Centerbrook, Connecticut, 1874-1983
Tool Types: Others
Identifying Marks: Various arrangements and combinations of the maker name, city, state, and "TOOLSMITHS." Alternately, "G P/Co" in a shield outline.
Remarks: While they advertise as 'manufacturers,' they may have, in fact, only been dealers. The only known patent to Goodnow is a washer cutter. By 1898, the Goodnow patent washer cutter was being sold by Luther H. Wightman & Co., their successor. The 1882 catalog lists them as "importers, manufacturers and dealers in tools of all kinds for machinists, pattern-makers, carvers, model makers, amateurs, cabinet-makers, jewelers, etc. and materials of all kinds" (Nelson 1999; Goodnow & Wightman [1882]). The catalog does, in fact, contain planes for sale, including Bailey's patent planes. A note in the beginning notes that they make "models and small experimental Machines of all kinds, to order."
References:
Goodnow & Wightman. [1882] n.d. *Price list of Goodnow & Wightman, importers, manufacturers and dealers in tools of all kinds for machinists, pattern-makers, carvers, model makers, amateurs, cabinet makers, jewelers, etc. Boston, Mass.* The Early Trades & Crafts Society and the Mid-West Tool Collector's Association.
Links: http://www.davistownmuseum.org/bioGoodnowWightman.html

Goodspeed & Wyman
Winchendon, Massachusetts, 1826-1876-
Tool Types: Carpentry Tools, household Tools, Lathes, Sewing Machines, Butter Churns, Spool/Bobbin Machines, and Saws
Remarks: G. N. Goodspeed and Harvey Wyman started this company, sometimes adding "Machine Co." The barrel stave cylinder saw they produced was their own invention. At some point after 1876 the name changed to G.N. Goodspeed Co. and later the Goodspeed Machine Co., which persists today (Nelson 1999).

Gray, John
Kingston, Massachusetts, -1849-
Tool Types: Axes, Chisels, and Other Edge Tools
Identifying Marks: J.GRAY/CAST STEEL/KINGSTON (Nelson 1999)

Greenfield Tool Co.
Greenfield, Massachusetts, 1851-1883-
Tool Types: Clamps, Marking

No. 476 Turning Chisel

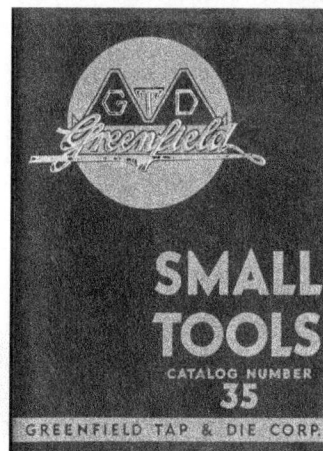

GTD Greenfield

SMALL TOOLS
CATALOG NUMBER 35
GREENFIELD TAP & DIE CORP.

Gauges, Wood Planes, and Other Hand Tools

Identifying Marks: GREENFIELD TOOL Co/GREENFIELD MASS (versions with both straight and curved lines)

Remarks: While best known for its planes, this company, which succeeded the Conway Tool Co. with Alonzo Parker as its agent, manufactured a variety of marking, cutting and slitting gauges and clamps and was reported as working until 1883 making an "iron plane gauge," patented 26 July 1887 by Edward B. Shapardson. This product was still being sold in 1905 (Nelson 1999). This was an extremely prolific company, producing a wide variety of hand tools that can be found in circulation today.

References:
Greenfield Tool Co. [1854] n.d. *Price List of Joiners' Bench Planes and Moulding Tools Manufactured by the Greenfield Tool Company.* Fitzwilliam, NH: Ken Roberts Publishing Co.
Greenfield Tool Co. [1872] 1978. *Illustrated Catalogue and Invoice Price List of Joiners' Bench Planes, Moulding Tools, Handles, Plane Irons, &c., Manufactured by the Greenfield Tool Company.* Fitzwilliam, NH: Ken Roberts Publishing Co.

Greenlee Bros./Greenlee Tool Co.
Chicago and Rockford, Illinois, 1876-1927-
Tool Types: Carpentry Tools
Identifying Marks: ROCKFORD/GREENLEE/ILL. U.S.A.; ROCKFORD ILL/GREENLEE/G (in a diamond)
Remarks: This company was founded by the Greenlee twins, Robert Lemuel and Ralph Stebbins, who were born in PA and moved to Chicago in 1859 to work for the Machine Roller Co. In 1876 they perfected a "hollow chisel" and in 1881 a power ripsaw. "& Co." was added to their name in 1890; in 1904 they moved to Rockford. They dealt primarily in power tools, not offering hand tools until 1910. They acquired the Reliance Edge Tool Works of Youngston, OH, in 1910, the Rockford Bit Co. of Kokomo, IN, in 1916, and the Jackson Mfg. Co. of Jackson, OH, in 1918 (Nelson 1999). Products of note include the Greenlee Hollow Chisel Mortiser, a combination four-sided chisel and rotating bit that allowed square holes to be made in wood; a Greenlee Tie Machining Car, a mobile railroad tie milling machine which gained import in the post-civil war western expansion; and a Self-feed Power Ripsaw, a wildly popular power saw that proved both safer and more effective than existing equivalents. After 1910 they became one of America's most prolific manufacturer of edge and other small hand tools.
References: Greenlee Bros. & Co. 1940. *Greenlee Mortising and Boring Tools.* Rockford, IL: Greenlee Bros. & Co.
Links: http://www.greenlee.com/Company Info/history.shtml -- The modern day Greenlee's site (now Greenlee & Textron), including company history.
http://www.owwm.com/MfgIndex/detail.aspx?id=403 – OWWM entry on Greenlee Bros.
http://www.davistownmuseum.org/bioGreenlee.html

Gregg, Mahlon
Rochester, New York, 1854-1870-
Tool Types: Cooper Tools, Draw Knives, Edge Tools, Agricultural Tools, and Leather Tools
Identifying Marks: M.GREGG/ROCHESTER N.Y.
Remarks: Gregg was part of Gregg & Hamilton from 1866 to 1867 (Nelson 1999).

Griffin, Edmond
E. Bridgewater, Massachusetts, -1849-
Tool Types: Edge Tools (Nelson 1999)

Griffith & Co.
W. Cambridge, Massachusetts, -1849-
Tool Types: Saws
Remarks: This company is unconnected to Charles Griffiths & Co. of Boston as evidenced by directory listings (Nelson 1999).

Griswold & Co., George G.
Chester and Clinton, Connecticut, 1857-1858-
Tool Types: Augers and Bits
Identifying Marks: GEO.G.GRISWOLD & Co.
Remarks: There are some disparities in directory listings regarding whether they were in Clinton or Chester and how early or late they worked. Patents included an auger making process from April 1, 1856 and a November 29, 1864 patent for a grindstone, both listing his home as Chester, though one 1857-1858 directory lists this company in Clinton (Nelson 1999).

Griswold, Charles L.
Chester, Connecticut, 1855-1884
Tool Types: Augers, Awls, Bits, and Other Edge Tools
Remarks: Charles is sometimes confused with George G. Griswold but it is unknown whether they were in any way related. His patents included an auger from May 30, 1865; a gimlet from November 26, 1872; and a gimlet handle patent from September 23, 1873 (Nelson 1999).

Hammacher, Schlemmer & Co.
New York City, New York, 1885-1900-
Tool Types: Adzes, Axes, Pliers, Rules, Wood Planes, and Wrenches
Identifying Marks: H.S.&Co NY; Various configurations of the maker's name, city name or initials, and sometimes a street address.
Remarks: Albert Hammacher and William Schlemmer ran a hardware store, marking a variety of tools. It is unclear whether Schlemmer worked for Hammacher & Co. before becoming a partner in 1885. An existing William Schlemmer & Co. hardware business in 1871 seems to make it likely that he was independent. Some tools they sold with unknown makers include nipper pliers patented 25 September 1893 (invalid date) and called "Mediden" and a piano tuning wrench patented by "Mueller," 3 October 1899 (Nelson 1999). Hammacher, Schlemmer & Co. was a very successful hardware store. By 1912 it had the largest catalog of any hardware dealer on the east coast, totaling 1,112 pages. Their product line eventually included the first pop-up toaster, electric razor, answering machine and microwave oven. They are in business today, carrying lines of giftware.
References:
Hammacher, Schlemmer & Co. n.d. *Piano, Organ and Violin Tools, Catalog No. 142*. Bath, NY: Martin J. Donnelly Antique Tools.
Links: http://www.davistownmuseum.org/bioHammacher.html

Hammond, Charles
Philadelphia, Pennsylvania, 1869-1908-
Tool Types: Adzes, Axes, Hammers, Hatchets, and Other Edge Tools
Identifying Marks: C.HAMMOND/PHILADA (sometimes curved, sometimes on the same line)
Remarks: Hammond added "& Son" to his name, possibly not until after 1900. Brand names include MECHANICS PRIDE on hammers. The mark is sometimes misread "O. HAMMOND" (Nelson 1999).

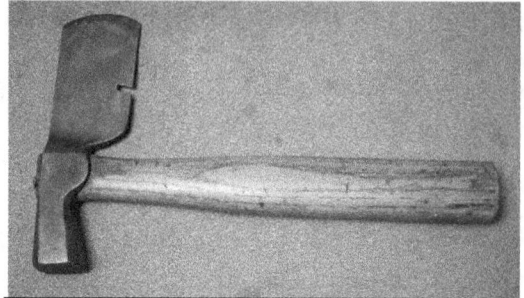

A Hammond hatchet, Davistown Museum MIV Collection ID# 70209T1.

Hannum, Caleb W.
Chester Village (later renamed Huntington), Massachusetts, -1849-1855-
Tool Types: Axes and Chisels
Identifying Marks: C.W.HANNAM/CHESTER VILLAGE; C.W.&J.HANNUM/HUNTINGTON,MA.
Remarks: Caleb W. (May 16, 1810-November 24, 1868) was the song of Caleb Hannum. The "J." in the mark probably refers to his brother John or Joseph (Nelson 1999).

Hardy, David P.
Hebron, New Hampshire, 1872
Tool Types: Edge Tools (Nelson 1999)

Hardy, Ephraim L.
Brookline and Hollis, New Hampshire, 1821-1870
Tool Types: Axes, Coopers' Tools, Draw Knives, and Shaves
Identifying Marks: HARDY
Remarks: Ephraim L. Hardy (October 14, 1801-November 28, 1870) moved from Hollis to Brookline circa 1840. One source says he marked tools E. HARDY but no such tools have been found (Nelson 1999).

Harlow, C. C.
Bridgewater, Massachusetts, -1869-1875-
Tool Types: Augers and Metal Planes
Identifying Marks: C.C. HARLOW MAKER BRIDGEATER,MASS.
Remarks: Products included Russell Phillip's patent plow planes and hollow augers, some of which were marked by Babson & Repplier, possible marketers (Nelson 1999).

Harmon & Co., Bronson
N. Bennington, Vermont, -1848-1854
Tool Types: Rules and Squares
Remarks: Harmon may have worked under his own name prior to 1848 (Nelson 1999).

Harmon, John W.
Boston, Massachusetts, 1860-1907

Tool Types: Levels, Yardsticks, Spirit Levels, Telescopic Sighting Levels, and Rules
Identifying Marks: J.W.HARMON/BOSTON,MASS
Remarks: The telescopic sighting level was made under patents from November 23, 1880 (Harmon's) and January 23, 1883. The former was also used by Grade Level Co. (Nelson 1999).

Hart, L.
Farmington, Connecticut, -1823-
Tool Types: Tinsmiths' Shears (Nelson 1999)

Hart, William
Portsmouth, New Hampshire, 1757-1809
Tool Types: Rules, Scales, Survey and Navigational Instruments, Scales, Gauging Rods, Telescopes, and Other Scientific Instruments
Identifying Marks: WM:HART/FECIT
Remarks: Hart died January 13, 1812 (Nelson 1999).

Hartford Tool Co.
Hartford, Connecticut, 1883-1890
Tool Types: Augers, Dies, Reamers, Threading Tools, Lathe Tools, Machinist Tools, and Others
Remarks: This company was formed by former Pratt & Whitney employees E.G. Parkhurts, M.D. Pratt, J.E. Woodbridge, and Frederick N. Gardner and may have been a Pratt and Whitney subsidiary. They may have marked tools the H.T. Co. patents include a drafting compass from September 28, 1886 by Gardner, later assigned to Pratt & Whitney. They went out of business in 1890 but continued to sell inventory until 1892 (Nelson 1999).

Harvey, H. H. & Co.
Augusta, Maine, and Boston, Massachusetts, 1872-1914
Tool Types: Blacksmithing Tools, Ice Tools, and Stone-working Tools
Identifying Marks: H.HARVEY/AUGUSTA.ME./MANUF'R
Remarks: The "& Co." is inconsistently used. The factory was in Augusta while headquarters and distribution was in Boston (Nelson 1999).
References:
Harvey, H.H. 1896. *H. H. Harvey's special illustrated catalogue and price list for 1896 – 7. Granite, marble, and soft stone workers', blacksmiths' and contractors', hammers and tools, manufactured by him at Augusta, Maine.* Publisher unknown.
Links: http://www.davistownmuseum.org/bioHarvey.html

Haselton, Rufus B.
Groton and Contoocock, New Hamsphire, 1847-1875-
Tool Types: Calipers, Rules, and Squares
Identifying Marks: R.B.HASELTON/GORTON,N.H. (sometimes with an eagle)
Haselton & Son, Rufus B.
Groton, New Hamsphire, -1877-1879
Tool Types: Calipers, Rules, and Squares
Haselton, Hermon R.

Contoocock, New Hamsphire, 1880-1939
Tool Types: Calipers and Rules
Remarks: Rufus B. Haselton, sometimes recorded as Heselton or Hazelton, worked with his son, Hermon, from 1877-1879 but it is unknown how deeply his son was involved. Rufus moved to Contoocock in 1871. Hermon used his father's mark throughout his career—rules marked Contoocock are generally attributed to Hermon while the ones marked Groton are more likely Rufus's (Nelson 1999).

Hathaway, Braddock D.
New Bedford, Massachusetts, -1836-1873-
Tool Types: Axes, Carpenter Tools, Coopers' Tools, and Other Edge Tools
Remarks: Hathaway and his two sons were blacksmiths who may or may not have made wooden planes (or if another unrelated B.D. Hathaway did) (Nelson 1999). The Hathaway family was one of New Bedford's most prolific shipsmiths and edge toolmakers. See *Hand Tools in History* volume 7 for an illustration of a Hathaway tool (Brack 2008a, 37).

Hawes, Silas
Shaftsbury, Vermont, 1814-1828
Tool Types: Squares
Remarks: Markers other than Silas Hawes (most notably Eagle Square Co.) marked their squares with variations of "HAWES PAT." or similar markings. Hawes held a patent on a steel square from 1819 and a May 11, 1814 file patent though it is unknown whether he ever produced the latter (Nelson 1999). Hawes is generally credited with the invention of the framing square. He was the first of a prolific community of Vermont framing square makers.
Links: http://www.bobvila.com/HowTo_Library/The_Carpenter_s_Square-Hand_Tools-A2046.html

Hawkins, William S.
New York City, New York, 1852-1871-
Tool Types: Axes, Coopers' Tools, Draw Knives, Shipsmiths' Tools, and Other Edge Tools
Identifying Marks: HAWKINS/N.Y.
Remarks: Hawkins forged shipsmiths' tools (Nelson 1999).

Hawks, Loomis & Co.
Bennington, Vermont, -1854-1859
Tool Types: Squares
Remarks: This company succeeded Rufus W. Bangs after an 1852 flood destroyed his shop and were absorbed by Eagle Square Co. in 1859 along with several other local steel square makers (Nelson 1999).

Hay-Budden Mfg. Co.
Brooklyn, New York, 1890-1931-
Tool Types: Anvils (Nelson 1999)
Remarks: This company consisted of James Hay and Frederick C. Budden. By 1905, the company claimed to have 100,000 anvils in circulation and they remain prized among blacksmiths.
Links: http://www.fholder.com/Blacksmithing/anvil.htm

Heald, Addison
Hudson and Milford, New Hampshire, 1868-1873
Identifying Marks: ADDISON HEALD/MILFORD.N.H. (name line curved, star between lines)

Heald & Son, Addison
Milford, New Hampshire, 1873-1906
Identifying Marks: A.HEALD&SON/MILFORD NH and variations, sometimes curved, sometimes without the city and state
Tool Types: Coopers' Tools, Shaves, and Wood Planes
Remarks: Addison Heald (February 25, 1817-January 18, 1895) worked in Nashua as A. Heald & Co. circa 1856, as Warren & Heald circa 1860, alone from 1868 to 1873, and then with his son Daniel Milton Heald (born in Ohio January 9, 1852, died October 30, 1929), who continued using the name after Addison's death in 1905. Daniel had patents on a plane iron holding and adjusting device from November 19, 1878 (Nelson 1999).

Heald, Paul
Atkinson, New Hampshire, -1856-1860-
Tool Types: Edge Tools
Remarks: Heald moved from working as a machinist to specializing in edge tools in 1856 (Nelson 1999).

Hedge, Lemuel
Windsor, Vermont, -1830
Tool Types: Rules
Remarks: Hedge was part of Hedge & Ayers in 1813, Pomeroy & Hedge circa 1815 (possibly cabinet makers, possibly tool and rule makers), then moved to Brattleboro, Vermont, where he patented a rule joint on April 22, 1835 used by Morton Clark & Co., and later worked for E.A. Stearns & Co., Morton Clark's successor (Nelson 1999).

Heebner & Sons
Worcester and Lansdale, Pennsylvania, 1840-1926
Identifying Marks: HEEBNER'S; HEEBNER & SONS LANSDALE , PA.
Tool Types: Agricultural Implements and Machinery
Remarks: David S. Heebner (1810-1900), a farmer by trade, took out a loan against his farm in 1840 to begin production of agricultural equipment. His sons, Isaac D., Josiah D., and Jacob D. were all active in the firm, William being the sole owner from 1887. His early threshing machines were well-received by the relatively primitive farming communities in Worcester. They were made to order and took roughly six weeks apiece to construct. In 1862, Isaac and Josiah Heebner joined on and began manufacturing two-wheeled mower/reapers under a Ball patent. By 1877, the firm was shipping horse powers and threshers as far away as Nova Scotia and Prince Edward Island, with inquiries from South America, Bulgaria and Turkey. By 1882, Heebner & Sons sold fifty percent more machinery than the entire business done by all their local competitors put together. The "Little Giant" thresher was especially popular in Maine.
References:

Blase, Francis Jr. 1984. *Heebner & Sons, Pioneers of Farm Machinery in America*. Hatfield, PA: Hatfield Publishing Company.
Links: http://www.davistownmuseum.org/bioHeebner.html

Heller & Bros.
Heller Bros.
Heller Brothers
Heller Tool Co.
Newark, New Jersey, 1866-1955-
Identifying Marks: A standing horse, sometimes shown by a farrier with a rasp; MASTERENCH; MASTERWRENCH; H & B; variations of the horse and farrier mark with the Heller name (sometimes Heller Wagon Co.)
Tool Types: Blacksmith Tools, Files, Hammers, Wrenches, and Others
Remarks: This company used "Heller & Bros." and "Heller Bros." interchangeably from 1866 to 1899, Heller Brothers Company from 1899 to 1955, and Heller Tool Company from 1955 on. It was originally founded by three Heller brothers, Peter J., Lewis B., and Elias G. Jr. as a successor to their father, Elias Heller Sr. This company specialized in farriers' tools (Nelson 1999). America's most prolific manufacturers of farriers' rasps and blacksmiths' hammers.
Links: http://www.davistownmuseum.org/bioHellerBros.html

Hjorth, William
Jamestown, New York, 1896-1903-
Tool Types: Drills, Pliers, and Wrenches
Identifying Marks: Wm HJORTH JAMESTOWN N.Y; HJORTH DRILL//PAT JUNE 11, 96
Remarks: Hjorth patented and produced a wrench from December 15, 1896, a Drill from June 11, 1896 (invalid date), and a pair of "wrench-like" pliers from September 8, 1903. In 1903, he added "& Co." to his name (Nelson 1999).
Links: http://www.alloy-artifacts.com/other-makers-p2.html -- Photo of Hjorth pliers

Hobart, George W. L.
Brookline, New Hampshire, 1855-1881
Tool Types: Edge Tools

Remarks: Hobart was an edge tool blacksmith (Nelson 1999).
Hobbs, C. E.
Barre, Vermont, -1887-1890-
Tool Types: Bushing Hammers and Stone-working Tools
Remarks: Hobbs had a partner named McDonald. An S. E. Hobbs also made bushing hammers—this may have been the same person, a successor, or completely unrelated (Nelson 1999).
Hoe & Co., R.
New York City, New York and Boston, Massachusetts, 1828-1969-
Tool Types: Planes, Saw Tools, Saws, and Others
Identifying Marks: R.HOE & CO./NEW YORK; MANUFACTURED BY/-/R.HOE & CO/NEW YORK.N.Y. (first

line curved upward)

Remarks: Robert Hoe was born in England in 1784 and died in 1833. He was succeeded by his son Richard, who ran the company from 1838 to 1909. In 1843, its name was also recorded as "The Hoe Printing Press & Saw Mfg. Co." Hoe & Co. was America's largest producer of circular saw blades. Harvey Peace's father managed the Hoe saw business for a period and Harvey worked there from 1849 to 1861 (Nelson 1999).

References:

White, Frank G. Messrs. Hoe & Co's printing press manufactory. *The Chronicle* of the Early American Industries Association, Inc. (accessed August 23, 2007) http://findarticles.com/p/articles/mi_qa3983/is_200009/ai_n8917187

R. Hoe & Co. [1855] n.d. *R. Hoe & Co., Manufacturers of Patent Ground Warranted Cast-Steel Saws*. Long Island, NY: Early Trades & Crafts Society.

Links: http://www.davistownmuseum.org/bioHoe.html

Holway, Seth W.
N. Sandwich, Massachusetts, -1860-1871-
Tool Types: Edge Tools
Remarks: Holway lived with Harriet N. Howes, Lewis Howes's widow, and may have taken over the Howes business (Nelson 1999).

Hoole Machine & Engraving Works
New York City and Brooklyn, New York, 1832-1911
Tool Types: Engraving Patterns
Identifying Marks: HOOLE/MACH.&ENG.WK'S./BROOKLYN, N.Y.
Remarks: While the "Est" date is 1832, this name and the Brooklyn address may be post-1900. The Hooles were America's most prolific bookbinding toolmakers, including John R. Hoole and his son, William E. Hoole (Nelson 1999). Prints owing to Hoole design can be found in innumerable books from the 19th and 20th centuries.

References:

Hoole Machine and Engraving Works. [1911] 1985. *Hoole Machine and Engraving Works*. The Special Publications Committee, Midwest Tool Collectors Association.

Links: http://www.davistownmuseum.org/bioHoole.html

Hope & Co.
Providence, Rhode Island, 1868
Tool Types: Engraving Machines
Identifying Marks: "HOPE & Co. PROV. R.I." on one side and "SPRING STEEL" on the other.
Remarks: Hope & Co. made "Engraving machines (Nelson 1999). Very few examples of this short-lived company's endeavors exist.
Links: http://www.davistownmuseum.org/bioHope.html

Hopkins, Richard Henry
Chesterfield and Hinsdale, New Hampshire, 1855-1873
Tool Types: Augers, Bits, Chisels, Draw Knives, Hatchets, Spinning Wheels, and Other Household

Tools

Remarks: Hopkins worked for Hopkins & Pierce from 1868 to 1870, as a part of Wilder & Hopkins from 1870 to 1873, and as a part of Howe & Hopkins, possibly only working alone circa 1855 in Chesterfield. He had a patent on an auger from June 21, 1870 (Nelson 1999).

Hough, Isaac J.
Middletown, Connecticut, -1856-1858-
Tool Types: Tinsmith Tools
Identifying Marks: I.I.HOUGH/MIDDLETOWN.C
Remarks: Hough also may have worked as Hough & Co. (Nelson 1999).

Hovey, William
Boston, Massachusetts, 1830-1833-
Tool Types: Plane Irons
Identifying Marks: CAST STEEL/*W.HOVEY*/WARRANTED (top and bottom lines curved)
Remarks: Hovey had a patent on the plane iron making process used from March 10, 1830, and owned the Mill Dam Foundry, sometimes using its name in lieu of his own (Nelson 1999).

Howard & Co., E.
North Bridgewater and Boston, Massachusetts, -1849-
Tool Types: Cobblers' Tools, Coopers' Tools, Draw Knives, Hammers, Leather Tools, and Other Edge Tools
Identifying Marks: E.HOWARD & CO./N.BRIDGEWATER; E.HOWARD/N.BRIDGEWATER
Remarks: Howard & Fisher and this company are both listed as edge and shoe tools and are probably related. Manter & Blackmer are the successors of both E. Howard and E.S. Morton but whether Howard and Morton worked together at any point is unknown. It is possible another E. Howard & Co. made cooper's crozes (and possibly other tools) in Boston; there may have been multiple E. Howards working in North Bridgewater (Nelson 1999).

Howard, Leonard D.
St. Jonsbury, Vermont, -1867-
Tool Types: Bevels
Remarks: Howard patented a bevel on November 5, 1867, that Star Tool Co. made and reportedly marked with his name, a box opening tool on November 7, 1869 (invalid), and another bevel from February, 1871 (though it's unknown whether the latter two were ever made) (Nelson 1999).

Howe, Joel
Medford, Massachusetts, -1864-1868-
Tool Types: Hammers and Hatchets
Identifying Marks: JOEL HOWE-PATENT-CAST STEEL
Remarks: Howe had a July 8, 1834 patent on a shingling hatchet/hammer that he may or may not have also produced (Nelson 1999).

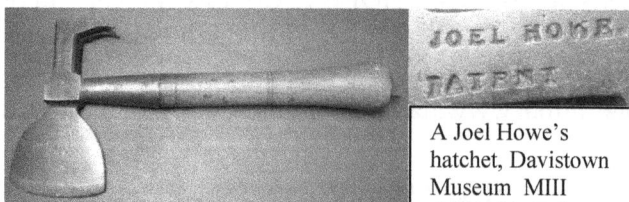

A Joel Howe's hatchet, Davistown Museum MIII Collection ID# TCC2011.

Howes, Lewis
North Sandwich, Massachusetts, -1849-1859
Tool Types: Axes, Chisels, and Other Edge Tools
Identifying Marks: L.HOWES/SANDWICH
Remarks: Howes (September 16, 1812-January 7, 1860), was succeeded by Seth W. Holway (Nelson 1999).

Hubbard & Curtiss Mfg. Co.
Middletown Connecticut, 1871-1874-
Tool Types: Chisels, Rules, Machinist Tools, and Other Edge Tools
Remarks: This company succeeded Warwick Tool Co. and may not have used "Mfg. Co." in ads and marks. A dollar sign was sometimes used in the trademark (Nelson 1999).

Hubbard Hdw. Co.
Middletown, Conencticut, -1868-1874-
Tool Types: Chisels, Rules, Shaves, and Others
Identifying Marks: HUBBARD HARDWARE CO./MIDDLETOWN, CT.; HUBBARD H.W. Co./MIDDLETOWN.CONN (sometimes without second line) (Nelson 1999)

Humphrey Machine Co.
Keene, New Hampshire, -1874-1920
Tool Types: Lathes, Rules, Saws, Log Calipers, Axe Handles, Band Saws, and Machinist Tools
Identifying Marks: J.HUMPHREY/KEENE.N.H.; John Humphrey & Co. (paper label)
Remarks: This company succeeded John Humphrey & Co. Humphrey's name sometimes appeared alone on this company's markings (Nelson 1999).

Humphreysville Mfg. Co.
Seymour, Connecticut, 1852-1904-
Tool Types: Augers and Bits
Identifying Marks: HUMPHREYSVILLE MFG. CO. (sometimes curved)
Remarks: This company was owned by N. Sperry, whose name was sometimes used on their augers and bits (Nelson 1999).

Hunt, Warren
Douglas, Massachusetts, and St. Stephen, New Brunswick 1836-1892
Tool Types: Axes
Identifying Marks: W.HUNT DOUGLAS; W.HUNT/MFD BY/DOUGLAS AXE MFG. CO.
Remarks: Warren worked with his father as Oliver Hunt & Co. from 1815 to 1830, as Warren Hunt & Co. from 1830 to 1836, and his name continued to be used for marking Douglas Axe Mfg. Co. axes from 1836 to 1892. He held a patent on an axe testing machine from September 2, 1856 (Nelson 1999).

Hyde & Co., Isaac Perkins
Southbridge, Massachusetts, 1875-1985
Tool Types: Cutlery, Hammers, Leather Tools, Shaves, and Others

Identifying Marks: I.P.HYDE/SOUTHBRIDGE,MASS

Remarks: Hyde worked as Superintendent of the Theodore Harrington Knife factory from 1872 to 1874 before making this company to specialize in mill knives for cobbling the upper part of shoes. The brand name DIAMOND appeared sometime after 1900, followed by Dexter and Russell-Harrington in the later 1900s. They produced a shoe shave patented by Albert E. Johnson on July 30, 1867 (Nelson 1999). Many specimens of Hyde & Co. shoe knives have been recovered and sold by or are available at the Liberty Tool Co.

References:
Crane, Ellery Bicknell. (1907). *Historic Homes and Institutions and Personal Genealogical and Personal Memoirs of Worcester County, Massachusetts.* The Lewis Publishing Company, New York, NY.

Hynson Tools & Supply Co.
St. Louis, Missouri, 1851-1920

Tool Types: Coopers' Tools and Wood Planes

Identifying Marks: HYNSON/EST./TOOLS & SUPPLY Co/1851/ST. LOUIS (top and bottom line curved, third line widening from the center); I. HYNSON TOOL & SUPPLY CO. ST. LOUIS

Remarks: Augustus R. Hynson first worked for Hall & Hynson in 1851, which is assumed to provide the "EST. 1851" mark (Nelson 1999). Hynson marketed largely to coopers and was the first to produce a barrel heater.

References:
Hynson Tool & Supply Company. [1903] n.d. *52 annual catalogue: Hynson Tool & Supply Company.* Mid-West Tool Collectors Association and the Early American Industries Association.

Links: http://www.davistownmuseum.org/bioHynson.html

Irwin Auger Bit Co.
Wilmington, Ohio, 1885-1991-

Tool Types: Augers and Bits

Identifying Marks: IRWIN; THE IRWIN BIT; STRAIT LINE (with arrow through it), and variations including the name, city, and/or state

Remarks: Charles Irwin ran this company and owned a half interest in William Dimitt's October 21, 884 auger bit that Irwin later improved and patented on April 19, 1887. They began to produce tools other than augers and bits after 1900 (Nelson 1999). Along with the C. E. Jennings Co., Irwin was America's most prolific manufacturer of wood auger bits.

Links: http://www.irwin.com/irwin/consumer/jhtml/irwinHistory.jhtml

Ives, William A.
New Haven, Hamden, and Wallingford, Connecticut, 1868-1971

Tool Types: Augers, Bits, Braces, and Handles

Remarks: Hamden produced tools and used techniques based on 12 or more patents he received between 1868 and 1887. He sold under the names Ives, W.A. Ives, W.A. Ives & Co., and, after 1900, W.A. Ives Mfg. Co. (Nelson 1999).

Jackson, S. Robert
Watertown, New York, 1910-?

Tool Types: Levels

Identifying Marks: DAVIS & COOK/WATERTOWN N.Y.; DAVIS & COOK; MANUFAC'D BY DAVIS & COOK * WATERTOWN N.Y. (sometimes in a circle)

Remarks: Hynson appears to have primarily supplied Davis & Cook with levels and supplies and he notes in the catalog in our collection that he has succeeded them (Nelson 1999; S. Robert Jackson n.d.).

References:

S. Robert Jackson. n.d. *The Level You Need.* Martin J. Donnelly Antique Tools.

Links: http://www.davistownmuseum.org/bioJackson.html

Jackson & Tyler
Baltimore, Maryland, circa 1880

Tool Types: Machinists' Tools, Planing and Saw Mills, Blacksmiths, Pattern Makers, Model Makers, Cabinet Makers, Piano-Forte Makers, Carvers, Molders, and Carpenters' Tools, and Amateurs' supplies

Identifying Marks: Reseller of Baxter Steam Engines, Richardson & Bros., Reynolds & Co., Union Stone Co., Johnson & Bro., Morse Twist Drill Co., Darling, Brown & Sharpe, Prentiss & Co. and Lowell Wrench Co. among others.

Remarks: This company's catalog appears to indicate that they were retailers with no product line of their own. The catalog nonetheless remains a valuable resource for placing dates on certain patented items (Nelson 1999).

Links: http://www.davistownmuseum.org/bioJacksonTyler.html

Jennings & Co., Charles E.
New York City, New York;
Factories in Yalesville, Chester,
and New Haven, Connecticut, and
Port Jervis, New York, 1878-1923

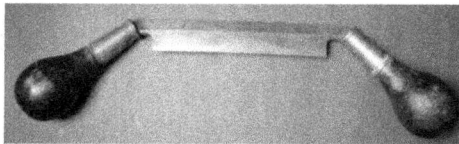

A small C. E. Jennings drawshave, Davistown Museum IR Collection ID# 31808SLP17.

Tool Types: Augers, Bits, Chisels, Draw Knives, Levels, Metal Planes, Saws, and Wood Planes

Identifying Marks: C.E.JENNINGS & CO/N.YORK; CHAS.E.CHANNINGS & Co; J (on an arrowhead); C.E.JENNINGS & CO./NEW YORK.U.S.A. (name line curved)

Remarks: While it is possible Charles E. Jennings worked under his name a few years before adding the "& Co.," it is obvious that the company grew out of acquiring other companies as evidenced by its factories' locations and numerous brand names. In 1901, the company took credit for the following marks: C.E. JENNINGS & CO.; C.E. JENNINGS; JENNINGS & GRIFFIN MFG. CO.; L'HOMMEDIEU; MERRILL & WILDER; WATROUS & CO.; NOBLE'S MFG. CO.; CLARK TOOL CO.; BRATTLEBORO TOOL CO.; PASSAIC MFG. CO.; E.H. TRACY; GEO. S. WILDER; PLINY MERRILL; HINSDALE MFG. CO.; and EXCELSIOR MFG. CO., all

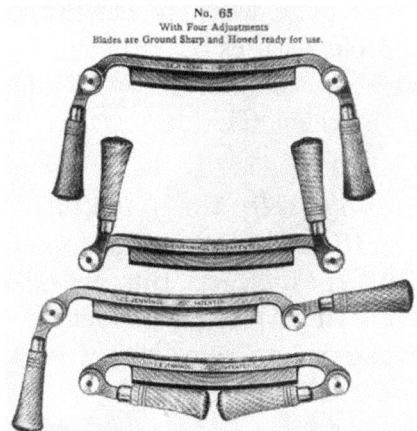

Folding drawknife illustrated in a 1985 reprint of a C. E. Jennings & Co. catalog.

manufactured at the L'Hommedieu Tool Works in Chester, CT. They took credit for the following saw brands: WM. B. ASTEN; JOSEPH HARRIS; THE ORIENT; EXCELSIOR SAW CO.; IMPERIAL; GENEVA SAW WORKS; THE TRANSVAAL; CLARK & CO.; MERRILL'S FAULTLESS; CLARK'S COMBINATIN; J. DOUBLEDAY; HORTON'S; NEW YORK SAW CO.; A.G. MORTON; S. MORRELL; KING; HOWARD & CO.; CLARK'S FARMERS; J.I. SEE; BRIGHTON; J.&G. MFG. CO.; THE MAGNETIC; NEW CENTURY and GRIFFIN'S. ARROW HEAD and LONDON SPRING were also marks they used (Arrowhead appears twice on the cover of the catalog in the Davistown Museum collection) (Nelson 1999). While part of Jennings & Griffin (1885-1900 and on), Jennings also stamped his tools "MERRILL & WILDER".

References:

Jennings, C. E. & Co. Jennings, Charles E. and Griffin, Francis B. 1985. *Price list. no. 13: C. E. Jennings & Co., manufacturers of C. E. Jennings' Arrowhead high grade tools.* Westborough, MA: The Stanley Publishing Co.

Links: http://www.mwtca.org/OTC/ar000021.htm -- Article in *The Gristmill* including notes on C.E. Jennings

http://www.davistownmuseum.org/bioJennings.html

Jennings & Griffin Mfg. Co.
Hinsdale, New Hampshire, and Yalesville, Connecticut, 1883-1900

Tool Types: Augers, Bits, Cutlery, and Other Edge Tools

Identifying Marks: JENNINGS & GRIFFIN/MFG.CO.; J.&G. MFG. CO.

Remarks: This company formed when George S. Wilder was succeeded by Charles E. Jenning and Francis B. Griffin. Wilder continued to run the Hinsdale plant. The initial relationship was between this company and C.E. Jennings & Co., but by 1901 this was only a brand name that persisted at least until 1932. Other brands used included the National Tool & Mfg. Co. and the Sherman Saw Works (Nelson 1999).

References: Jennings, C. E. & Co. Jennings, Charles E. and Griffin, Francis B. 1985. *Price list. no. 13: C. E. Jennings & Co., manufacturers of C. E. Jennings' Arrowhead high grade tools.* Westborough, MA: The Stanley Publishing Co.

Gillespie, Charles Bancroft and Curtis, George Munson. (1906). *An Historical and Pictoral Description of the Town of Meriden, Connecticut and Men Who Have Made It.* Journal Publishing Co., Meriden, CT.

Jennings Mfg. Co., Russell
Deep River and Chester, Connecticut, 1853-1944

Tool Types: Augers, Bits, and Chisels

Identifying Marks: Variations of the name and initial, sometimes with Mfg. Co., sometimes in a circular pattern, sometimes with patent dates and/or model numbers

Remarks: Russell's death was reported both 1885 and March 8, 1888. He was a partner in Stephen Jennings & Co, succeeding Stephen closely associated with the C. E. Jennings Co. Unlike C. E. Jennings, which made all types of hand tools, including drawknives, Russell Jennings specialized in the auger bits still frequently found today in Jennings three drawer wooden bit boxes. He held the January 30, 1855 patent on the famed "Jennings Pattern" bit made by a number of companies, as well as auger-related patents from July 3, 1866 and July 31, 1866. The Chester plant was built in 1865 and both the Chester and Deep River plants ran until 1890 with an office remaining in Deep

River until 1902. Stanley bought them out in 1944 and continued to use the name until 1960 (Nelson 1999).

References:
Jennings, Russell. (1899). *The Russell Jennings Manufacturing Company Catalog.* Reprinted by Astragal Press, Lakeville, MN.
Links: http://www.sydnassloot.com/Brace/RJennings.htm

Jones & Co., Solomon A.
Hartford, Connecticut, 1838-1841
Tool Types: Bevels, Marking Gauges, Rules, and Squares
Identifying Marks: S.A.JONES&CO./HARTFORD-CON.
Remarks: See Willis Thrall & Son (Nelson 1999).
Links: http://www.davistownmuseum.org/bioSAJones.html

Keen Kutter
St. Louis, Missouri, 1870-1940-
Tool Types: Axes, Bits, Chisels, Forks, Levels, Planes, Saw Tools, Squares, and Wrenches
Identifying Marks: Variations of KK, the company name, E.C.SIMMONS, patent dates, and/or a shield
Remarks: This was a brand of the E.C. Simmons Hardware Co. (Nelson 1999) and one of the more collectable of 20th century tool marks.
References:
Heuring, Jerry and Heuring, Elaine. (1990). *E.C. Simmons Keen Kutter collectibles: An illustrated price guide.* Second edition. Collector Books, P. O. Box 3009, Paducah, KY 42001.
The Standard-Simmons Hardware Co. *Simmons mail order want book: E. C. Simmons Keen Kutter cutlery and tools.* The Standard-Simmons Hardware Co., Toledo, OH. Reprinted in 1989 by R. L. Deckeback, Royal Oak, MI.
Sellens, Alvin. (2004). *Keen Kutter Pocket Knives.* Schiffer Publishing, Atglen, PA.
Sellens, Alvin. (2004). *Keen Kutter Planes.* Schiffer Publishing, Atglen, PA.
Simmons Hardware Company. (1930). *E.C. Simmons Keen Kutter cutlery and tools.* Simonds Steel and Saw Co., Fitchburg, MA.
Simmons Hardware Company. (no date). *E. C. Simmons Keen Kutter cutlery and tools.* 21st catalog. Reprinted in 1971 by American Reprints, St. Louis, MO.
Walter, John. (1989). *Antique and collectable Keen Kutter hand tools: 1989 price guide.* The Tool Merchant, Akron, Ohio.
Links: http://www.thckk.org/ -- Keen Kutter Kollector's Klub
http://www.oldandsold.com/articles/article034.shtml -- Collector's Guide to Keen Kutter

Kellogg, James
Amherst, Massachusetts, 1835-1867
Tool Types: Wood Planes
Identifying Marks: J.KELLOGG/AMHERST.MS
Remarks: James was part of Kellogg, Fox & Washburn until 1839, Kellogg & Fox from 1839-1840 and J. Kellogg & Son from 1865-1867. William Kellogg, his son, continued to use his mark after his retirement in 1867 (Nelson 1999). Kellogg's first company was purchased from Eli Dickinson and

became a successful manufacturer of planes. At one point, a portion of Amherst was called "Kelloggville" and was occupied by two of his factories; producing 150 to 200 planes a day; they were often unable to fill all the orders they received. In 1886, the dam supplying power for the factories was washed away and production remained idle for several years.

References: Carpenter, Edward Wilton and Morehouse, Charles Frederick. *The History of the Town of Amherst, Massachusetts*. Excerpt on James Kellogg http://books.google.com/books?id=p_95jmEyb58C&pg=PA294&lpg=PA294&dq=%22james +kellogg%22+wood+planes&source=web&ots=D8g0iCipgs&sig=XMdY1xXAZrCAY8xx_ AvPZ1pPBa8

Kelly Axe Co., William C.
Louisville, Kentucky, and Alexandria, Indiana, 1874-1909
Tool Types: Axes
Remarks: This company operated in Kentucky until circa 1890 when they moved to Indiana, changing their name to Kelly Axe Mfg. Co. They were sometimes recorded as W.C. Kelly & Co. and used the brands FULTON CLIPPER, FULTON SPECIAL, KELLY HAMMER, KELLY CROSSCUT, KELLY PERFECT AXE, THE WORLD KELLY, KELLY STANDARD, and W.C. KELLY FLINT EDGE. This company becomes hard to track after 1900, having been bought out by the American Fork & Hoe Co. whose name was changed to True Temper and was bought out by Barco Industries in 1987 (Nelson 1999).

Kendall & Vose
Windsor, Vermont, 1885-1886-
Tool Types: Shaves
Remarks: This company consisted of Elton P. Kendall and Ambrose S. Vose, who had a patent on a "Windsor Beader" from September 15, 1885 and a try/bevel square from July 6, 1886. The former was produced by this and other companies, the latter may never have been made (Nelson 1999).

Keuffel & Esser Co.
New York City, New York, and Hoboken, New Jersey, 1867-1962-
Tool Types: Levels, Rules, Survey and Drafting Equipment, Transits, and Other Scientific Instruments
Identifying Marks: K+E; K-E; KEUFFEL & ESSER CO.,N.Y.; EXCELSIOR (all sometimes with patent dates, brand names, and others)
Remarks: Wilhelm J.D. Keuffel (July 21, 1838 to October 1, 1908) and Herman Esser (December 30, 1845-April 16, 1908) incorporated this company in 1889 in Hoboken, New Jersey, as a New York City company. They produced under numerous patents, including a slide rule by Edwin Thacher from November 1, 1881, and many issued to Willie L.E. Keuffel (May 10, 1861-May 5, 1952), a nephew of Wilhelm who joined as head of manufacturing in 1884 (Nelson 1999).
References: Stoll, Cliff. (May 2006). When Slide Rules Ruled. *Scientific American*.
Keuffel & Esser Company. (1976). Calendar: *Early American engineers: The early craftsman, his tools, his creations and his design*. K&E Co., Long Island City, NY.
Links: http://www.mccoys-kecatalogs.com/ -- Keuffel & Esser Catalogs
http://www.sphere.bc.ca/test/ke-sliderule.html -- On K&E Slide Rules
http://www.surveyhistory.org/keuffel_&_esser1.htm -- On K&E Transits

Kimball, Caleb Jewett
Milford, Wilton, and Bennington, New Hampshire, 1841-1914-

Tool Types: Axes, Cutlery, Draw Knives, Hoes, Knives, Leather Tools, and Shaves

Remarks: Caleb Jewett Kimball (1817-1896) made draw knives, hoes and shaves using the mark C. J. KIMBALL. His solo working dates are from 1841 to 1872. He started in Milford, NH then moved to Wilton, NH from 1849 to 1851 and finally to Bennington, NH. In 1873, he started working with his son, George Edward Kimball (1842- 1913) as C. J. KIMBALL & SON. Their tool line expanded to include axes, cutlery, knives and leather tools. In 1894, Caleb retired and the company became the Caleb Jewett Kimball Co. in which two of his younger sons were partners (Fred Hastings Kimball b. 1857, d. 1917 and Charles Herbert Kimball b. 1848, d. 1912) along with William H. Odell. In 1894, Wilbur Webster of E. Jaffrey, NH sold his cutlery, knives and leather tools business to the C. J. Kimball Co. (Nelson 1999).

"In the late 1880s the Kimball's started manufactured drawknives for E. C. Simmons, the large hardware firm of St. Louis, MO, which used the brand name KEEN KUTTER." (Smith 1997, 6).

"In August of 1914 Fred Hastings Kimball, at that time president of the firm and only surviving son, announced that the machinery and tools would be moved to the Walden Knife Co. in Walden, N.Y., a company also largely owned by the E. C. Simmons Co. The Kimball firm continued to manufacture drawknives until 1915 when they completed their move to New York." (Smith 1997, 6)

References:

Smith, Roger K. (1997). *Caleb Jewett Kimball: Edge tool maker: Bennington, NH.* The Fine Tool Journal 47(1): 5-7.

Links: http://www.kimballfamily.com/Bios/cjk.htm
http://www.mainememory.net/bin/Detail?ln=16813
http://www.davistownmuseum.org/bioKimball.html

Kinsley Iron & Machine Co., Lyman
Canton, Massachusetts, 1854-1891-

Tool Types: Axes, Crowbars, Tire Benders, Coachmaking Tools, Shovels, and Others

Remarks: Lyman Kinsley succeeded his father Adam and his partner, Johnathan Leonard, who worked in Canton from 1788 to 1854 (but not as toolmakers). This company made a number of non-tool iron items. Ames shovel company had controlling interest by 1863 (Nelson 1999).

Klein & Sons, Mathias
Chicago, Illinois, 1885-1969

Tool Types: Knives, Pliers, and Wrenches

Remarks: The name was changed to "& Sons" in 1885, then changed to Klein Tools Inc. circa 1970 and was still being managed by Mathias's descendents in 1994. Prior to 1900, most of their tools were for servicing telegraphs (Nelson 1999). After 1900, they became one of America's prolific makers of pliers, wire cutters, and other hand tools.

References:

Leonard, John W. and Marquis, Albert Nelson. (1911). *The Book of Chicagoans.* A.N. Marquis and Company, Chicago, IL.

Currey, Josiah Seymour. (1918). *Manufacturing and Wholesale Industries of Chicago.* Thomas B. Poole Company, Chicago, IL.

Mathias Klein & Sons. (1969). *Klein Tools Catalog #112*. Mathias Klein & Sons, Chicago, IL.
Links: http://www.highbeam.com/doc/1G1-18608691.html

Kraeuter & Co.
Newark, New Jersey, 1879-1931-
Tool Types: Calipers, Chisels, Hammers, Leather Tools, Machinist Tools, Pliers, Wrenches, Plumbing Tools, and Others
Identifying Marks: KRAEUTER; KRAEUTER & CO./NEWARK NJ USA
Remarks: August Kraeuter formed this company after working with Heuschkel, Kraeuter & Co. from 1866 to 1871 and Foerster & Kraeuter from 1871 to 1879, though Kraeuter & Co. cites 1860 as its establishment date (Nelson 1999).

Lang & Jacobs
Boston, Massachusetts, 1884-1890
Tool Types: Coopers' Tools
Identifying Marks: LANG & JACOB/BOSTON, MASS
Remarks: The found marked tools are both coopers' hoop drivers. The "Jaco" in the mark is inconsistent with that depicted in the catalog (Nelson 1999).
References:
Lang & Jacobs. [1884] n.d. *Catalogue and price list of Lang & Jacobs' head quarters for coopers' supplies & cooperage stock, Boston, Mass. Coopers' tools & truss hoops a specialty.* Long Island, NY: The Early Trades and Crafts Society.
Links: http://www.davistownmuseum.org/bioLangJacobs.html

L'Hommedieu
Chester, Connecticut, 1809-1849-
Tool Types: Augers and Bits
Identifying Marks: L'HOMMEDIEU
Remarks: L'Hommedieu got his start producing an auger based on a July 31, 1809 patent, later patenting a machine for making augers with R.N. Watrous on July 24, 1838. L'Hommedieu Hardware Co. was listed in Wallingford, Connecticut from 1881 to 1884. C.E. Jennings & Co. advertised in 1901 that they were the sole producers of L'Hommedieu augers and bits (Nelson 1999).

Lambert, George H.
Cambridge, Massachusetts, -1849-1851-
Lambert, Mulliken & Stackpole
Boston, Massachusetts, 1852-1855
Identifying Marks: LAMBERT,MULLIKEN/&/STACKPOLE/BOSTON, MASS. (with eagle)
Tool Types: Levels and Others
Remarks: Lambert worked for Lambert, Mulliken & Stackpole in Boston from 1852 to 1855. They made spirit levels and plumbs and were succeeded by Mulliken & Stackpole (Nelson 1999).

Lamson & Co., Ebenezer G.
Windsor, Vermont, 1850-1877

Tool Types: Cutlery, Guns, Gun Making Machines, Needles, Handles, Household Tools, Machinist Tools, Scythe Snaths, and Others

Remarks: Lamson began making scythe snaths and cutlery, acquired a Robbins & Lawrence Co. gun factory in 1861 while he was part of the Lamson & Goodnow Mfg. Co., but kept the two companies separate. Guns are sometimes marked "L.&G." or "L.G.&Y." (supposedly Lamson, Goodnow and Yale—Yale's association is unknown). Lamson had a patent on a May 10, 1870 machinist's square by David M. Moore but it is unknown whether it was produced (Nelson 1999).

Lamson & Goodnow Mfg. Co.
Shelburne Falls, Massachusetts, Windsor, Vermont, 1851-1982
Tool Types: Augers, Bits, and Cutlery
Remarks: See Ebenezer G. Lamson & Co.

Landers, Frary & Clark
New Britain and Meriden, Connecticut, and New York City, New York (sales office), 1865-1955
Tool Types: Apple Parers, Cleavers, Edge Tools, Household Tools, Ice Tools, Knives, Food Mills, Juicers, Meat Presses, Vegetable Slicers, Scales, Coffee Mills, Glue Pots, and Others
Identifying Marks: L.F.&C.
Remarks: Brand names used by this company include UNIVERSAL, DOMESTIC, SAMUEL LEE, and COLUMBIA. They made items patented June 10, 1873 by A. Turnbull & R.L. Webb; May 1886; October 12, 1897, and April 18, 1899. A "Handy Family Glue Pot" had an 1872 patent and an invalid April 16, 1873 patent. They bought out Meriden Cutlery Co. in 1866 and Humason & Beckley Mfg. Co. in 1912 (Nelson 1999).

Langdon Mitre Box Co.
Millers Falls, Massachusetts, -1876-1882-
Tool Types: Metal Planes and Saw Tools
Identifying Marks: LANGDON MITRE BOX CO./MILLERS FALLS,MASS. (also with "LANGDON" separated)
Remarks: This company and Millers Falls Co. made a shoot board and plane marked with a September 19, 1882 patent date. The company made miter boxes that came with saws (but not the saws themselves). In 1876 they merged with the Millers Falls Co., though the name continued to be used, including on a box branded ACME, patented August 6, 1895 (Nelson 1999).

Leighton, William, W.
Auburn and Manchester, New Hampshire, -1849-1885
Tool Types: Axes, Chisels and Edge Tools
Identifying Marks: LEIGHTON AUBURN N.H.; LEIGHTON / MANCHESTER
Leighton & Co., William W.
Manchester, New Hampshire, 1854
Tool Types: Edge Tools
Leighton & Lufkin
Auburn, New Hampshire, 1856-1860
Tool Types: Edge Tools

Leighton & Son
Manchester, New Hampshire, 1885
Leighton, Charles O.
Manchester, New Hampshire, -1880-1882-
Remarks: Andrew P. Leighton (1793-1882) fathered William P. Leighton (1815-1885) and they worked together in William W. Leighton & Co. Leighton & Son consisted of William P. and his son Charles O. Leighton, who also worked alone. Leighton & Lufkin consisted of William W. Leighton and Jacob Lufkin (1825-1872), who probably worked as part of Underhill, Leighton & Lufkin. William W. Leighton was a prolific and widespread edge toolmaker, working with Underhill, Brown & Leighton circa 1849, Underhill & Leighton circa 1852, W.W. Leighton & Co. circa 1854, Leighton & Lufkin 1856 to 1860, Amoskeag Axe Co. from 1866 to 1879, Underhill Edge Tool Co. circa 1881, for the Manchester Locomotive Works from 1881 to 1883, for S.C. Forsaith & Co. in 1884, probably as part of Leighton & Son circa 1885, and probably as part of Underhill, Leighton & Lufkin (Nelson 1999). According to the Leighton family genealogy, William Leighton is a descendant of Captain William Leighton, originally of Kittery, Maine, born December 26, 1815, and married Susan Hall of Auburn, NH in 1844. He worked as foreman in the Amoskeag Axe Works for eight years and of Underhill Edge Tool Manufactury in Nashua for five. (Jordan 1885, 56).
References:
Jordan, Tristam Frost and Parsons, Usher. (1885). *Leighton Genealogy*.

Little, Charles S.
New York City, New York, -1846-1872-
Identifying Marks: C.S. LITTLE/59 FULTON St NY; C.S.LITTLE&CO/NEW YORK
Little, Charles E.
New York City, New York, -1874-1891-
Tool Types: Carpentry Tools, Copper's Tools, Cutlery, Edge Tools, and Stoneworking Tools
Remarks: Charles S. Little is the Little from Osborn & Little who continued the business. Charles E. succeeded Charles S. (Nelson 1999).

Lockport Edge Tool Co.
Lockport, New York, 1860-1870
Tool Types: Axes, Wood Planes, and Other Edge Tools
Identifying Marks: LOCKPORT/EDGE TOOL CO./LOCKPORT N.Y
Remarks: This company was run by Daniel and Jonas Simmons. Their planes may have been made by someone else (Nelson 1999).

Lovejoy & Webster
Bristol, New Hampshire, -1824-1849-
Tool Types: Axes, Draw Knives, and Edge Tools
Identifying Marks: A.LOVEJOY
Remarks: Lovejoy lived in Alexandria prior to 1823 and may or may not have made tools there. He was a part of Lovejoy & Webster from 1829 on, working as a blacksmith until 1865 but ceasing to specialize in tools after 1849 (Nelson 1999).

Lowell Wrench Co.

Worcester, Massachusetts, 1869-1967

Tool Types: Vises and Wrenches

Identifying Marks: LOWELL WRENCH/Co/PAT.AUG 10/1875/WORCESTER,MASS (curved into an oval)

Remarks: This company was formed by John E. Sinclair and Milton P. Higgins to make D.M. Moore's patented ratchet wrenches, though it didn't have its own shop until circa 1895. Moore's origin in Lowell is the origin of the company name. Sinclair worked at the Worcester Polytechnic Institute until circa 1877 when the institute ceased commercial activities. Higgins went on to found the Norton Co. and Sinclair received a number of patents for Jeweler's vises, bench vises, strap wrenches, and improved ratchet wrenches which were eventually made on a subcontract basis by a Mr. Ballard and a Mr. Pollard. The name changed to the Lowell Corp. in 1967, which persists today (Nelson 1999).

Lowentraut, Peter

Newark, New Jersey, -1869-1894-

Tool Types: Calipers, Dividers, Hammers, Leather Tools, Shaves, Wrenches, Plumbers' Tools, Box Scrapers, Punches, and Others

Identifying Marks: P. LOWENTRAUT/NEWARK N.J.; P.L./MFG.CO. (in a diamond)

Remarks: Lowentraut was using "Mfg. Co." at the end of his name by 1905. His best known product is a wrench brace, patented December 4, 1895 by S.J. Johnston of Leesburg, Virginia (Nelson 1999).

Lufkin Rule Co.

Cleveland, Ohio, and Saginaw, Michigan, 1885-1967

Tool Types: Rules

Identifying Marks: Variations of the maker name and city/state, the maker name only or "Lufkin" with various brand names, model numbers, etc.

Remarks: Originally E.T. Lufkin in the Lufkin Board & Log Rule Mfg. Co. in Cleveland, this company's name was changed when bought out by four members of the Morley Bros. Co. in Saginaw (Nelson 1999). Edward Lufkin owned the patent on a "Board Measure," April 7, 1874, which appeared in the 1890-1891 catalog. America's largest manufacturer of measuring tools for logging and timber harvesting, Lufkin measuring tools are also commonly encountered in modern workshops and collections.

References:

The Lufkin Rule Co. (1888) *Lufkin Measuring Instruments, Exerpts from Trade catalogues, 1888 to 1940, Documentary and Arrangement by Kenneth D. Roberts.* Cleveland, OH: Clark-Briton Printing Co.

No author. (1982). Lufkin's History. *Cooper Canada Newsletter.* 4(4). Reprinted by Kenneth D. Roberts Publishing Co.

The Lufkin Rule Co. (1888). *Lufkin Compendium of Early Catalogs.* Reprinted by Kenneth D. Roberts, 1976.

Links: http://www.skowheganwoodenrule.com/history.htm
http://www.davistownmuseum.org/bioLufkin.html

Machinists Tool Co.
Providence, Rhode Island, circa 1868
Tool Types: Lathe Tool Holders
Identifying Marks: MACHINISTS TOOL CO. PROV RI PATENTED MAY 26, 1868
Remarks: This company is one of the many tool manufacturers of the area at the time. Currently, the Davistown Museum collection includes a lathe tool holder, patented May 26, 1868. It appears to be drop-forged iron.
Links: http://www.davistownmuseum.org/bioMachinistool.html

Mack & Co.
Rochester, New York, 1875-1940
Tool Types: Coopers' Tools, Wood Planes, and Other Edge Tools
Remarks: W.R. and R.L. Mack were partners with David R. Barton & Co. until they took it over and changed the name, later acquiring the D.R. Barton Tool Co., continuing to use its "D.R. BARTON -1832" trademark (Nelson 1999).

Mann Co., David M.
Lincoln, Massachusetts
Tool Types: Engine Rules
Remarks: An engine rule made by the David W. Mann Co. was probably used by the L. S. Starrett Co. for checking their calibrations on precision tools such as Vernier calipers up until the mid-1950s. During the late 1950s and early 1960s these manually operated engine rules became obsolete due to the advent of new photoengraving techniques.

The following description came with the tools pictured: "manually operated grid ruling engine, used for checking scales." "This instrument consists of a base with supports for an overhead bridge slide. Beneath the bridge a precision case and stage motion is bolted to the base casting. The bridge runs perpendicular to the precision stage thus a tool mounted on the bridge motion can scribe lines substrates born by the stage. Movement of the bridge (ruling stroke) is 150 mm. This is controlled by turning a hand wheel attached to a 10 lead screw - 1 mm piten giving 10 mm motion per turn. Precision stage 200 mm travel metered by a metric lead screw. Intervals are hand set using a dial graduated in 1000ths of a millimeter." "This is a prototype instrument, rough finished and the bridge is fabricated from several pieces. This instrument was not intended for sale so was very rudiment in construction. The automatic feature consists of a fabricated cross head motion for the bridge and a pawl arm to turn a rotature plate attached to the precision stage was 8 inches travel metered by a 20 thread per inch lead screw." "Base, 500 mm travel 'scale checker'. This instrument provides a stage approx. 500 mm long on supporting glass scales for inter comparator micro-processing."

The Davistown Museum owns one David W. Mann Co. engine rule. It is located in the Banks Garage next door to the museum. Its description is in the IR Collection listing. Two other engine rules are for sale at the Liberty Tool Co.

References:

Palmer, Christopher. 2002. *Diffraction Grating Handboook.* Rochester, NY: Richardson Gratings.
Semiconductor Magazine. Great Moments in Our Industry Become Defining Moments in Information Technology. *Semiconductor Magazine* 1(7).
Stowsky, Jay. 1987. The Weakest Link: Semiconductor Production Equipment, Linkages, and the Limits to International Trade, BRIE Working Paper #27. Berkeley Roundtable on the International

Economy (BRIE), University of California, Berkeley.
Links: http://gratings.newport.com/library/handbook/chapter3.asp - Christopher Palmer's
Diffraction Grating Handbook.
http://brie.berkeley.edu/publications/WP%2027.pdf
http://www.davistownmuseum.org/bioMann.html

Mann, Harvey
Bellefonte, Pennsylvania, 1834-1870-
Tool Types: Axes
Remarks: William Mann Sr. fathered Harvey Mann (July 2, 1804 – June 4, 1870) and William Jr.,
who worked together from 1825 to 1833 as William & Harvey Mann, later moving to Bellefonte and
working alone. Harvey had axe patents from June 3, 1862 and August 17, 1869. His nephew, J.
Fearon Mann, apparently operated the Harvey Mann axe factory, later joining American Axe & Tool
Co, possibly using his father's name (Nelson 1999).

Marble Safety Axe Co.
Gladstone, Michigan, -1898-1911
Tool Types: Axes, Hatchets, Guns, and Knives
Identifying Marks: MARBLES/GLADSTONE, MICH. USA
Remarks: W.L. Marble, the holder of an 1898 hatchet patent, is assumed to be the Marble of this
company. In 1911 they changed their name to Marble Arms & Mfg. Co., which made a hatchet
based on an 1883 patent, indicating a possible earlier starting date (Nelson 1999). This company
began making guns in 1908, including interchangeable barrel and stock break-action designs.
Marble's axes and edge tools are among the most sought after of all 20th century American
toolmakers.
References: Sawyer, Charles Winthrop. (1920). *Firearms in American History*. The Cornhill
Company, Boston, MA.
Links: http://www.marblesoutdoors.com/index.html
http://www.knifeworld.com/ahuyeofsahub.html -- A Hundred Years of Safety Hunters

Maydole, David
Norwich and Lebanon, New York, 1828-1877
Tool Types: Hammers and Edge Tools
Identifying Marks: MAYDOLE; D.MAYDOLE
Remarks: David Maydole (January 27, 1807 – October 14, 1892) was a prolific and difficult to
track individual whose hammers were not always marked by the company under which he was
working. From 1828-1830, he probably worked with his brother David as a blacksmith. In 1830 he
worked for Gardner & Abbott in Lebanon, NY, then working with a David Abbott (possibly the
same Abbott) from circa 1831 to 1833. In 1834 he went back to Eaton and it is unclear what he did
until 1840 when he formed Maydole and Ray in Norwich, New York until 1847, when he invented
(but did not patent) the adze-eye hammer, a lasting industry standard. In 1845, he had created his
own company, Maydole Co., which inducted N.B. Hale as a partner from 1851 to 1854. In 1861,
David changed the name of his company to David Maydole & Co. and worked with his son-in-law,
Charles H. Merritt. The business was sold in 1931 and the name was changed to Maydole Tool Co.
(Nelson 1999).

References:
Parton, James. (1917). *A captain of industry: The story of David Maydole: Inventor of the adz-eye hammer: To which is added a catalog of the principal varieties of hammers made by The David Maydole Hammer Company.* The David Maydole Hammer Co., Norwich, NY.
Links: http://www.usgennet.org/usa/ny/county/chenango/1898-m.htm -- Biography of Maydole

McKnight, G. L.
Worcester, Massachusetts, 1867
Tool Types: Calipers and Dividers
Remarks: McKnight made a combination divider and caliper he patented April 30, 1867 and May 28, 1867 (Nelson 1999).

Mead, Charles L.
Brattleboro, Vermont, 1861-1863
Tool Types: Levels and Rules
Remarks: Charles L. Mead (January 31, 1833-August 19, 1899) invested in E.A. Stearns & Co. in 1857 and advertised as the successor in January 1861, then sold the business to Stanley in 1863, worked for them after the Civil War, and was President from circa 1884 to 1899. He had three patents assigned to Stanley from January 20, 1885, April 10, 188, and (with Justus A. Traut) February 28, 1888 (Nelson 1999).

Meriden Cutlery Co.
Meriden, Connecticut, 1855-1870
Tool Types: Cutlery and Knives
Remarks: At least two Clark brothers, later involved in the Higganum Mfg. Co., were involved in this company. One source says they worked up to 1870 while another says they were bought out by Landers, Frary & Clark in 1866 (Nelson 1999). One of America's most prolific manufactures of household cutlery.

Merrick, J.
Springfield, Massachusetts 1845-46
Tool Types: Adjustable Wrenches
Remarks: Recent research by Herb Page and others suggests the J. Merrick pattern monkey wrench is the prototype of the Coes Adjustable wrench. Its production may be earlier than 1845 and run possibly from the 1820s.

Merrick, Solyman
Springfield, Massachusetts, 1834-1835-
Tool Types: Hole Punches and Wrenches
Identifying Marks: MERRICK'S/PATENT/SPRINGFIELD
Remarks: Solyman Merrick had wrench patents from April 18, 1834 and August 1835, at least one of which was made by Stephen C. Bemis & Co. It is unknown whether he ever produced them himself (Nelson 1999). Solyman is generally credited with the invention of the first adjustable wrench, as well as the hole punch.
Links: http://ourpluralhistory.stcc.edu/industrial/innovators.html

Merrill & Wilder
Hinsdale, New Hampshire, 1860-1901
Tool Types: Chisels, Draw Knives, and Plane Irons
Identifying Marks: Merrill & Wilder (sometimes with a Buck's head)
Remarks: This company consisted of Pliny Merrill and his nephew George S. Wilder. Wilder later worked alone and was absorbed by the Jennings & Griffin Mfg. Co. in 1883, who began using the Merrill & Wilder mark again, producing many more tools than the original partners. 1860 is an approximate date as they were never listed in a directory together (Nelson 1999).

Merrill, Pliny
Hinsdale, New Hampshire, -1844-1868-
Tool Types: Chisels and Draw Knives
Remarks: Pliny Merrill (sometimes recorded "Pliney") was Pardon-Haynes Merrill's brother and George S. Wilder's uncle. He was a founder and blacksmith who specialized in making tools, working as P. Merrill & Co. circa 1856, in Merrill & Wilder circa 1860, and having C. E. Jennings produce his tools by 1901 under his name. Wilder was still calling his chisels "Merrill chisels" but it is likely he made them on his own (Nelson 1999).
Links: http://www.mwtca.org/the-gristmill/sample-articles/96-notes-on-new-england-edge-tool-makers-i-edge-tool-makers-hinsdale-nh-1840-1900.html

Merritt, Charles H.
S. Scituate, Massachusetts, 1850-1893
Tool Types: Wood Planes
Remarks: Charles H. Merritt was a part of Tolman & Merritt from 1864 to 1880 (Nelson 1999).

Merritt, James
South Scituate and Hanover, Massachusetts, -1870-1878-
Tool Types: Wood Planes
Identifying Marks: J.MERRITT/HANOVER/MASS.
Remarks: Merritt's move from S. Scituate to Hanover was sometime between 1871 and 1877 (Nelson 1999).

Millers Falls Company
Millers Falls, Massachusetts, -1870-1931
Tool Types: Anvils, Bits, Boring Machines, Braces, Drills, Handles, Levels, Metal Planes, Saws, Shaves, Vices, and Others
Identifying Marks: Variations and combinations of "M.F. CO.," "MILLERS FALLS TOOLS," the company name, city or state, patent dates, model numbers, etc.
Remarks: This company was formerly the Levi J. Gunn and Charles H. Amidon Co. In 1872

A typical Miller Falls egg beater drill, Davistown Museum IR Collection ID# 112400T1.

they dropped the Mfg. to become Miller's Falls Co., which lasted from 1872 to 1931 (Nelson 1999). One of the most prolific of New England's late 19th and 20th century toolmakers, their bit braces and hand drills are commonly encountered in workshops and collections (FFLTC).

References:
Millers Falls Co. (April 15, 1878). Catalog: *Millers Falls Co., Millers Falls, Mass., No. 74 Chambers Street, NY*. New York City, NY: Millers Falls Co. Reprinted in 1992 by Philip J. Whitby
Millers Falls Co. (1887). *Catalog*. Reprinted in 1981 by Astragal Press.
Reeder, Layla. Development of modern bench planes: Millers Falls - the "Buck Rogers" planes. *Tool Talk*.
Links: http://oldtoolheaven.com/ -- Information and photographs by Randy Roeder.
http://rosetools.bizland.com/id82.html
http://www.davistownmuseum.org/bioMillersFalls.htm

Mix & Co.
Cheshire, Connecticut, 1820
Tool Types: Chisels and Draw Knives
Remarks: This could be one or more makers as marks on chisels include C.I. Mix & Co., T.I. Mix with no city and a draw knife marked Cheshire (Nelson 1999). Mix edge tools, especially timber framing tools, are occasionally encountered by the Liberty Tool Co. (FFLTC).

Monhagen Saw Works
Middletown, New York, -1860-
Tool Types: Saws
Remarks: A catalog from this company is dated 1860 and lists a number of patented cast steel ground saws, including circular and handsaws.
References:
Monhagen Saw Works. (1860). *Illustrated Price List.* The "Press" Printing Establishment, Middletown, NY.

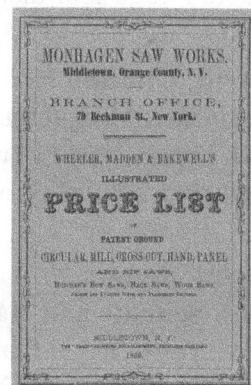

Morris, Ezekiel
Little Falls and Baldwinsville, New York, 1835-1869
Tool Types: Axes, Chisels, Draw Knives, and Other Edge Tools
Identifying Marks: Merrill & Wilder (sometimes with a Buck's head)
Morris, Henry D.
Baldwinsville, New York, 1869-1879
Tool Types: Axes and Edge Tools
Morris, John Roseberry
Jewel City, Kansas, 1898-1905-
Tool Types: Fencing Tools and Pliers
Remarks: Ezekiel (February 29, 1804-October 17, 1869) moved from Little Falls to Balsdwinsville in 1850 after being a part of Windsor & Morr. Henry D. Morris, his son, succeeded him, after working on his own. Henry had an axe head shaping machine patent from January 11, 1870 and, possibly, a similar 1869 patent. In 1879 he moved to California but it is unknown whether he worked there. John R. Morris (May 17, 1863-September 22, 1914) was born in Pennsylvania. His patents included a combination fencing tool patented July 19, 1898, March 7, 1899, and March 14, 1905.

The 1905 patent was produced by Marshalltown Drop Forge after he joined them in 1907 but it is unknown whether he made it prior to 1907. He was president of Waterloo Drop Forge Co. by 1911 (Nelson 1999).

Morse Twist Drill & Machine Co.
New Bedford, Massachusetts, 1865-1990
Tool Types: Bits, Dies, Drill Chucks, and Drills
Identifying Marks: MORSE; M.T.D.&M.Co.
Remarks: The Morse Twist Drill & Machine Company was founded by Stephen A. Morse and located in New Bedford, MA from 1864 to 1990. They made bits, dies, drill chucks and drills. They used the marks MORSE and M.T.D.&M.Co. Their name was also recorded as Morse Twist Drill & Mfg. Co. (Nelson 1999).

The following information is from a *New Bedford Sunday Standard-Times* 1964 article: "Stephen Morse developed the idea of creating a twisted drill consisting of two parallel spiral grooves with a straight cutting edge. Prior to this, drills were made from a flat piece that was pointed and sharp. He began manufacturing drills in October of 1861 with a small shop. His original patent, No. 38119, is dated April 7, 1863." You can see this patent on the DATAMP website (http://www.datamp.org/displayPatent.php?number=38119&type=UT).

Listing of companies purchased by Morse:

1871	**American Standard Tool Company**	Danbury, CT	Listed also on Morse Cutting Tools website history page
1871	**Standard Tool Company**	Newark, NJ	Sayer lists this company (looks like it may be an error as it seems to combine the one above and below).
--	**Manhattan Fire Arms Company**	Newark, NJ	Listed on Morse Cutting Tools website history page
1874	**New York Tap and Die Company**	Bridgeport	Listed also on Morse Cutting Tools website history page
1879	**New York Twisted Drill Company**	Brooklyn	
1885	**Rockford Twist Drill and Patent Company**	Rockford, IL	
1897	**T & B Tool Company**	Danbury, CT	Listed also on Morse Cutting Tools website history page

In 1923 the **Van Norman Machine Company** of Springfield, MA, purchased the Morse company. In 1964, **Universal American Corp.** was the parent company.

The following information is from Sayer's *New Bedford,* pg. 253-6: "Stephen Morse's original shop of 1861 was in East Bridgewater. He moved to New Bedford in June of 1864 because he was able to acquire enough interested capital to set up a new shop there." According to the Morse Cutting Tools website: "The current owners, a group of American investors, purchased the company from a Scottish manufacturing concern, and are committed to upholding the Morse reputation for high-quality, American-made cutting tools." America's most prolific manufacturer of machinists' twist drills.
References:

No author. 1872. The Morse twist-drill. *The Manufacturer and Builder* 4(4): 78-79.

No author. 1893. The Morse Twist Drill Co. and its products. *The Manufacturer and Builder* 25(12): 269.

Morse Twist Drill & Machine Co. 1935. *Machinist's practical guide: The Morse Twist Drill and Machine Co.: Incorporated 1864: Makers of twist drills, reamers, milling cutters, taps, dies, sockets, gauges, chucks, machinery and machinists' tools.* New Bedford, MA: Morse Twist Drill & Machine Co.

No author. 1964. 100 years of leadership in metal cutting. *The New Bedford Sunday Standard-Times*, June 21.

Sayer, William L. Ed. 1889. *New Bedford: Massachusetts: Its history, industries, institutions, and attractions.* The Board of Trade.

Links: http://www.flickr.com/photos/nbwm/sets/72157612474032637/ -- New Bedford Museum's flickr gallery of Morse factory photographs

http://www.davistownmuseum.org/bioMorseTwist.html

Morss, Joab
Philadelphia, Pennsylvania, -1867-1879
Tool Types: Levels, Saws, and Squares
Remarks: Joab Morss (January 8, 1840-February 13, 1879), son of Thomas L. Morss, belonged to Disston & Morss, having married a niece of Disston's. He held four tool-related patents but is not known to have produced on his own (Nelson 1999).

Newman, Andrew W.
Roxbury, Massachusetts, -1847-1852-
Tool Types: Adzes and Axes
Identifying Marks: UNDERHILL/CAST STEEL/ROXBURY/NEWMAN
Remarks: Newman was a blacksmith making cooper tools whose relation to the Underhill name is unknown (Nelson 1999).

Nicholson File Company
Providence 1, Rhode Island, 1864-1972
Tool Types: Files and Rasps
Identifying Marks: Variations of full name and city/state, "Nicholson," brand names, patent dates, etc.
Remarks: William T. Nicholson founded this company after inventing a file cutting machine, but in actuality was its agent Vice President. A series of buyouts followed, including the New American File Co. in 1890, the Great Western File Co. circa 1893 and the McClellan File Co. circa 1898. After 1900, numerous other companies were acquired, including M. Buckley & Co.; Eagle File Co.; Kearny & Foot; Arcade File Co.; J. Barton Smith Co.; Globe File Mfg. Co.; Mechanic's Star File Mfg. Co.; Toronto File Works; and G&H Barnett Co. Some of the company brand names persisted after the buyouts. Nicholson acquired file patents on 22 September 1864, 10 January 1865, 11 September 1866, 12 June 1877, 12 February 1878 and 4 June 1878. 18 February 1876 was patent-marked by the company (later declared invalid) and 1 January 1878. In 1972, the company was bought out by Cooper Industries, though the brand name persists (Nelson 1999). Nicholson File Company published a treatise on files and rasps, included in the bibliographical listing below.

America's largest producer of hand files.

References:

Nicholson File Company. 1878. *A treatise on files and rasps: Descriptive and illustrated: For the use of master mechanics, dealers, &c. in which the kinds of files in most common use, and the newest and most approved special tools connected therewith, are described -- giving some of their principal uses. With a description of the process of manufacture, and a few hints on the use and care of the file.* Reprinted by the Early American Industries Association, 1983.

Nicholson File Company 1945. Nicholson Files and Rasps and X.F. Swiss Pattern Files. Providence, RI: Nicholson File Company.

Nicolson File Company. 1956. *File filosophy and how to get the most out of files (-being a brief account of the history, manufacture, variety and uses of files in general.)* Twentieth Edition. Providence, RI: Nicolson File Company.

Links: http://www.cooperhandtools.com/brands/nicholson_files/index.cfm -- Files currently available from Nicholson under Cooper Tools.

http://www.rihs.org/mssinv/Mss587.htm

http://www.davistownmuseum.org/bionicholson.html

North Bros. Mfg. Co.
Philadelphia, Pennsylvania, 1880-1946

Tool Types: Combination Tools, Drills, Household Tools, and Screwdrivers

Identifying Marks: CROWN; YANKEE; Configurations of the company name including city/state, patent dates, brand names, etc.

Remarks: From 1880-1887, S.G. and R.H. North ran an iron foundry under the name "North Bros." In 1887, they incorporated, adding Mfg. Co. to their name. They bought out the American Machine Co. in 1892 and the Shepard Hdw. Co. of Buffalo, NY in 1893, adding ice cream freezers, ice shaves, egg beaters, flooting machines and other household tools to their offered inventory, including a fluting machine patented 2 November 1875 by Herman Albrecht and a similar tool with a 3 July 1887 patent. Their screwdrivers were based on an 1895 patent of Zachary Furbish as well as a 2 November 1897 patent and several post-1900 patents. They continued to use the American Machine Co. "CROWN" brand. A combination tool marked NORTH MFG. CO./PHILAD. is attributed to this company (Nelson 1999). The "YANKEE" brand was apparently used by Stanley Rule & Level Co. after they bought out North Bros. Mfg. Co. America's most prolific manufacturer of push drills. See Stanley Level & Rule Co.

References:

North Bros. Manufacturing Co. [1908] 1985. *"Yankee" Tool book describing, with illustrations, some up-to-date labor saving tools. More especially: ratchet screw drivers, spiral screw drivers, automatic or push drills, breast and hand drills etc., etc.* Philadelphia, PA: O.M. Ramsey and Philip Whitby of the Midwest Tool Collectors Association.

North Bros. Manufacturing Co. [1912] 1988. *Catalogue of "Yankee" tools ice cream freezers etc., etc.* The Mid-West Tool Collectors Association.

Links: http://www.sydnassloot.com/Brace/Northb.htm

http://www.davistownmuseum.org/bioNorth.html

http://www.davistownmuseum.org/bioFurbish.htm -- to find more information on "Yankee" screwdrivers

North, Edmund
1855-1856-
North, Jedediah
-1810-1824
Identifying Marks: J.NORTH
North, Jedediah & Edmund
-1824-1854
Identifying Marks: J. & E. NORTH/ BERLIN/ CONN
North, Levi
-1782-
Berlin, Connecticut
Tool Types: Tinsmith Tools and Others
Remarks: Levi made tools and nails and may have been related to Jedediah and Edmund North. Edmund (1797-1874) is likely the reported E. North and E. North Mfg. Co. from this era and worked as part of J.&E. North from 1824 to 1854; his brother Jedediah (1789-1855) died the next year. This cooperative effort was also reported as J.&E. North Mfg. Co. There also was an E. North reported working 1816 to 1823 (Nelson 1999).

Northfield Knife Co.
Northfield, Connecticut, 1858-1929
Tool Types: Cutlery, Shears, Razors, and Knives
Remarks: This company also used the brand name UN-X-LD and served for agents of Frary Cutlery Co. They bought out American Knife Co. of Thomaston, Connecticut in 1865, and Excelsior Knife Co. of Torrington, Connecticut in 1885, and were subsequently bought out themselves by Clark Bros. Cutlery in St. Louis, Missouri in 1929 (Nelson 1999).

North Wayne Tool Co.
Hallowell and Oakland, Maine, 1879-1969
Tool Types: Axes, Farm Tools, Scythes, and Others
Identifying Marks: LITTLE GIANT (possibly only after 1900); NO. WAYNE TOOL CO/OAKLAND.ME.U.S.A.; NORTH WAYNE TOOL CO./OAKLAND, MAINE
Remarks: Charles W. Tilden, Joseph E. Bodwell and Wiliston Jennings founded this company and moved to Oakland circa 1900--their original location has been reported as Hallowell, West Waterville, and Wayne. (Nelson 1999) Special thanks to John P. Miller, President of the Wayne Historical Society for furnishing us with a copy of Kallop's history of the North Wayne Tool Co. This text gives a comprehensive history of this company and the many names it used and people who were involved with it. It states, "In the *Maine Register* the tool company's presence in North Wayne is first noted in 1881, when it is identified at Bodwell & Harvey, edge-tools. Not until the year following and thereafter is the published listing identified by company name. Whatever was the reason for the initial listing with personal rather than company name, it nevertheless leads to speculation on the complex role of William Harvey in these various transactions, and his emergence as an apparently equal partner with Joseph R. Bodwell." (Kallop 2003, 77). "At the opening of the new century the numerous firms manufacturing edge tools during much of the 19th century were reduced to three; the Dunn Edge Tool Company, Emerson & Stevens, and the American Axe and Tool Company. The last was to be out of business soon after the century began, leaving only two,

but in 1907 they were joined by the King Axe Company. With an earlier existence as King & Messer, the company continued under its new name until 1922 when it was sold to others, then some twenty years later was resurrected and survived for a brief time as King Axe and Tool Company. With a far shorter lifetime is identified in 1906 still another newcomer to the list -- William Harvey & Sons." (Kallop 2003, 109). "In 1904, sharing a page with four others whose business addresses are in either Hallowell or Gardiner, is the North Wayne Tool Company. Identified as Manufacturers of Agricultural Edge Tools, the company's products are named under the heading Specialties: *C. C. Brooks' Bread Knives, Corn Hooks, Hay Knives and Hoes. C. C. Brooks' little Giant Scythes. C. C. Brooks' Be Ve Be Scythes. H. S. Earle's Little Giant Grass Hooks. H. S. Earle's Corn Knives. Hand Made Axes of all Patterns. Lefavour's Favorite Weeders."* (Kallop 2003, 110). The most prolific of all Maine agricultural edge toolmakers, this company is one of the few Maine toolmakers reported in this section. For a comprehensive listing of Maine toolmakers before 1900, see volume 10, *Registry of Maine Toolmakers* (Brack 2008d).

References:
Kallop, Edward L. Jr. 2003. *A history of the North Wayne Tool Co. manufacturers of axes, corn hooks, scythes and hay knives.* Wayne, ME: Wayne Historical Society.
Links: http://www.davistownmuseum.org/bioNorthWayne.html

Norton Emery Wheel Co.
Worcester, Massachusetts, 1885-1990
Tool Types: Grinders, Grindstones, Emery Wheels, Machinists' Tools, and Whetstones
Identifying Marks: NORTON COMPANY WORCESTER MASS.U.S.A.
Remarks: This company used to be F.B. Norton & Co. and used patents from February 6, 1877; February 7, 1881 (invalid), July 11, 1882, and a "Walker's Patent." They bought out Pike Mfg. Co. in 1932 and were later bought out by Saint Gobain circa 1990 (Nelson 1999). America's most prolific manufacturer of grinding and emery wheels and stones.
Links: http://www.ind.nortonabrasives.com/data/aboutus/About_Us.asp?SEQ=90

Ohio Tool Co.
Auburn, New York and Columbus, Ohio, 1823-1920
Tool Types: Augers, Axes, Bits, Chisels, Clamps, Draw Knives, Metal Planes, Plane Irons, Shaves, Vises, and Wood Planes
Identifying Marks: Various configurations of the name; COLUMBUS; NEW YORK; SCIOTO

Remarks: This company was formed in 1823 though the name was not used until incorporation in 1851. The Columbus branch used prison labor from the Ohio State Penitentiary from 1841 to 1880. In 1893, they merged with Auburn Tool Co. until all operations moved to Charleston, West Virginia in 1914. Peter Hayden and George Gere were involved both with tool production and as officials of the company after its incorporation (Nelson 1999). Along with the Auburn Tool. Co's "Star," Ohio Tool Co. also owned and used New York Tool Co.'s "Thistle" brand (FFLTC).

References:
Ohio Tool Company. [ca. 1900] 1981. Catalog: *High-grade mechanics' tools.* Fitzwilliam, NH: Reprinted in by Ken Roberts Publishing Co.
Ohio Tool Company. [1910] 1976. *Ohio Tool Company established 1823: Catalogue no 23.*

Lancaster, MA: Roger K. Smith.
Links: http://pages.friendlycity.net/~krucker/OhioTool/history.htm -- Contains some historical information on Ohio Tool Co.
http://www.davistownmuseum.org/bioOhio.html

Oldham, Joshua
New York City, New York, 1867-1887
Tool Types: Chisels, Files, Knives, Rules, and Saws
Identifying Marks: His name, W. BRINDSWORTH; J. THOMAS; J. ARMITAGE
Remarks: Oldham is believed to have manufactured his own saws but may have simply dealt in other products. He had a 5 October 1880 patent on a saw (Nelson 1999). The catalog in the Davistown Museum collection includes an essay on saws, their history and use back to Talus's invention in ancient Greece.
References: Oldham, Joshua. [1887] 1978. Catalogue and price list: *Joshua Oldham, manufacturer of saws, machine knives, &c. of every description.* The Early American Industries Association.
Links: http://www.davistownmuseum.org/bioOldham.html

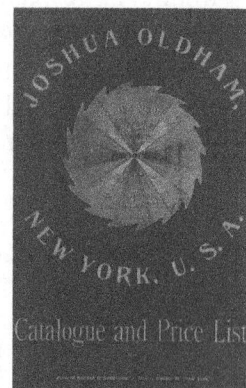

Orr, Hugh
Bridgewater, Massachusetts, 1738-1760-
Orr, Robert
Sutton and Bridgewater, Massachusetts, -1776-
Tool Types: Axes, Edge Tools, Hammers, Scythes, and Shovels
Remarks: Hugh Orr (January 2, 1715-December 6, 1798). Robert was Hugh's son and worked at the Springfield Arsenal by 1804 (Nelson 1999). The Orr's were also well known gunsmiths.
References: Brack, H. G. 2008a. *Art of the Edge Tool.* Hulls Cove, ME: Pennywheel Press.

Osborne & Co., Charles Samuel
Newark, New Jersey, 1826-1992-
Tool Types: Augers, Dividers, Hammers, Leather Tools, Levels, and Others
Identifying Marks: C.S. OSBORNE &CO./NEWARK, N.J. (sometimes without city line, sometimes with "EST. 1826" or a patent date); C.S.O. & CO.
Remarks: Osborne succeeded William Dodd after marrying his daughter. the "Est. 1826" date refers to the maker prior to Dodd. In 1906, the company moved from Newark to

An Osborne leather-working hammer, Davistown Museum IR Collection ID# 70209T2.

Harrison, NJ, where it is still operated by the Osborne family. Charles had a brother, Henry Frank

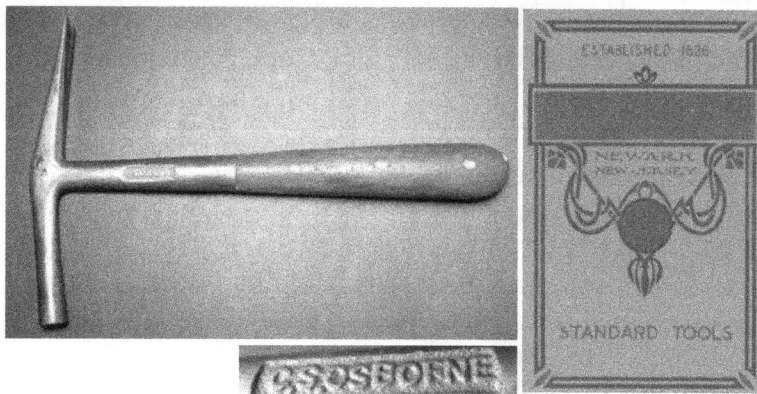

Osborne, who worked with the company prior to setting up his own in 1876. Patents include a washer cutter attributed to Kirkland October 11, 1875 (invalid date) and a leather cutting gauge, patented August 1, 1876, repatented July 17. 1877. In 1960, they acquired Mound Tool Co. (Nelson 1999). This company is one of many mass-producers of the area at the time and America's most prolific manufacturer of leather-working and leather-cutting hand tools.

References:
Osborne, C. S. & Co. [1911] 1976. Catalog: *Established 1826: C. S. Osborne & Co., Inc.: Standard tools.* The Early American Industries Association and the Early Trades and Crafts Society and the Mid West Tool Collectors.

Links: http://www.davistownmuseum.org/bioOsborne.html

Osborne, Henry Frank
Newark, New Jersey, 1876-1905
Tool Types: Dividers and Leather Tools
Identifying Marks: H.F. OSBORNE/NEWARK-NJ; H.F. OSBORNE/TRADE OPT MARK/NEWARK N.J.
Remarks: Henry was Charles S. Osborne's brother and worked for his company prior to starting his own company, then sold it out to them in 1905. The OPT in his maker's mark stands for "Osborne Patent Tools." He produced a leather draw gauge patented by E.G. Latta of Friendship, New York, on February 8, 1881 (Nelson 1999).

Page, George
Keene, New Hampshire, 1825-1838
Tool Types: Awls, Bits, Carpenter Tools, Chisels, Drills, and Others
Remarks: Page was a machinist with patents on a pump from 1833, a mortise machine from 1836, chisels from August 14, 1833 and July 7, 1835, a countersink from November 17, 1836, and a geared drill from May 8, 1838. It is unknown whether he made anything other than the mortise machine or pump. He also made screw gimlets with Everett Newcomb (Nelson 1999).

Parker Co., Charles
Meriden, Connecticut, 1854-1926-
Tool Types: Vises
Identifying Marks: C. PARKER/MERIDEN, CT/PAT'D DEC 17, 1867
Remarks: Parker was listed in Meriden as Parker, C., C. Parker, Chas. Parker, Charles Parker, Parker & Co., C. Parker Co., and others, and used the brand names VULCAN and ECLIPSE. Vice patent dates included November 26, 1867 and June 20, 1854 (Nelson 1999). One of the most prolific of bench vice manufacturers (FFLTC).

Peace, Harvey W.
New York City and Brooklyn, New York, 1861-1890
Tool Types: Saws
Identifying Marks: HARVEY W. PEACE/BROOKLYN N.Y./PATD DEC 21, 1869
Remarks: Harvey Peace (August 10, 1831-September 21, 1907) was born in England and moved from New York City to Brooklyn in 1863. His shop was called the Vulcan Saw Works and used the brand VULCAN. He also may have used the brands or marks J.D. DARLINGTON, C.H. GOBLE

DIAMOND BRAND, B.P. BALDWIN, NEW YORK SAW CO., WILKINS, and H.P. WARREN. The company name had "Co." or "& Co." added but was omitted in marks. He used a number of patents, including W.H. Hankin & C. Tinney's from July 3, 1883; F.A. Buell's from July 17, 1883; A. Sloan's from October 16, 1883; A. Boynton's from November 22, 1887; and others from May 12, 1874; January 19, 1875; August 29, 1876; November 18, 1879; and March 15, 1881 He was acquired by Disston in 1890 and sold out to National Saw Company in 1891 (Nelson 1999).

References:
Brundage, Larry. (March 1993). Harvey W. Peace's fan club. *The Chronicle*. 46(1). pg. 19.
Cope, Ken. (March 1993). Harvey W. Peace, sawmaker. *The Chronicle*. 46(1). pg. 18-9.

Pearce, Jonathan W.
Fall River, Massachusetts, and Providence, Rhode Island, -1840-1879
Tool Types: Edge Tools and Wood Planes
Identifying Marks: J.W.PEARCE/FALL-RIVER (with starburst figure); J.W.PEARCE/PROV.R.I. (name line curved)
Remarks: Jonathan's last name was also recorded "Pirce" and "Pierce." His move from Fall River to Providence occurred in 1852 (Nelson 1999).

Peavey Mfg. Co.
Brewer, Oakland, Bangor, and Eddington, Maine, 1857-todate
Tool Types: Edge Tools and Wood Planes
Remarks: This is one of numerous Peavey Co. names utilized by the famous Peavey clan of logging tool manufacturers. So many Peaveys made tools that a complete compilation of their numerous 19th century activities is, as yet, unavailable. Peavey Mfg. Co. is listed among Yeaton's ax-makers in his directory *Axe Makers of Maine*. From 1900 to 1918, the company was in Bangor, then in Brewer from 1918 to 1923, then in Oakland from 1927 to 1965. The Peavey Manufacturing Company is now located in Eddington, Maine. The company claims to have been in operation since 1857.

References:
Brack, H. G. (2008d). *Registry of Maine Toolmakers*. Hulls Cove, ME: Pennywheel Press.
Yeaton, Donald G. 2000. *Axe makers of Maine*. Rochester, NH: Unpublished, Donald Yeaton, 51 Strafford Rd., 03867-4107.
Links: http://peaveymfg.com/history.html
http://www.mainemade.com/members/profile.asp?ID=1958
http://www.davistownmuseum.org/bioPeavey.htm

Peck & Co., A. G.
Cohoes, New York, 1876-1904
Tool Types: Adzes, Axes, Edge Tools, Hatchets, and Picks
Remarks: This company succeeded M.H. Jones & Co. with Jones as a partner. Brand names include Empire Tool Works, SUPERB, SUPERIOR, CHAMPION BLADE, ROYAL, BOSTONIAN, and MANHATTAN CLIPPER. Peck's Edge Tool is thought to be a variation of the name (Nelson 1999).

Peck Stow & Wilcox Co.
Southington, Connecticut, 1870-1950
Tool Types: Bits, Braces, Chisels, Dividers, Draw Knives, Hammers, Household Tools, Machinists'

Tools, Screwdrivers, Tinsmith Tools and Wrenches, and Others

Identifying Marks: Various combinations of the name, "P.S.&W. CO.," "PEXTO," City/state, patent dates, often Plantsville, CT or Cleveland, OH

Remarks: This company was formed from a merger of Peck, Smith & Co; the S Stow Mfg. Co.; and the Roys & Wilcox Co. Other minor incorporated companies included AW. Whitney & Sons; J.E. Hull & Co; Woodruff & Wilcox; Hart, Vliven & Mead Mfg. Co.; Cheshire Edge Tool Co.; and Johns & Co. Finally, they were bought out by Billings & Spencer in 1950, though the name persisted. Household tools included a variety of meat cutters, choppers and grinders, sausage stuffers and coffee mills, some of which were branded "LITTLE GIANT." Patent dates include March 15, 1859; January 10, 1860; March 9, 1869; April 19, 1892; and September 3, 1895. Also under the company, Amos Shepard had a December 30, 1884 patent on a brace, James H. Culver had an October 11, 1887 patent on calipers, Henry Smith had a July 1, 1873 wire gauge patent. A set of Ellrich saws was also patented. Other reported patent dates include March 20, 1888, November 20, 1888, July 15, 1890, and August 26, 1890 (Nelson 1999). Among the most prolific (top 10) producers of hand tools in the early 20th century.

References:

Page, Herb. (Fall 2008). Reach for the wrench: The wrenches of Peck, Stow & Wilcox Co. *Fine Tool Journal*. 58(2). pg. 20-3.

Links: http://www.alloy-artifacts.com/peck-stow-wilcox.html
http://www.davidrumsey.com/detail?id=1-1-26594-
1110095&name=Peck,+Stow+&+Wilcox+factories
http://www.sydnassloot.com/Brace/PSW.htm
http://www.roseantiquetools.com/id192.html
http://www.davistownmuseum.org/bioPeck.html

Peck & Co., Seth
Southington, Connecticut, Circa 1835
Tool Types: Tinsmith Tools
Identifying Marks: SETH PECK & CO
Remarks: Reports on this maker are confusing due to conflicting information concerning tinsmith toolmakers named Peck in this town. His birth has been reported as late as 1816, but he has been reported working as early as 1819 and as late as 1867. His partner or partners, if he had any, may have been Edward Converse, Romeo Lowrey, Orrin Peck, Noble Peck, and/or Wyllys Smith. It is generally agreed that Peck, Smith & Co. succeeded him (Nelson 1999).

Peck, Smith & Co.
Southington, Connecticut, -1860-1870
Tool Types: Drills, Tinsmith Tools, and Wrenches
Identifying Marks: PECK SMITH & CO/SOUTHINGTON, CT (sometimes without &); MADE BY PECK SMITH MFG CO/SOUTHINGTON CT
Remarks: Orrin Peck, Wyllys Smith, and Benjamin Seward were principals in this company, which succeeded Seth Peck (possibly with him still in the company) and later merged into Peck, Stow & Wilcox Co. in 1870 (Nelson 1999).

Pierce, Benjamin
Chesterfield and Spofford, New Hampshire, 1851-1882

Tool Types: Augers, Bits, Braces, Spinning Wheel Heads, and Household Tools

Identifying Marks: B. PIERCE; Benj.Pierce & Co.

Remarks: Benjamin Pierce worked for Richardson & Huggins as a salesman prior to succeeding them and also served as the superintendent of a chisel factory, probably Pliny Merrill's, in Hinsdale, NH, from 1852 to 1865. The Currier Brothers succeeded him. "Benj.Pierce & Co." was used on paper labels marked "Chesterfield Factory, N.H." or "Spofford, N.H." (Nelson 1999).

Links:
http://www.accessgenealogy.com/scripts/data/database.cgi?file=Data&report=SingleArticle&ArticleID=0012645

Pillsbury, M. M.
Napanoch and New York City, New York, Circa 1878

Tool Types: Adzes, Axes, Edge Tools, Hatchets, and Picks

Remarks: This company succeeded Napanoch Axe & Iron Co. It is possible he moved to New York City, but that location could have merely been a sales office (Nelson 1999).

Plumb, Fayette R.
Philadelphia, Pennsylvania, 1888-1964

Tool Types: Axes, Blacksmithing Tools, Chisels, Hammers, Hatchets, Picks, Railroad Tools, Stone-working Tools, and Bolo Knives

Identifying Marks: FAYETTE R PLUMB, PLUMB, Plumb (scripted on an anchor), ARTISAN'S CHOICE, BLUE GRASS, O-V-B (Our Very Best), DIAMOND EDGE, KNOCKER, SERVALL, PHILA. TOOL CO, QUAKER CITY, HOME THRIFT, POWER STROE, CHAMPION, BOY SCOUT and AU-TO-GRAF.

Remarks: Plumb started off as part of Yerkes & Plumb but began working alone at some point between 1856 and 1897; 1887-1888 seems most likely. The anchor brand became prominent around 1890. (Nelson 1999) During the First World War, they became prominent manufacturers of trench tools (including bolo knives, picks, and hand axes) for the Allied forces. Among the top ten most prolific 20th century toolmakers.

References:
The North American. (1891). *Philadelphia and Popular Philadelphians Illustrated*. The North American, Philadelphia, PA.

Links: http://www.workshopoftheworld.com/richmond_bridesburg/plumb.html -- Information on production for the Allied Forces.
http://www.time.com/time/magazine/article/0,9171,853600,00.html?iid=chix-sphere -- Article on a lawsuit against "Plomb" tools, a rival manufacturer.
http://www.workshopoftheworld.com/richmond_bridesburg/plumb.html
http://query.nytimes.com/gst/abstract.html?res=9E06E5D9163DE733A2575AC0A9679C946497D6CF – Article on Plumb's bizarre death
http://www.yesteryearstools.com/Yesteryears%20Tools/Plumb%20Co..html
http://www.davistownmuseum.org/bioPlumb.html

Pomeroy, A. H.
Hartford, Connecticut, -1886-
Tool Types: Cutlery, Lathes, Saws, and Others
Identifying Marks: A.H. POMEROY
Remarks: It is unclear whether Pomeroy made any of his products or simply distributed them (Nelson 1999). Testimonials in Pomeroy's catalog indicate that he did business as far away as St. John, NB, Canada and Eagle Pass, Texas.
References:
A. H. Pomeroy. 1886. *Illustrate Catalogue of Scroll Saws, Lathes, Fancy Woods, Clock Movements, Pocket Cutlery, Mechanics' Tools, &c. &c.* MA: Pilgrim Publishers.
Links: http://www.davistownmuseum.org/bioPomeroy.html

Poole, Williams & Co.
Windsor, Vermont, and New Britain, Connecticut, 1884-1886-
Tool Types: Shaves
Identifying Marks: Poole, Williams & Co./Windsor, Vt.
Remarks: This company, consisting of Lawrence V. Poole and Orlando E. Williams, was one of several companies making a scraper called a "Windsor beader." They used a September 15, 1885 patent by Elton P. Kendall and Ambrose S. Vose of Windsor. They also used November 28, 1884, March 10, 1885, June 2, 1885, and March 10, 1885 patents (Nelson 1999).

Porter Co., H. K.
Boston, Massachusetts, 1888-1900
Tool Types: Bolt Cutters and Others
Remarks: The successor of Porter & Wooster, they moved from Boston to Everett in 1900 and changed their name to H. K. Porter Inc., remaining in business through 1990 under that name. Brand names included HKP, HYPOWA (both probably used after 1900), EASY and NEW EASY on bolt cutters patented April 6, 1880; January 18, 1881; August 9, 1881; and October 18, 1892. They bought out the Disston Saw Works in 1955 and sold it to Sandvik of Sweden in 1975 (Nelson 1999).

Porter, H. S.
Fairlee and Thetford, Vermont, -1846-1886-
Tool Types: Axes, Edge Tools, and Scythes
Remarks: This could be one or more people: a Herman Porter was listed in Thetford from 1846 to 1855, a Hammond S. Porter was listed in Thetford in 1849, an H.S. Porter was listed in Fairlee in 1872, and an H. Porter was listed in Fairlee from 1885 to 1886 (Nelson 1999).

Pratt & Whitney Co.
West Hartford, Connecticut, 1860-1966
Tool Types: Bevels, Dies, Drill Chucks, Drills, Lathes, Machinists' Tools, Taps, and Others
Identifying Marks: P.&.W. Co.
Remarks: Francis A. Pratt (1827 -1902) and Amos Whitney (1832-1920) made gunmaker's tools, drop hammers, screw machines, bolt cutters, gang drills, gear cutters, knurling tools and so on, including a protractor patented to Ambrose Swazey on 13 July 1875, possibly a wrench patented by

John J. Grant on 1 February 1876, a compass patented by Frederick Gardner on 28 September 1886, and a caliper attachment patented by Bengt M. Hanson on 19 December, 1899. Other patent dates include 31 August 1875 and 10 August 1897--another patent-holder's name was apparently Woodbridge (Nelson 1999). One of America's most important and prolific 19th century machine tool manufacturers.

References:
Pratt & Whitney Co. 1950. *Pratt & Whitney Co. Small Tools.* West Hartford, CT: Niles-Bement-Pond Co.

Links: http://www.davistownmuseum.org/bioPrattWhitney.html

Prentice, James
New York City, New York, -1846-1883
Tool Types: Bevels, Rules, Scales, Measuring Tapes, Protractors, and Other Scientific Instruments
Remarks: James Prentice (January 2, 1812-August 25, 1888) immigrated in 1842 and was succeeded by James Prentice & Son in 1883. The latter company was only known to have been a manufacturer of optical equipment (Nelson 1999).

Prentiss Vise Co.
Watertown and New York City, New York, 1872-1948
Tool Types: Anvils, Clamps, Dies, Taps, Vices, Drill and Wrench Attachments, and Others
Identifying Marks: PRENTISS VISE CO./NEW YORK (sometimes with brand names); P.V. CO. NY.
Remarks: This company used the brands MAGIC, BULL DOG, RAPID TRANSIT, MONARCH, REX, GIPSY, STAR, HANDY, YANKEE, and ECLIPSE. They also used patent holder names as brands, including SHEPARD, BINGHAM, BLAKE, and LEW Watertown was the factory, New York City was their sales office (Nelson 1999) (FFLTC).

Providence Tool Co.
Providence, Rhode Island, 1845-1883
Tool Types: Augers, Bits, Edge Tools, Plane Irons, Gunsmith Tools, and Others
Identifying Marks: Providence Tool Co.; PROV TOOL CO
Remarks: Wing H. Taber and Thomas H. Abbott were the owners of this company at some point and were succeeded by the Rhode Island Tool Co. (Nelson 1999).

Putnam Machine Co.
Fitchburg, Massachusetts, 1854-1912
Tool Types: Levels and Machinists' Tools
Identifying Marks: Putnam Machine Co./FITCHBURG MASS
Remarks: Their products included a gear tooth layout scale patented by John Putnam on May 8, 1877, and possibly an iron level patented by Edward E. Webb on December 7, 1886. Levels marked "Fitchburg Level Co." are probably from this company (Nelson 1999).

Putnam, John & Salmon W.
Fitchburg, Massachusetts, 1836-1854
Tool Types: Rules, Steam Engines, and Steam-Powered Machine Tools

Remarks: John Putnam (1810-1888) and Salmon W. Putnam (1815-1872) were possibly the first commercial steel machinists' rule producers and became Putnam Machine Co. in 1854 (Nelson 1999).

Quinnipiac Malleable Iron Co.
New Haven, Connecticut, -1857-
Tool Types: Braces, Wrenches, Drill/Bit Stocks, Wagon/Carriage Hardware, and Gun Parts (Nelson 1999)

Reading Hardware Company
Reading, Pennsylvania, 1868-1910-
Tool Types: Household Tools, Apple Parers, Food Choppers, and Tobacco Knives
Identifying Marks: MADE ONLY BY THE READING HARDWARE CO./READING. PA. U.S.A. (in a circle with "78" in a shield at center); R.H. CO./READING PA US/PATENTED NOV 14 1876 (from nail-grabbing tongs)
Remarks: Patent dates used by this company included July 22, 1873 and December 10, 1872, by W.A.C. Oakes of Reading, Pennsylvania, under the name CENTENNIAL. They also used patents from May 5, 1868; May 3, 1875; October 19, 1875; November 14, 1876, May 22, 1877, and in 1878. Their food chopper was similar to one patented by Athol in 1865. Their tobacco cutter was called a "Standard Tobacco Knife" but whether "Standard" was a brand name or just part of the product name is unclear (Nelson 1999).

Reynolds, Henry C.
Manchester, New Hampshire, 1855-1877
Tool Types: Axes, Edge Tools, and Handles
Remarks: Henry C. Reynolds (1829-1877) is not known to have worked alone, but worked for the Blodgett Edge Tool Mfg. Co. and Amoskeag Ax Co. circa 1855 to 1877 and was part of Benjamin H. Piper & Co. 1867 to 1877. His patents included axes from August 1, 1865 and August 6, 1867 and Amoskeag's combination hatchet design may have been his (Nelson 1999).

Richardson, Charles Fred
Athol, Massachusetts, 1883-1904
Tool Types: Calipers, Levels, and Scientific Instruments
Identifying Marks: C.F. RICHARDSON ATHOL MASS
Remarks: Charles Richardson's father, Nathaniel, owned a machine shop from around 1835 to 1883 but was never known to make tools. This was the shop in which L.S. Starrett began making combination squares in 1878. Charles and his brother George H. worked under their father until his death when Charles bought out George. His products included a spring caliper patented by Frederick Thomas March 22, 1881, the rights to which he later sold to J. Stevens Arms & Tool Co. and a sighting/grade level patented in 1887, obtaining a patent for such a tool himself on October 27, 1896. Circa 1895, he added "& Son" to the company name and sold out to Goodell-Pratt and L.S. Starrett in 1904 and 1905, respectively (Nelson 1999).
References:
Crane, Ellery Bicknell. (1907). *Historic Homes and Genealogical and Personal Memoirs of Worcester County, Massachusetts*. The Lewis Publishing Company, New York, NY.

Tuholski, Robert. (1999). *Images of America: Athol, Massachusetts*. Arcadia Publishing, Dover, NH.

Ring & Co., E. & T.
Worthington, Massachusetts, 1840-1849
Tool Types: Wood Planes
Identifying Marks: E&T RING&CO/WORTHINGTON.MS
Remarks: Probably the brothers Elkanah and Thomas Ring (Nelson 1999). Hand planes marked "Ring" are commonly encountered in New England collections (FFLTC).

Rixford, O. S.
East Highgate, Vermont, 1812-1889
Tool Types: Axes, Handles, Scythes, and Whetstones
Identifying Marks: RIXFORD; EBONY
Remarks: O. S. Rixford was the youngest son of one Luther Rixford and evidently succeeded him around 1838. He built a foundry to make stoves around 1865 and employed roughly 30 people at the time.
References:
Rixford, O. S. Circa 1887. *O. S. Rixford's scythes and axes East Highgate, Vermont.* Boston, MA: T. O. Metcalf & Co., Printers.
Links: http://www.rootsweb.com/~vermont/FranklinHighgate.html -- Highgate township history including notes on O.S. Rixford
http://www.davistownmuseum.org/bioRixford.html

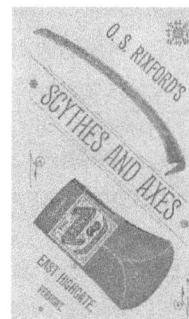

Robbins & Lawrence Co.
Windsor, Vermont, -1855-1861
Tool Types: Machine Tools and Gun- and Pistol-Making Machinery
Remarks: This company was succeeded by E. G. Lamson & Co. (Nelson 1999). The Robbins & Lawrence factory is the current location of the American Precision Museum, of which the editor is a long time member. Robbins & Lawrence is America's most important manufacturer of gun-making machinery and a key player in the evolution of the American system of mass production. Numerous references to Robbins & Lawrence are contained in the historic surveys and manuscripts cited in our bibliographies.
Links: http://www.americanprecision.org/
http://www.aaawt.com/html/firearms/f99.html

Robinson & Co., I. J.
St. Johnsbury, Vermont, -1872-1877-
Tool Types: Bevels and Squares
Remarks: Robinson and H. Fairbanks had bevel patents from April 9, 1872 and March 11, 1873 (Nelson 1999).

Rowland Saw Works Anchor Brand
Rowland & Co., William
-1802-1856-

William & Harvey Rowland
1849-1870
Philadelphia, Pennsylvania
Tool Types: Saws

Remarks: The DATM indicates that there are some conflicting reports. This family made saws in Philadelphia, Pennsylvania between 1802 and 1870.

William Rowland Sr. (b. 1780 d. 1857) is reported as working between 1802 and his death in 1857. He is reported as running William Rowland & Co. in 1856. This company apparently was also known as the Rowland Saw Works. He used the brand name ANCHOR BRAND at some time in his career. It is not known when the "& Co." was added to the name of the company. One source reports it as being used only until 1851, but it is found on an 1856 catalog.

William Rowland Jr. is the son of William Rowland Sr. He has "solo" working dates from 1835 to 1849. Apparently, prior to 1835 he worked in his father's business. In 1849, he became part of William & Harvey Rowland. The sources for these dates conflict with one showing this partnership starting in 1835 and another listing only William Jr. in 1837. Yet another source shows William Jr. working alone beyond 1851.

Willam & Harvey Rowland: This partnership existed from 1849 to 1870, when the company was acquired by Disston. It is not clear if Harvey Rowland was William Jr.'s brother, son, or other relative.

DATM (Nelson 1999) lists the following other Rowlands who worked in the Philadelphia area: B. Rowland& Co. of Philadelphia made shovels and related tools (scoops, draining/ditching tools) 1876. Jonathan Rowland made shovels in 1838, location unknown. T. Rowland and T. Rowland's Sons of Cheltenham, PA made shovels 1894 - 1901 then merged into Ames Shovel & Tool Co. Thomas & Benjamin Rowland of Philadelphia made shovels in 1836 (Nelson 1999).

The Davistown Museum has now received the following information about Rowlands in Philadelphia: "My father (William Rowland) was from Philadelphia and he spoke of the factory that his family owned. The factory manufactured leaf springs. My father told me that they used to manufacture farm implements before they went into leaf springs."

Links: http://www.davistownmuseum.org/bioRowland.htm

Roys & Co., Franklin
Berlin, Connecticut, 1840-1849-
Tool Types: Tinsmith Tools, Silversmith Tools, and Others

Identifying Marks: F.ROYS & CO./BERLIN

Remarks: This company, which overlapped the start of the Roys & Wilcox Co., consisted of Franklin Roys, Noah C. Smith, Benjamin Wilcox, and Benjamin F. Savage (Nelson 1999).

Roys & Wilcox Co.
East Berlin, Connecticut, -1845-1870
Tool Types: Calipers and Tinsmith Tools

Identifying Marks: ROYS & WILCOX/EAST BERLIN.CT; ROYS & WILCOX Co/EAST BERLIN Ct; ROYS & WILCOX/BERLIN CT (with eagle)

Remarks: This company, which merged into Peck Stow & Wilcox in 1870, initially consisted of Franklin Roys, Edward Wilcox, Elisha Norton, and Samuel Wilcox. Roys also worked with Josiah Wilcox in Wilcox & Roys of North Greenwich, Connecticut, and with Benjamin Wilcox in another

"F. Roys & Co." The relationship(s) between all of these Wilcoxes is unknown. This company made calipers patented by Noah C. Smith on September 14, 1869 (Nelson 1999).

References:
Herndon, Richard, and Burton, Richard. (1897). *Men of Progress*. Alfred Mudge & Son, Boston, MA.
Connecticut State Library. (1908) *Public Documents of the State of Connecticut: 51ˢᵗ Annual Report*. Hartford Press and The Case, Lockwood & Brainard Company, Hartford, CT.
Links: www.tintinkers.org/files/tool_list.pdf
http://dunhamwilcox.net/bios/wilcox.htm

Russell & Erwin Mfg. Co.
New Britain, Connecticut, 1846-1919-
Tool Types: Axes, Hammers, Hatchets, Household Tools, Levels, Saws, and Wrenches
Identifying Marks: R.&E.MFG.CO.
Remarks: This company consisted of Henry E. Russell and Cornelius B. Erwin, who added "Mfg. Co." in 1851. They succeeded a number of other lock and hardware makers who had been in operation since 1835, continuing to make a number of household tools and remark other manufacturers' tools, often using the brand RUSSWIN after 1886. This name was used specifically on a wrench also showing patents from October 30, 1900, January 12, 1901, and February 7, 1901. Tools they made themselves included pressing irons and stands, a sausage stuffer patented by O.W. Stowe of Plantsville, Connecticut, on July 6 1858; a "Hale" meat cutter, and a number of Nathaniel Waterman's patented kitchenware. Tools they sold but did not necessarily make included a level patented March 12, 1878, a Baxter patent wrench, hammers, axes, and hatchets with the name G.B. Germond, and saws. Their company catalog also lists shovels made by Ames and others who were independent manufacturers or brand names (Nelson 1999). From their 1877 price catalog, "The Russell & Erwin Manufacturing Company originated in 1839 when H.E. Russell, Cornelius B. Erwin, and Frederick T. Stanley formed a partnership to produce locks and builders' hardware, under the name of Stanley, Russell & Company. When Mr. Stanley withdrew from the partnership in 1840, Smith Matteson and John H. Bowen were added, changing the name of the company to Matteson, Russell & Company. In 1846, with the death of Mr. Matteson and the expiration of the partnership terms, the company's name changed to Russell & Erwin. In 1851 the partnership was reorganized as a joint stock company and was from that time known as Russell & Erwin Manufacturing Company, until its merger with P. & F. Corbin in 1902. Cornelius Erwin served as president of the company from 1851 until his death in 1885. The company is best known as the pioneer of the wrought steel lock industry."
References:
Camp, David Nelson. (1889). *History of New Britain*. William B. Thompson & Company, New Britain, CT.
Russell & Erwin Mfg. (1877). Co. *Price List And Descriptive Catalogue Of Hardware Made By The Russell & Erwin Manufacturing Co*. Russell & Erwin Manufacturing Co., New Britain, CT.
Links: http://www.brooklynmuseum.org/opencollection/objects/2227/Doorknob

Russell & Co., John
Greenfield and Deerfield, Massachusetts, 1832-1865
Tool Types: Axes, Knives, Hammers, Hatchets, Household Tools, Levels, Saws, and Wrenches

Identifying Marks: J.RUSSELL & CO./GREEN RIVER WORKS

Remarks: It is difficult to track Russell's movements since he worked under several names at several MA locations. Francis Russell was an early partner of his, though for how long is unclear. Though he started in Deerfield in 1832, his most famous factory was at Greenfield, where he produced edge tools and knives using the mark "GREEN RIVER" as well as John Russell Mfg. Co. One source shows he also remained in Deerfield until 1864 with John Russell Mfg. Co. operating from 1870-1872. John Russell Mfg. Co. later became John Russell Cutlery Co. sometime around 1884 and ran at least until 1936. BARLOW jack knives were made from 1875 to 1920 under various names. Wiley & Russell Mfg. Co also produced tools under the "GREEN RIVER" brand name, suggesting an unexplored connection (Nelson 1999). Green River knives, particularly bowie knives, were prominent in the "Old West" scene and were known for their rugged durability. Green River Works continues to manufacture Bowies today, which are widely available in cutlery and hunting stores. John Russell is well known as one of the most important innovators of the American system of manufacturing and is frequently mentioned in many of the bibliographic citations in this volume. See Brack (2008a) for additional comments on Russell.

References:

Brack, H. G. 2008a. *Art of the Edge Tool*. Hulls Cove, ME: Pennywheel Press.
Merriam, Robert. 1976. *The History of the John Russell Cutlery Company, 1833-1936*. Greenfield, MA: Bete Press.

Links: http://home.att.net/~mman/JRussellCo.htm
http://preservationgreenfieldma.org/places.html
http://www.jstor.org/pss/1496598
http://www.davistownmuseum.org/bioRussel.html

Sanborn, David Page
Littleton, New Hampshire, and Worcester, Massachusetts, -1841-1865
Identifying Marks: Variations of "D.P. SANBORN" and "LITTLETON" or "WORCESTER/MASS."

Sanborn, Francis Davidson
1873-1875-

Tool Types: Carpenter Tools, Household Tools, Wood Planes, Hammers, Handles, Rules, Saw Tools, Moulding Tools, Churns, and Vises

Remarks: David Page Sandborn (1810-1871) was James Dow's son-in-law, who was a brother-in-law of Franklin J. Gouch, who was Francis David Sanborn's father and Minot Weeks's father-in-law. David Page may have been working in New Hampshire or Massachusetts in the early 1840s, but by 1845 he was in Worcester, probably working alone, as part of Sanborn & Gouch, and probably as Sanborn & Co. Circa 1850, he moved back to Littleton, taking his son as a partner in 1866 and possibly working in Sanborn & Weeks. Francis (1834-1880) worked with his father circa 1866 and as part of Sanborn & Weeks until the death of his brother-in-law Minot in 1873. He produced planes and other tools like rules, wooden mallets, handles, saw horses, bench screws, cheese presses, washing machines, rolling pins, potato mashers, and butter pats while working with others but made solely woodenware alone (Nelson 1999).

Sandusky Tool Co.
Sandusky, Ohio, 1869-1926

Tool Types: Axes, Clamps, Hammers, Hoes, Metal Planes, Picks, Plane Irons, Screwdrivers, Shaves, Wood Planes, and Others

Identifying Marks: SANDUSKY TOOL CO/OHIO (in straight lines or scrolled double curve); SANDUSKY TOOL CO/SANDUSKY OHIO (name curved); OGONTZ TOOL CO.

Remarks: Aside from planes, this company may have solely been a distributor for some items. Patented planes include one by Cyrus Kinney in 1855, two by Ellis H. Morris on 8 November 1870 and 21 March 1871, and one by Harmon Vandbuskirk on 30 November 1869. They were bought by American Fork & Hoe Co. in 1926. (Nelson 1999) Sandusky produced a line of semi-steel planes incorporating an alloy of 85% gray iron, 10% steel, and 5% Mayari iron (probably from Cuban ore). A prolific late 19th century maker of wood planes, including some fancy plow planes, which are now collector's items.

References:

Schwer, Wilbert. 1993. The Sandusky Tool Company story. *The Chronicle* 46(3): 57-68.

Sandusky Tool Co. [1877] n.d. *Illustrated List of Planes, Plane Irons, etc. The Sandusky Tool Co.* Fitzwilliam, NH: Register Team Printing Establishment & Ken Roberts Publishing Co.

Links: http://www.davistownmuseum.org/bioSandusky.html

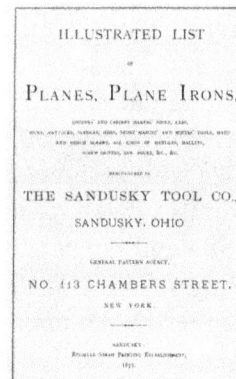

Sargent & Co.
Hartford and New Haven, Connecticut, 1869-1926

Tool Types: Axes, Clamps, Hammers, Hoes, Metal Planes, Picks, Plane Irons, Screwdrivers, Shaves, Wood Planes, and Others

Identifying Marks: SARGENT and variations

Remarks: Sargent & Company was formed in 1858 when three companies, located in different states and involving three different Sargent brothers (Joseph B., George & Edward) were combined. The companies were J. B. Sargent & Co., Sargent & Brother, and Peck & Walter. In 1863 a new, larger location was purchased in New Haven, Connecticut. Prior to that the company had been located in Hartford with sales offices in New York City. Sargent & Co. was incorporated in October of 1864 and in 1866 a fourth brother, Harry, joined the company. His involvement was short-lived though. By the end of the 1960s Sargent hand tools had been completely eclipsed by their production of locks and builders hardware. In 1967 Sargent & Co. became a division of Walter Kidde & Co., Inc. and the name was changed to Sargent Manufacturing Company (Lamond 1997, 10). The Sargent & Co. copied many of the Stanley tool designs and, after the Stanley Company, was the second most prolific manufacturer of malleable iron hand planes and other tools.

References:

Lamond, Thomas C. 1997. Sargent spokeshaves: A brief company history and an overview of their spokeshaves. *The Fine Tool Journal* 47(2): 10-12.

Sargent & Co. [n.d.] 1975. *Wood bottom and iron planes.* Lancaster, MA: Roger K. Smith.

Sargent & Co. 1993. *The Sargent tool catalog collection: A reprint of the Sargent tools illustrated in*

the Company's 1894, 1910, and 1922 catalogs, Forward by Paul Weidenschilling*. Mendham, NJ: The Astragal Press.

Sargent & Co. *Duralumin and Steel Carpenter Squares*. New Haven, CT.

Links: http://www.sargentlock.com/index.asp -- Sargent: an ASSA ABLOY group company still in business today.

http://www.rostratool.com/default.htm -- Sargent Quality Tools is also still in business today.

http://www.davistownmuseum.org/bioSargent.htm

Sawyer Tool Co.
Athol, Fitchburg and Ashburnham, Massachusetts, 1894-1915

Tool Types: Bevels, Levels, Machinists' Tools, Rules, and Screwdrivers

Identifying Marks: SAWYER TOOL MFG. CO./ASHBURNHAM MASS (sometimes without the city/state)

Remarks: The "Mfg." in the title was not always used. The company was formed by Burnside E. Sawyer. The move from Athol to Fitchburg occurred in 1898 and to Ashburnham in 1912. In 1915, the name changed to Almond Mfg. Co. They produced screwdrivers and screwdriver bits for use in SAMSON brand braces, likely based on a 9 May 1899 patent belonging to Sawyer and William Arnott. Further productions include a screwthread gauge (patented to Sawyer, September 21, 1897), a surface gauge (patented to Sawyer, May 9, 1899), a bevel (patented to Carl G. Osteman, July 25, 1893) which was also manufactured by L.S. Starrett and, possibly, the S.A. Woods Machine Co., and a tap wrench (patented to Henry B. Keiper, December 14, 1897). See also Almond Mfg. Co. (Nelson 1999).

Links: http://www.americanartifacts.com/smma/advert/ay223.htm -- A device containing a piece manufactured by Almond Tool Co.

http://www.davistownmuseum.org/bioSawyer.html

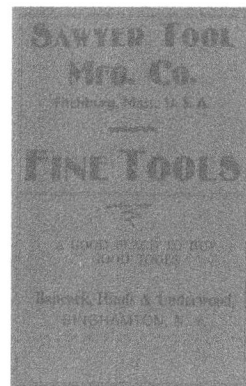

Sawyer Watch Tool Co.
Fitchburg, Massachusetts, 1867-1881-

Tool Types: Calipers, Dividers, and Watchmaker Tools

Remarks: An S. Sawyer, probably Sylvanus Sawyer, the president of this company in 1881, held divider and caliper patents from April 9, 1867 and July 7, 1868. This company is apparently unconnected and unrelated to Sawyer Tool Co. (Nelson 1999).

Sayre, L. A.
Newark, New Jersey, 1884-1916-

Tool Types: Dividers, Hatchets, Hoes, Household Tools, Leather Tools, Pliers, and Saw Tools

Identifying Marks: L.A.SAYRE & SON/NEWARK,N.J.; L.A.SAYRE & CO/NEWARK,NJ (first line curved)

Remarks: Sayre's name is also reported as Sayer and sometimes appears with "Co.," "& Co.," and "& Son." He held a patent on an apple parer from January 29, 1884 under the brand WAVERLY and may have marked and sold tools made by others (Nelson 1999).

Schollhorn Co., William S.
New Haven, Connecticut, 1873-1913-

Tool Types: Dividers, Pliers, Punches, Eyelet Setters, and Others
Identifying Marks: Variations of the full name or W.S.S.CO., sometimes with city, state, patent dates, patent holder's name, etc.
Remarks: William Schollhorn (1834-1890) had patents on a pair of dividers with a pencil attachment from June 17, 1873 and a divider shared with F. P. Pfleghar from January 9, 1886. The latter were made under the brand EXCELSIOR. This company also made many pliers and dividers patented by William A. Bernard on May 6, 1890; July 19, 1892; October 24, 1899; August 7, 1900; January 1, 1901; April 2, 1907; and November 6, 1907 (Nelson 1999).

Sears, Roebuck & Co.
Chicago, Illinois, -1893-todate
Tool Types: Numerous
Identifying Marks: CRAFTSMAN
Remarks: This company marks a variety of tools with their name and began to use the brand CRAFTSMAN after 1900. Richard Sears, a watch company owner in Minneapolis, Minnesota, was operating as early as 1886. He hired Alvah C. Roebuck, a watchmaker, and opened a branch in Chicago in 1887. By 1893, they were using their corporate name and had moved to Chicago, publishing their first catalog covering more than watches in 1896 (Nelson 1999).

Seymour & Co., William N.
New York City, New York, -1828-1861-
Tool Types: Marking Gauges, Metal Planes, Squares, and Wood Planes
Identifying Marks: W.N.SEYMOUR & CO./NEW-YORK (first line curved, sometimes without state)
Remarks: This company is listed as early as 1828 but a trade is not specified until after 1842 when they are listed as a hardware dealer. While they are reported as makers of machinist's try squares, they may have only marked and resold them, as with most of their tools including wooden planes, some of which seem to be from the Greenfield Tool Co. The metal planes found could be from prior to 1842 and appear to be piano maker's iron miter planes (Nelson 1999).

Shepardson & Co., H. S.
Shelburne Falls, Massachusetts, 1855-1876-
Tool Types: Augers, Awls, Bits, Braces, Chisels, and Farm Tools
Remarks: H. S. Shepardson held patents on a brace from March 1, 1870, a countersink from March 8, 1864, and a gimlet handle from August 12, 1873 (Nelson 1999).

Siegley Tool Co., Jacob
New York City, New York, and Wilkes Barre, Pennsylvania, 1878-1905
Tool Types: Metal Planes, Plane Irons, Wood Planes, and Other Planes
Identifying Marks: SIEGLEY; J. SIEGLEY; JACOB SIEGLEY; SSS; STS; SBS (on plane irons)
Remarks: Jacob Siegley (1846-January 19, 1937) had plane patents from December 6, 1878; July 1, 1879; August 16, 1881; January 2, 1883; March 11, 1884, February 10, 1891; and December 5, 1893. His company moved from New York to Pennsylvania in 1883 and sold it to Stanley Rule & Level in 1905, who continued to use his name into the 1920s, using different letters in the markings to signify iron type and thickness (Nelson 1999).

Silver & Deming
Salem, Ohio, 1851-1900-
Tool Types: Bits, Boring Machines, and Drills
Remarks: E. W. Silver and J. Deming are assumed to be the principals of this company. Their date of establishment is listed at 1854, probably when the partnership began, and apparently began to use the name "Silver Mfg. Co." starting with an appearance in a catalog from 1900. They were credited with inventing the first bits with reduced shanks, allowing larger bits to fit into smaller chucks. E.W. Silver and J. Deming had a patent with E. W. Fawcett on September 23, 1873 for a hollow auger. They produced wheelwright's hub borers patented July 25, 1851, and blacksmith style drill presses from 1854, 1864, and 1868 (Nelson 1999). Their drill bits are still frequently encountered, suggesting additional production well into the 20th century. Silver & Deming bits are highly regarded by many customers of the Liberty Tool Co. (FFLTC).

Simmons & Co., Daniel
Cohoes, Albany, and New York City, New York, 1834-1860
Tool Types: Adzes, Axes, Hammers, and Hatchets
Identifying Marks: SIMMONS/MOSS; D.SIMMONS; D.SIMMONS & CO. COHOES-N.Y.
Remarks: The Albany and New York City branches were likely just sales offices for the Cohoes factory. This company was reportedly succeeded in 1860 by Weed, Becker & Co. One source says Daniel Simmons and Jonas Simmons (possibly his brother) formed the Lockport Edge Tool Co., also in 1860. MOSS may be a brand name or partner's name. The D.SIMMONS mark could be prior to the addition of "& Co." to the company name (Nelson 1999).

Simmons Hdw. Co., Edward Campbell
St. Louis, Missouri, 1869-1940
Tool Types: Numerous
Identifying Marks: Variations of SIMMONS, E.C. SIMMONS, S.H. Co., sometimes with city/state; KEEN KUTTER
Remarks: This company is also known as E.C. Simmons & Co. and was a hardware wholesaler that marked and marketed countless tools. While not a tool maker itself, this company did acquire controlling interest in Walden Knife Co. in the 1890s. They were bought out by A.F. Shapleigh Hdw. Co. in 1940. KEEN KUTTER was their best brand, but other brands included ARKANSAS on splitting wedges, AXTEL on horse rasps, BAY STATE on hatchets, BLACK JACK on assorted tools including a wrench patented January 14, 1896, BLACKSNAKE on saws, BLUE BRAND or B.B. on squares, BULL DOG on braces and tool sets, CHIPAWAY on axes, planes, hammers and others, CLAY BANK on shovels, COLUMBIA on saws, CUMBERLAND on axes, DEFIANCE on saws, DELMAR on coffee mills, DRIVEWELL on punches, DUCK BILL on wrenches, ESSEX on wrenches, EUREKA on rakes, FARMER BOY on scythe stones, FAST MAIL and W.M.FINCH on saws, HERCULES on wheelbarrows, HOWARD on axes, KEYSTONE on forks, KLICKER, KLINCHER on pliers, KORN KRUSHER on corn mills, KIGHTNIN, LONE STAR on hoes, H.M. MEIER on saws, MESQUITE on axes, MOGUL on shovels, MONARCH on saws, NEVER SLIP on wrenches, OAK LEAF, OHIO BOY on rakes, OHIO FALLS on axes, OZARK on splitting wedges, POLAR on ice saws, RED JACKET on hatchets, RED LINE on rules, RED TOP on scythes, ROYAL on axes, RUN EASY on plows and lawnmowers, SMILEY'S on saws, SURE GRIP on

vices, braces and wrenches, SWIFT on saws and saw tools, TRUE BLUE on axes, UTILITY on saws and wrenches, VANGUARD on saws, WINNER on lawn mowers, WOOD PECKER and ZULU on axes, and WM. ENDERS (Nelson 1999).

Links: http://www.thckk.org/history/simmons-hdwe.pdf

Simonds Saw Mfg. Co.

Fitchburg, Massachusetts, and Chicago, Illinois, -1890-1904

Tool Types: Files, Knives, Saws, and Saw Tools

Identifying Marks: THE SIMONDS SAW (in a banner); THE/SIMONDS/MFG./CO./PAT.DEC.27.1887 (on a saw handle rivet)

Remarks: This company uses an 1832 establishment date but was the successor to A. Simonds & Son, not using this name until the 1880s. A number of other names like Simonds Saw Works, Simonds Mfg. Co., Simonds Saw Co, and others, confuses the dating of this company as it is unclear whether or not they were related. They used a number of brand names including BAY STATE IROQUOIS, KING PHILIP, MOHAWK, OSCEOLA, PONTIAC, and SIOUX. By 1904 they were working out of Lockport, NY; Seattle, WA; Philadelphia and Pittsburg, PA; Vancouver, BC; Montreal, QU; Toronto, ON; St. John, NB; New Orleans, LA; Portland, OR; San Francisco, CA; Los Angeles, CA; and several locations in England. Patents include a saw swage dated February 22, 1881, a dove tail saw dated December 27, 1887, and a band saw guide dated August 7, 1894 (Nelson 1999).

References: Simonds Saw and Steel Company. (1937). *Woodworking saws and planer knives: Their care and use*. Simonds Steel and Saw Co., Fitchburg, MA. IS.

Simonds Saw and Steel Company. (1937). *The circular saw: A guide book for filers, sawyers and woodworkers*. Simonds Steel and Saw Co., Fitchburg, MA. IS.

Simonds Saw and Steel Company. (1937). *How to file a cross-cut saw*. Simonds Steel and Saw Co., Fitchburg, MA. IS.

Simonds Saw and Steel Company. (1937). *The cross-cut saw*. Simonds Steel and Saw Co., Fitchburg, MA.

Links: http://www.simonds.cc/company/history17.php?menu=../mnu/mnuCompanyHistory

Smith Machine Co., H. B.

Smithville, New Jersey -1892-

Tool Types: Carpenter Tools and Grinders

Remarks: "H. B. Smith, founder of the H. B. Smith Machine Co. (1847) of Smithville, New Jersey, pioneered the use of cast iron in woodworking machinery. Unlike his competitors, he used all iron construction in his major machines from the very beginning. Even so, some minor Smith machines were built with wooden frames to meet the needs of smaller shops. ...William C. Bolger pointed out its importance in *Smithville: The Result of Enterprise* (1980): 'The statement that 'It is all iron' is significant, both in terms of the man and his future career as a manufacturer of woodworking machines. His reputation in the machine business was due as much to his use of iron as to the designs he patented.'" (Batory 2004, 11).

References:

Batory, Dana Martin. 2004. Cast in iron. *Fine Tool Journal* 54(1): 8-13.

Links: http://www.njhm.com/irongrave1.htm - New Jersey's History Mysteries: The Iron Grave
http://www.owwm.com/MfgIndex/detail.asp?ID=766 - Old Woodworking Machines:

Smith, Otis A.
Rockfall, Connecticut, 1884-1917
Tool Types: Bits, Combination Tools, Levels, Marking Gauges, Metal Planes, Saw Tools, Wrenches, and Others
Identifying Marks: OTIS A. SMITH/ROCKFALL, CONN (sometimes with patent dates or other marks)
Remarks: Otis Smith (1836-1923) made planes patented by Amos Fales on March 7, 1882, April 1, 1884, and August 31, 1886, and a gasket cutter from October 24, 1865. He also made an adjustable countersink bit, a wire splicer, and guns. A Savage & Smith that made a gasket cutter patented 1865 in Vermont is not known to be related (Nelson 1999).

Snell Mfg. Co.
Fiskdale and Sturbridge, Massachusetts, -1850-1905-
Tool Types: Augers, Awls, Bits, Boring Machines, and Plane Irons
Identifying Marks: Variations of the maker's name, sometimes with Mfg., the city and state, "MANUF'D BY," or patent dates
Remarks: This company probably succeeded Snell & Bros., though it is unknown when (Nelson 1999). A prolific maker of augers and bits.

Standard Tool Co.
Athol, Massachusetts, Circa 1880-1905
Tool Types: Household Tools, Levels, Machinist Tools, Rules, Squares, and Wrenches
Identifying Marks: STANDARD TOOL CO/ATHOL, MASS (first line curved); CHAPLIN'S PATENT/REISSUE MAY 4 1880
Remarks: This company was a subsidiary of Athol Machine Co. and had the same factory and officers that specialized in machinist tools. The mark with Chaplin's May 4, 1880 patent was on a rule in a combination machinists' square; the patent was originally issued on May 8, 1866 to Orril R. Chaplin. Other patents used by Standard Tool Co. include patents belonging to Stephen H. Bellows from September 16, 1873, June 8, 1880; November 12, 1881 (invalid); March 11, 1884; April 15, 1884; June 3, 1884; May 21, 1889; May 13, 1890; May 11, 1897; and October 18, 1898. They also used David E. Woolsen's patent from October 7, 1879, Thomas Frederick's invalid patent from March 22, 1880, William A. White's from April 1, 1884, and John P.B. Wells's patents from November 22, 1887; July 22, 1890, and January 9, 1900. Their brand names included BOSS, BAY STATE, and sold AMERICAN choppers that were probably manufactured by Athol Machine Co. Laroy S. Starrett bought both Athol Machine Co. and Standard Tool Co. in 1905, continuing to use Athol Machine Co. but abandoning this component of the company due to his legal disputes with it over patent rights (Nelson 1999).

Stanley & Co., Augustus
New Britain, Connecticut, 1854-1857
Tool Types: Rules
Identifying Marks: *A.STANLEY*NEW BRITAIN,CONN.*

Remarks: The brief history of Augustus Stanley & Co., a rule-making company is complicated by the fact that the Stanley family had been producing hardware in New Britain since 1831, using a series of other names before becoming the Stanley Works in 1852. The Augustus Stanley & Co. was formed by brothers Augustus and Timothy Stanley with Thomas Conklin, an earlier rule maker from Bristol, Connecticut, at the same time they acquired Seth Savage's rule business in Middletown, Connecticut. In 1857, they merged with Hall & Knapp to form the Stanley Rule & Level Co. (Nelson 1999). Numerous citations in the bibliographies that follow this appendix discuss the pivotal event in the history of the Stanley company; i.e. the evolution of Augustus Stanley's rule manufacturing company to the Stanley Rule and Level Co. in 1857.
References: Camp, David N. (1889). *History of New Britain, Connecticut.* William B. Thompson & Company, New Britain, CT.
Links: http://www.nbim.org/

Stanley Works
New Britain, Connecticut, 1852-1920
Tool Types: Hardware
Identifying Marks: STANLEY
Remarks: At the same time that Augustus Stanley's rule company became the Stanley Rule and Level Company in 1857, the Stanley Works was continuing to manufacturing a wide variety of hardware in an associated group of buildings in New Britain. The Stanley Works was finally incorporated into the Stanley Rule & Level Co. in 1920. The Stanley Works was probably America's largest producer of hinges, hasps, and other commonplace hardware, which can still be found in New England workshops and collections.

Stanley Rule & Level Co.
New Britain, Connecticut, 1857-
Tool Types: Rules
Identifying Marks: *A.STANLEY*NEW BRITAIN,CONN.*

PRICE LIST. 1898
STANLEY
RULE AND LEVEL COMPANY.
NEW BRITAIN,CONN.U.S.A.
107 CHAMBERS ST.
NEW YORK.

Remarks: "The Stanley family had been making hardware in New Britain from 1831 on; they used a series of other names before they became the Stanley Works in 1852. In 1854, brothers August and Timothy Stanley and Thomas Conklin (an earlier rule maker in Bristol, CT) formed [the August Stanley & Co.] ...concurrently, they acquired the rule business of Seth Savage, Middletown, CT. In 1857, this company merged with Hall & Knapp as the Stanley Rule and Level Co."

Henry Stanley was the first president of Stanley Rule & Level Co. Henry was concurrently the president of the Stanley Works, a maker of hardware, which maintained a separate corporate identity from Stanley Rule & Level Co. until 1920 when they merged. "The S. R. & L. Co. continued to expand its product line by acquiring other companies making tools they wanted to add and to expand their market volume by acquiring competitive companies. Their major pre-1900 acquisitions were: Hill & Crum, Unionville, CT; Charles L. Mead (successor to E.A. Stearns & Co.), 1863; Bailey, Cheney & Co., 1869; Leonard Bailey & Co., 1878; Bailey Wringing Machine Co., 1880; R.H. Mitchell & Co., 1871; Upson Nut Co., 1893."

Patent rights acquired by S. R. & L. Co: Many of the approximately 44 issued to Leonard Bailey from 1855 to 1903 (Frank M. Bailey was a Stanley plane room foreman and had three patents assigned to Stanley); Nathan S. Clement, 19 March 1867 tool handle; A. Williams, combination gauge; W.T. Nicholson, levels; C.G. Miller, planes; G.A. Warren, planes; Dorn, planes; Justice Traut, multiple plane patents.

Atha Tool Co. may have been producing tools as early as 1875 in Newark, NJ. Buying out many competing tool and hammer makers, they were themselves purchased by the Stanley Rule & Level Co. in 1913, which retained its touch mark (Nelson 1999, 748-9).

The following history is excerpted from a 1937 *Tool Talks* publication by Stanley Tool: "The manufacture of "Bailey" Planes by Stanley marked a turning point in the Company's history. Other hand tools were soon added to the Stanley line-Mitre Boxes, Screw Drivers, Wood and Iron Levels, Bit Braces, Hand Drills, Hammers, Try Squares. With these new tools, The Stanley Rule & Level Co. produced the most complete line of woodworking tools in the world. Stanley's dominant position was recognized by carpenters and mechanics everywhere who turned over their problems and suggestions to the Company. This created a demand for specialized tools to perform certain jobs better than they could be done with regular size or style tools. Stanley responded by increasing its line to still greater proportions to include many more hand tools that helped craftsmen do better work. A search of the U. S. Patent Office would undoubtedly disclose that The Stanley Rule & Level Co. took out more patents during this period than any other industrial organization in the country. NEW COMPANIES ANNEXED: In the early years of this century the march of progress continued. In 1904 the George E. Wood Company, of Plantsville, Conn., manufacturers of "Hurwood" Screw Drivers was bought. The business was enlarged under Stanley leadership and Stanley "Hurwood" Screw Drivers became the biggest selling quality drivers in the world. Two other companies were purchased in 1913 and 1916. The products of these companies, Atha Tool Co., of Newark, N. J., and The Eagle Square Manufacturing Co., South Shaftsbury, Vt., brought handled hammers, sledges, wedges, anvil tools and carpenters' steel squares to the Stanley line. Today both these plants are busy producing hand tools as branch plants of the Stanley organization. To maintain leadership in the Canadian market, a tool plant was opened by Stanley at Roxton Pond, Quebec in 1906. Known in Canada as the Stanley Tool Company, Ltd., the Roxton Pond factory now makes 80 per cent of all the Stanley Tools sold in Canada. In 1920 The Stanley Rule & Level Co., for many years a full-grown organization merged with another New Britain firm, The Stanley Works."

References:

Aber, R. James. 1978. *Some notes on Gage planes*. Crafts of NJ meeting.

Astragal Press. 1989. *The Stanley catalog collection: 1855 - 1898: Four decades of rules, levels, try-squares, planes, and other Stanley tools and hardware*. Mendham, NJ: Astragal Press.

Blanchard, Clarence. 1997. The number one: Cute and useful. *The Fine Tool Journal* 46(4): 8-10.

Burdick, James M. ca. 1930. *History of the Bailey Plane business from 1869*. New Britain, CT: Internal history of the Stanley Rule & Level Company.

Heckel, David E. n.d. *The Stanley "forty-five" combination plane*. Charleston, IL: David E Heckel, 1800 McComb St.

Jacob, Walter W. 1998. The Stanley Rule & Level Company: Its historic beginning. *The Chronicle* 51(3): 80-84.

Jacob, Walter W. 1998. The Stanley Rule & Level Company: Charles L. Mead and the acquisition of E. A. Stearns. *The Chronicle* 51(4): 120-122.

Jacob, Walter W. 2000. Brace up for a bit of Stanley history part III: 1917 to 1958. *The Chronicle* 53(2): 62-65.

Jacob, Walter W. 2001. Stanley tapes measure the world part II. *The Chronicle* 54(1): 31.

Jacob, Walter W. 2001. Stanley tapes measure the world part III. *The Chronicle* 54(2): 75-80.

Jacob, Walter W. 2001. Stanley tapes measure the world part IV *The Chronicle* 54(3): 121-125.

Jacob, Walter W. 2002. The turn of the screw: The history of Stanley screwdrivers. *The Chronicle* 55(1): 31-34.

Jacob, Walter W. (December 2008). Stanley hand drills part VI: Defiance brand hand drills. *The Chronicle*. 61(4). pg. 166-70.

Lamond, Thomas C. 1996. The Bailey Tool Co. -- The Stanley Rule & Level Co. -- Edw. Preston & Sons, Ltd.: A circumstantial connection? or... whatever happened to the Defiance spokeshave line? *The Fine Tool Journal* 46(2): 11-12.

Pernis, Paul Van. 1998. Leonard Bailey's first planes. *The Fine Tool Journal* 48(1): 5-7.

Roberts, Ken. 1989. *The Stanley Rule & Level Company's combination planes featuring the development and use of the Miller, Traut, and Stanley 45 and 55 planes. Miller's patent combined plow, filletster and matching plane.* Mendham, NJ: Astragal Press.

Rodengen, J. L. n.d. *The legend of Stanley: 150 years of the Stanley Works.* Publisher unknown.

Sellins, Alvin. 1975. *The Stanley plane: A history and descriptive inventory.* South Burlington, VT: The Early American Industries Association.

Smith, Roger K. 1989. Transitional and metal planes: Stanley no. 18 and no. 19 knuckle-joint block planes, general information and type study. *Plane Talk* 13(1): 155-162.

Stanley, Philip E. 1984. *Boxwood & ivory: Stanley traditional rules, 1855 - 1975.* Westborough, MA: The Stanley Publishing Co.

Stanley, Philip E. 1985. *A concordance of major American rule makers.* Westborough, MA: Philip Stanley.

Stanley Tools. 1937. Stanley Tool history. *Tool Talks* 17. New Britain, CT: Stanley Tools.

Stanley Works. 1946. *Facts about tools: A message from Stanley.* New Britain, CT: Stanley Tools.

Stanley Rule & Level Co. n.d. *Stanley improved labor saving carpenters' tools including "Bailey" adjustable plane.* New Britain, CT: Stanley Rule & Level Co.

Stanley Rule & Level Co. [1859] 1975. *1859 Price list of boxwood and ivory rules, levels, try squares, sliding T bevels, gauges, &c., manufactured by the Stanley Rule and Level Company, also including the price list of boxwood and ivory rules manufactured by A. Stanley & Co., New Britain, Conn. Jan. 1855.* Fitzwilliam, NH: Ken Roberts Publishing Co.

Stanley Rule & Level Co. [1867] n.d. *Price list of U. S. standard boxwood and ivory rules, levels, try squares, gauges, handles, mallets, hand screws, &c. manufactured by the Stanley Rule and Level Company, New Britain, Conn., and Brattleboro, VT.* Bristol, CT: Ken Roberts Publishing Co.

Stanley Rule & Level Co. [1879] 1973. *Price list of U. S. standard boxwood and ivory rules, plumbs and levels, try squares, bevels, gauges, mallets, iron and wood adjustable planes, spoke shaves, screw drivers, awl hafts, handles, etc. manufactured by the Stanley Rule and Level Company, New Britain, Conn., U.S.A.* Reprinted in by Bristol, CT: Ken Roberts Publishing Co.

Stanley Rule & Level Co. 188?. *Bailey's patent adjustable bench planes and other improved carpenters' tools manufactured by the Stanley Rule and Level Company, New Britain, Conn.* New Britain, CT: Stanley Rule & Level Co.

Stanley Rule & Level Co. [1888] 1975. *Price list: Improved labor-saving carpenters' tools manufactured by the Stanley Rule and Level Co.* Fitzwilliam, NH: Ken Roberts Publishing Company.

Stanley Rule & Level Co. [1892] 1972. *Price list: Improved labor-saving carpenter's tools manufactured by the Stanley Rule and Level Co.* West Boylston, MA: H. C. Maddocks, Jr.

Stanley Rule & Level Co. [1898] 1975. *Price list of U. S. standard boxwood and ivory rules, plumbs and levels, try squares, bevels, gauges, mallets, iron and wood adjustable planes, spoke shaves, screw drivers, awl hafts, handles, etc. manufactured by the Stanley Rule and Level Co. New Britain, Conn., U.S.A.* Fitzwilliam, NH: Ken Roberts Publishing Company,

Stanley Rule & Level Co. [1900] 1983. *The Stanley bed rock: A new plane.* Port Angeles, WA: Bob Kaune.

The Stanley Rule & Level Co. [1909] 1975. Catalog: *Carpenters & Mechanics Tools: No. 102.* Lancaster, MA: Roger K. Smith.

The Stanley Rule and Level Plant. 1921. *"55" plane and how to use it.* New Britain, CT: The Stanley Rule and Level Plant.

Stanley Tools. n.d. *Read this before you use Stanley planes: A plane is no better than its cutter.* New Britain, CT: Stanley Tools.

Stanley Tools. n.d. *The Stanley catalog collection, 1855 to 1898: Four decades of rules, levels, try-squares, planes, and other Stanley tools and hardware.* Morristown, NJ: The Astragal Press.

The Stanley Works. [n.d.] 2002. *Stanley tools - in sets.* The Midwest Tool Collector's Association.

The Stanley Works. [1929] 1977. *Stanley tools for carpenters and mechanics: Catalog no. 129.* Lancaster, MA: Roger K. Smith.

The Stanley Works. 1955. *45 plane: Seven planes in one.* New Britain, CT: Stanley Tools.

Stanley. [n.d.] 2002. *Stanley tools.* The Midwest Tool Collector's Association.

The Stanley Rule and Level Company. [n.d.] 1973. *Combination planes: Historical development, patents and uses.* Fitzwilliam, NH: Ken Roberts Publishing Co.

Stanley. 1994. *Tool traditions catalog.* Phoenix, AZ: Stanley Mail Media, Inc.

Stanley. 1995. *Tool traditions catalog, Two different editions.* Phoenix, AZ: Stanley Mail Media, Inc.

Stanley. [n.d.] 2002. *Insert: Read this before you use: Combination plane no. 46.* The Midwest Tool Collector's Association.

Walter, John. 1988. *Antique and collectable Stanley planes: 1988 price guide.* Akron, Ohio: The Tool Merchant.

Walter, John. 2000. *Antique & Collectible Stanley tools: 2000 pocket price guide.* Marietta, OH: The Tool Merchant.

Walter, John. c. 1993. *Reproduction of a Stanley Tools newsletter: The Iron Age: Thursday, November 3, 1898: The making of the cast iron carpenters' plane.*

Links: http://www.stanleyworks.com -- This company is still in business.

http://www.supertool.com/StanleyBG/stan0a.html -- An important information source containing tons of information on Stanley Planes is on the web as: The Superior Works: Patrick's Blood and Gore.

http://www.roseantiquetools.com/id16.html -- A history of the Stanley Co. and descriptions of their tools has been created by Rose Antique Tools.

http://www.nbim.org/ -- The New Britain Industrial Museum has a history of The Stanley Works on their website.

Star Tool Co.
Middleton, Connecticut, 1867-1883
Tool Types: Bevels, Levels, Marking Gauges, and Squares
Identifying Marks: TOOL CO. (in a star), STAR BEVELS (in a star)

Remarks: Three companies with this name were simultaneously operating in Connecticut, Rhode Island and Vermont, confusing many historians and tool collectors. To further the ambiguity, this company made a bevel patented by Leonard D. Howard on 5 November 1867, who was living in St. Johnsbury, VT at the time. Rights to a marking gauge patented 21 April, 1868 by W. Brodhead of Meadville, PA and a caliper/divider patented 22 February, 1876 by Thomas McDonough were assigned, though it's unclear whether they ever made the latter. A Starr company produced squares during the same time and is possibly a typo or variation of this company (Nelson 1999).
Links: http://www.davistownmuseum.org/bioStar.html

Stark Tool Co.
Waltham, Massachusetts, 1862-1902-
Tool Types: Lathes and Other Tools
Remarks: This company, formed by John Stark, produced "fine tools" such as a spring chuck jeweler's lathe he patented in 1859 (possibly produced under his own name prior to the company's formation). His son, John Jr., succeeded him by 1902 (Nelson 1999).
Links: http://www.wade8a.com/history.htm
http://www.walthammuseum.com/
http://www.awco.org/Seminar2002/Machinery/damaskeeningmachines.htm#Engine

St. Johnsbury Tool Co.
St. Johnsbury, Vermont, 1870-1886-
Tool Types: Bevels and Squares
Identifying Marks: ST.JOHNSBURY TOOL CO.
Remarks: This company was founded by Isiah J. Robinson along with I. J. Robinson & Co. to make a double bevel and square combination tool patented in June, 1870. The distinction between the two companies is unknown (Nelson 1999).

Starrett Co., Laroy S.
Athol and Newburyport Massachusetts, 1880-1994-
Tool Types: Calipers, Dividers, Household Tools, Levels, Machinist Tools, Rules, and Squares
Identifying Marks: L.S. STARRETT CO./ATHOL.MASS (sometimes on one line and/or without "CO"); The L.S.S.Co./Athol,Mass.

An L.S.S. compass/divider, Davistown Museum IR Collection ID# 32708T46.

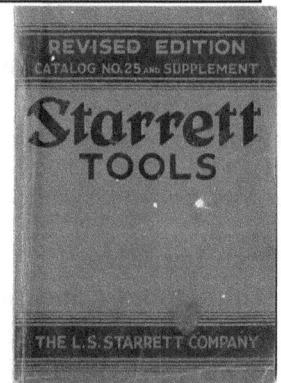

Remarks: Laroy (often incorrectly recorded "Leroy") S. Starrett (April 25, 1836-1922) invented and produced the HASHER, a meat chopping machine patented May 23, 1865. A few years later, he went to work for the Athol Machine Co., which produced it in 1868. Circa 1875 he was prompted to quit due to mounting legal disagreements over patents including a particular combination machinists' square he produced through the Richardson Machine Shop in 1877. In 1880, he won the lawsuit against Athol and formed his own company, buying out Charles P. Fay of Springfield, Massachusetts's caliper and divider stock, machinery, and patents in 1887. His wild success allowed him to buy out his old employer and legal foil,

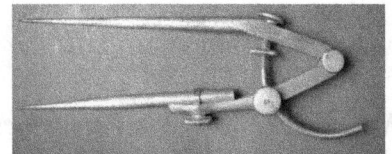

Athol Machine Co., in 1905. Starrett held over 100 patents, including a particularly significant micrometer patented July 29, 1890. He also produced tools with pre-1900 patents of Frederick A. Adams, Frank G. Lilja, Morris F. Smith, Burnside E. Sawyer, John D. Sloan, Edward C. Clapp, Carl G. Osteman, J.H. Cook, Patrick Kennelly, and George Thompson (Nelson 1999).

References:

Johnson, Laurence A. (1958). The hasher. *The Chronicle.* 11(4): 57-59, 64.

Starrett, L. S. (1895). *No. 13 catalogue and price list of fine mechanical tools, manufactured by L. S. Starrett Athol, Mass. U.S.A.* Reprinted by Bud Brown Publishing Co. in 1989, Reading, PA

L.S. Starret Company. (1968). *The Starret Book for Student Machinists.* Athol, MA: L.S. Starret Company.

Turner, Kenneth E. (1996). The Starrett wheels that led to more wheels. *The Chronicle.* 49(3): 69-71.

White, James T. (1922). *The National Cyclopaedia of American Biography, Volume XVIII.* James T. White & Co., New York, NY.

Links: http://www.starrett.com/ -- The L. S. Starrett Company's website
http://www.memorialhall.mass.edu/collection/itempage.jsp?itemid=4575 -- 1898 photograph
http://www.roseantiquetools.com/index.html -- A history of the L. S. Starrett Co. has been created by Rose Antique Tools
http://www.davistownmuseum.org/Inventoryofpictures/WebInfoStarrett.html -- Tools of the L. S. Starrett Co. in the Museum collection

Stearns & Co., Edward A.
Brattleboro, Vermont, and Springfield, Massachusetts, -1838-1863
Tool Types: Rules
Identifying Marks: E.A.STEARNS & CO
Remarks: Edward Stearns started as an employee of S. Morton Clark & Co. prior to buying them out and starting his own business circa 1838. He was succeeded by Charles L. Mead sometime between 1857 and 1861, who continued to use his name until 1863 when he sold out to Stanley Rule & Level Co. One source shows the company in Massachusetts in 1859 instead of Vermont, but this is as of yet unexplained (Nelson 1999).

Stearns & Co., Edward C.
Syracuse, New York, 1877-1941-
Tool Types: Augers, Clamps, Drills, Saw Tools, Shaves, and Wrenches
Identifying Marks: E.C.STEARNS & CO./SYRACUSE,N.Y. (sometimes on one line)
Remarks: Edward C. Stearns (?-1929) succeeded G. N. Stearns & Co. and is assumed to have been related to George N. Stearns. Avis Stearns

A Stearns butt gauge, photo compliments of the Liberty Tool Co.

Mead, most likely a daughter of George Stearns, was the Vice President and a principal of the company. They made the same saw vise as Seneca Falls Mfg. Co. patented by Wentworth on April 8, 1879 and a spoke shave patented March 27, 1900 by Herbert M. Coe. Their brand name MERIT TOOLS was used on scrapers, possibly only after 1900. They remained in business until circa 1956, but were not manufacturing tools by 1941 (Nelson 1999).

Stearns & Co., George N.
Syracuse, New York, 1864-1877
Tool Types: Augers, Box Scrapers, Farm Tools, and Shaves
Identifying Marks: G.N.STEARNS/PAT'D MAY 7, 78
Remarks: George Stearns used an Est. 1864 date for this company despite not having used "& Co." until circa 1870. The patent date above was used on augers and has not been found on record, though he did have patents on hollow augers from September 8, 1863 and August 27, 1872, as well as a spoke shave from December 13, 1870. Prior to 1864 when he became a machinist, he worked as a carriage maker. 1864 was the only year he was listed as working on mowers and reapers (Nelson 1999).

Stephens & Co., L. C.
Riverton, Connecticut, 1828-1901
Identifying Marks: STEPHENS & CO, RIVERTON CT (sometimes without city/state)
Tool Types: Bevels and Rules
Remarks: Lorenzo Case Stephens (1809-1871) began this company, possibly without "& Co.," and was succeeded by his son Deloss H. Stephens (1837-1919), who was bought out by Chapin-Stephens in 1901. It seems "L.C." was dropped from the company name after Lorenzo's death. They made a "No. 36" combination rule, level, and bevel tool patented by Lorenzo on January 12, 1858, which was later produced by Chapin-Stephens and Stanley (Nelson 1999).

Stevens & Co., Joshua
Chicopee Falls, Massachusetts, 1864-1903
Identifying Marks: STEVENS & CO.; J.STEVENS & CO/CHICOPEE FALSL, MASS; J.STEVENS A & T CO./CHICOPEE FALLS MASS
Tool Types: Bits, Calipers, Dividers, Levels, Guns, and Machinists' Tools
Remarks: This company changed its name to Stevens Arms & Tool Co. in 1886 (having always primarily produced guns), but still used "Joshua Stevens & Co." as late as 1898 in their catalogs. They were bought out by L. S. Starrett in 1903. They manufactured a number of calipers and dividers patented by Charles A. Fairfield on July 21, 1863; T.C. Page and George W. Hadley on February 8, 1870; Oliver D. Warfield on November 7, 1882, November 20, 1883, May 31, 1887, and April 10, 1888; Oscar Stoddard on August 27, 1872 and March 31, 1885; George M. Pratt on February 23, 1886; Joshua Stevens on October 30, 1883 and September 30, 1890; and Charles P. Fay on September 23, 1884, October 5, 1897, and January 10, 1899 (Nelson 1999).
References: Jendrysik, Stephen R. (2005) *Chicopee*. Arcadia Publishing, Mount Pleasant, SC.
Links: http://ugca.org/03mar/savage.htm -- The guns of Joshua Stevens & Arthur Savage

Stoddard, Oscar
Detroit, Michigan, 1872-1885-
Identifying Marks: PAT'D.AUG.27.1872/O.STODDARD (curved in an oval outline, arrowhead between the lines); PAT/APLD FOR//O.STODDARD
Tool Types: Calipers and Dividers
Remarks: Oscar Stoddard's patented dividers and calipers were made with extendable and replaceable tips by J. Stevens & Co. and possibly others, though it is unknown if Oscar ever produced them by such a design himself. He also had a March 31, 1885 patent date (Nelson 1999).

Stoney Brook Iron Works
Kingston, Massachusetts, 1805-1836
Tool Types: Augers and Others
Remarks: Seth Drew Jr., his brother-in-law Thomas Cushman, and Seth Washburn set up a works to do general blacksmithing and make ship's tools and supplies. C. Drew & Co. succeeded them (Nelson 1999).

Stortz, John
Philadelphia, Pennsylvania, 1853-1972
Identifying Marks: J.STORTZ & SON/PHILA (sometimes with full first name or without initial); J STORTZ/TOOLS/PHILA PA
Tool Types: Chisels, Coopers' tools, Drills, Knives, Stone-working Tools, and Others
Remarks: Stortz added "& Son" to his name at some point, at least by 1911 but possibly not before 1900. He specialized in slater, paver, brick layer, and cement worker tools, but also produced coopers' tools, race knives, oyster knives, caulking irons, and others (Nelson 1999).

Stow, Solomon
-1820-1847
Stow Mfg. Co., Solomon
-1853-1870
Southington and Plantsville (part of Southington), Connecticut
Identifying Marks: S.STOW MFG. CO./PLANTSVILLE CT (and variations)
Tool Types: Tinsmith Tools
Remarks: Solomon started out as a cabinet and clock maker who also produced brass gears and machine parts prior to concentrating on tinsmithing supplies and machines. He bought up Plant, Neal & Co. circa 1845, then his two sons joined him to form S. Stow & Sons circa 1847, Solomon Stow Mfg. Co. succeeded S. Stow & Co. and merged into Peck, Stow & Wilcox Co in 1870. Orson, Solomon's son, may have been involved and one of them had a November 12, 1867 patent on some sort of tinsmith machine (Nelson 1999).

Stratton Bros.
Greenfield, Massachusetts, 1869-1902
Identifying Marks: STRATTON BROTHERS/GREENFIELD/MASS (sometimes curved, with combined city and state line, or with an eagle)

An ebony level, Davistown Museum collection ID# 13102T1.

Tool Types: Levels

Remarks: Edwin A and Charles M. Stratton began making rifles for Springfield Armory before founding a level business in 1869. Edwin sold the business to his son-in-law, Raymond O. Stetson, in 1902, who may have continued to use the name up to 1908 (Nelson 1999). This company was later acquired by Goodell-Pratt. This company used Millers Falls for their sales agent initially.

Links: http://oldtoolheaven.com/history/history1.htm

Streeter, A. W.
Shelburne Falls, Massachusetts, 1855-1871-

Identifying Marks: A.W.STREETER/SHELBURNE FALLS MASS;
A.W.STREETER/SHELBURNE MASS

Tool Types: Braces

Remarks: A. W. Streeter put a January 23 1855 patent on his braces but the connection is unclear. He also marked them with the dates March 31, 1857, November 17, 1863, and January 8, 1867 (Nelson 1999).

Swan Tool Co., James
Seymour, Connecticut, 1877-1951

Identifying Marks: Configurations of Swan's name with or without "& Co.," city/state, patent dates, a swan figure

Tool Types: Augers, Awls, Bits, Boring Machines, Chisels, Draw Knives, Handles, and Screwdrivers

Remarks: Swan, who was born in Dumfries, Scotland, December 18, 1833, immigrated from Scotland in 1854 and worked at the Bassett Iron Works in Birmingham, CT, the Farrel Foundry & Machine Co in Ansonia, CT, and for Oliver Annes before buying Annes's business in 1877. The "Est. 1856" date is probably Annes's. He acquired

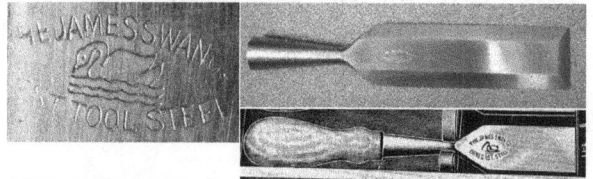

Socket firmer chisels, Davistown Museum IR Collection ID# 041709T3.

Douglass Mfg. Co. and changed its name, but apparently still used its name in 1894. Swan had patents on 20 August 1867 for a machine to make augers, 21 April 1868 for an augur handle, 9 June 1868 for auger bits, 14 July 1868 for an auger, 16 November 1869 for a bit/auger die, 14 December 1869 for an auger handle, 15 February 1870 for a machine to grind and polish bits, 15 March 1870 for an auger, 30 May 1870 for an auger, 27 June 1871 for an auger, 19 September 1871 for a hollow auger, 20 May 1873 for a machine to form lips on augers, 29 July 1873 for an auger, 27 June 1882 for a screwdriver, 12 June 1883 for an expansive bit, 25 December 1883 for an auger, 31 July 1885 for a draw knife, 11 May 1886 for a boring machine and 28 May 1894 for a hollow auger. Many of these patents were issued prior to the formation of his own company and may have been used by one or more of his previous employers. Upon Swan's death, the company passed to his son William, followed by his brother John, followed by James, Son of John. Upon its closing in 1951, it was owned by one R. S. Robie (Nelson 1999). The Davistown Museum has recently acquired a number of James Swan tools from a Boston hardware store which warehoused its inventory circa 1950. Many of these tools were in their original wrapping paper. Some were stamped "Douglass Co." or similarly with the James Swan stamp over it. Also of great interest were some ship auger bits stamped "Germany" and restamped with the James Swan logo, indicating that he was not only a manufacturer but an importer. In general, Swan-marked tools are frequently found by the Liberty Tool Co. The Swan company, along with the Stanley Rule & Level Co., are considered to be the last manufacturers of fine edge tools working at the end of the classic period of American toolmaking.

References:
Swan, James Company. (1911). *Illustrated catalog and price list of premium mechanics' tools manufactured by The James Swan Co.* Fitzwilliam, NH: Ken Roberts Publishing Company.
Pape, William Jamieson. (1918). *History of Waterbury and the Naugatuck Valley, Connecticut.* The S. J. Clark Publishing Company, New York, NY.
Links: http://www.davistownmuseum.org/bioSwan.html

Ten Eyck Mfg. Co.
Cohoes and New York City, New York, 1866-1880-
Tool Types: Adzes, Axes, Edge Tools, Hammers, Hatchets, Hoes, Picks, and Vises
Remarks: This company succeeded W. J. Ten Eyck & Co. and was sometimes recorded as Ten Eyck Axe Mfg. Co. The headquarters was in New York City but they had a factory in Cohoes and produced a Stevens Patent hand vise. It is possible they were succeeded either by Williams, Ryan & Jones in 1872 or Cohoes Axe Mfg. Co. The aforementioned Stevens vise was listed in an 1880 catalog under their name and found in an 1882 catalog with no reference to a maker. A. RIDER may have been one of its brands (Nelson 1999).

Thayer, John A.
Boston, Massachusetts, 1862-
Tool Types: Hammers and Others
Identifying Marks: THAYER'S PAT./JUNE 24 1862//C ("C" inside a diamond)
Remarks: The patent date on the mark was for a combination hammer, tack puller, screwdriver, rule, and more! The C in a diamond does not appear on all Thayer patent tools and is probably a maker's mark, though it is unknown whose (Nelson 1999).

Thompson, Francis M.
Greenfield, Massachusetts, 1868-1871
Tool Types: Braces and Marking Gauges
Remarks: F. M. Thompson is listed separately from Thompson Mfg. Co. in 1870 and a "Thompson" brace may have been produced by either. Francis M. also made a marking gauge. Together, he and J. W. Thompson patented braces on September 15, 1868 and February 23, 1869 (Nelson 1999).

Thompson, H.
Concord, New Hampshire, -1874-
Tool Types: Cutlery, Edge Tools, Knives, and Saws
Remarks: H. Thompson was listed as an edge toolmaker and was possibly the Thompson reported making saws in Concord. An H. Thompson from Concord who may or may not have been the same person had patents dated 1869 and 1871 for beef steak cutters and crushers that may never have been produced (Nelson 1999).

Thrall, Willis
Hartford, Connecticut, 1842-1860
Tool Types: Rules
Identifying Marks: Possibly "S A JONES & CO HARTFORD CON"
Remarks: Willis Thrall succeeded Solomon A. Jones & Co., adding "& Son" to the company name in 1860 (Nelson 1999).
Links: http://www.davistownmuseum.org/bioWillisthrall.html

Tinkham, Levi
Middleboro, Massachusetts, -1800-
Tool Types: Draw Knives, Planes, Slicks, and possibly Other Tools
Identifying Marks: L:TINKHAM
Remarks: Levi Tinkham (1766 - 1857) was a plane and edge toolmaker living in Middleboro, Massachusetts. A probable descendent of Ephraim Tinkham who immigrated to Plymouth, MA, as an indentured servant in 1629, the Davistown Museum has one Levi Tinkham slick and numerous Tinkham documents in its collection. Levi Tinkham was one of numerous toolmakers working in the Taunton River watershed area in the late 18th and early 19th centuries. A possible source of some of his iron was the Nemasket River forge at Middleboro (1692 to ±1800). His toolmaking operations were also in close proximity to the long established Leonard forge at Two Mile River in Taunton (1652 to 1777).

W. Sullivan has kindly given us a photograph of a Tinkham plane and states, "The nose of the plane is clearly marked 'L TINKHAM' (top line) and 'MIDDLEBORO' (bottom line). It's 13" long and 6" from the bottom of the body to the top of the tote. The body is 2 1/8" high and 2 5/8" across the profile. It has an off-set tote and a full original cutting iron. The iron is stamped 'E. BENNET' who I'm told was a local blacksmith."
Links: http://www.davistownmuseum.org/bioTinkham.htm

Toby, F. G.
Gr. Barrington and Mattapoisett, Massachusetts, -1849-
Tool Types: Edge Tools
Remarks: An 1849 directory listed a P. G. Toby in both cities but it is unclear whether they were the same person (Nelson 1999).

Tolman, Joseph Robinson
Hanover and Boston, Massachusetts, -1825-1849-
Tool Types: Wood Planes
Identifying Marks: J.R.TOLMAN/HANOVER/MASS. (sometimes without state and/or city)
Remarks: Tolman was probably making planes in South Scituate, Massachusetts, in the 1820s and 1830s and was in Boston by 1841, settling in Hanover by 1849. He fathered Thomas J. Tolman (Nelson 1999). Tolman was the foremost craftsman of spar planes for the shipwrights of the North River. His distinctive concave planes were widely circulated in the New England shipyards as evidenced by the numerous specimens received by the Liberty Tool Co. His touchmark sometimes accompanies that of Cumings of Boston on some spar and other planes.

Tomlinson, D.
Brookfield, Connecticut
Tool Types: Leather Tools
Identifying Marks: JD.TOMLINSON/PATENT
Remarks: Tomlinson's curriers' fleshing knife was patented July 6, 1820.

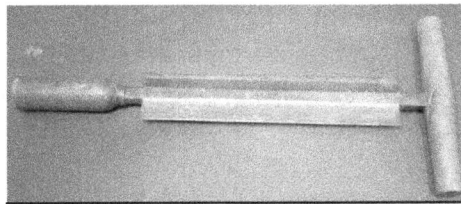

Curriers' fleshing knife, Davistown Museum MIII Collection ID# 62406T2.

Tower & Lyon
New York City, New York, and Glen Ridge, New Jersey, 1862-1902-
Tool Types: Chisels, Levels, Metal Planes, Planes, Screwdrivers, Wrenches, and Others
Identifying Marks: Variations of this name alone, with "MFR. BY" (and variations), cities, patent holders names and dates, "T&L," and others
Remarks: John J. Tower and Polhemus Lyon dealt both in tools they and others made, adding "& Co." to their name in 1902 and continuing at least until 1927 under that name. They began in New York and expanded to New Jersey at some point between 1898 and 1901, retaining the New York City location as a headquarters. Orril R. Chaplin of Boston's May 7, 1872 patent appears on their iron and wood planes (often as "CHAPLIN'S PATENT"). Their Arthur T. Goldsborough of Washington, DC planes patented September 11, 1883 and February 19, 1884 were marked with the brand name CHALLENGE. Their "Chaplin's Improved" planes (actually covered by Maschil D. Converse's February 14, 1899 patent) were branded CHAMPION. This brand was also used on screwdrivers covered by Morris's May 15, 1877 patent and on Iver Johnson Arms Co. guns. They also produced planes patented by Iver Johnson and Reinhard T. Torkelson on April 17, 1888. Some of the planes and other tools Tower & Lyon dealt in were made by Iver Johnson Arms Co. One of their combination levels had a wrench and was covered by Wood's June 14, 1887 patent and Byron Boardman's July 10, 1866 patent. They also made Clark's expansive bit, probably after 1902, and used the brands SAFETY and GEM on wrenches (Nelson 1999).

Towne, Snell & Co.
Snellville (part of Sturbridge), Massachusetts, 1841-1844
Tool Types: Augers and Bits
Remarks: This name was sometimes reported differently ("Towne Snell" and "Towne & Snell"). He apparently became part of Towne, Chaffee & Co. and Smith, Snell & Co. (Nelson 1999).

Trafton Bros.
1860-1874-
Trafton & Son, Alfred S.
-1879-1882
Trafton George A.
1883-1900-
Portsmouth, New Hampshire
Tool Types: Edge Tools
Remarks: Trafton Bros. consisted of Alfred S. and Timothy J. Trafton working as edge tool blacksmiths. Alfred worked with his son, George A. Trafton, as a blacksmiths and shipsmiths who made edge tools, under "Alfred S. Trafton & Son." George went on to work alone (Nelson 1999).

207

Traut & Hine Mfg. Co.
New Britain, Connecticut, -1862-1895-
Tool Types: Cutlery, Can Openers, Razors, and Household Tools
Remarks: Justus A. Traut was the first president of this company and other Trauts, assumed to be relatives, held office during and after his presidency. Henry C. Hine served as secretary from 1888 to 1919. Judd & North bought them out in 1925 (Nelson 1999).

Traut, Justus A.
New Britain, Connecticut
Tool Types: Metal Planes and Others
Remarks: Justus A. Traut worked for Stanley Rule & Level Co. and had a number of patents assigned to them, even after becoming a part of Traut & Hine Mfg. in 1888. His patent dates included November 18, 1862; May 9, 1872; October 22, 1872; March 4, 1873; August 5, 1873; February 7, 1875 (invalid); October 5, 1875; February 16, 1876 (invalid); April 4, 1876; April 18, 1876, January 16, 1877; July 30, 1878; September 2, 1879; January 8, 1884; March 11, 1884; October 4, 1884 (invalid); October 21, 1884; November 18, 1884; April 21, 1885; June 2, 1885; December 15, 1885; February 9, 1886, February 23, 1886; March 23, 1886, May 18, 1886, January 17, 1888; July 24, 1888; March 13, 1895; November 6, 1894; January 22, 1895; January 29, 1895; April 2, 1895; January 21, 1896; February 25, 1896; March 10, 1896; and October 12, 1897 (Nelson 1999). For an encyclopedia of additional information on this important plane designer, consult the numerous texts in the bibliographies that contain descriptions of the planes manufactured by the Stanley Rule and Level Co. after 1862.

Trenton Vise & Tool Co.
Trenton, New Jersey, -1870-1900-
Tool Types: Axes, Blacksmith Tools, Farrier Tools, Picks, Vises, and Others
Remarks: Some of their tools were marked "TRENTON" in a diamond (Nelson 1999).

Trimont Mfg. Co.
Roxbury (now part of Boston), Massachusetts, 1889-1920
Tool Types: Wrenches
Identifying Marks: PERFECT HANDLE/TRIMONT MFG. Co. ROXBURY, MASS.
Remarks: This company used the brand TRIMO (Nelson 1999). (FFLTC).
Links:
http://www.museumofamericanspeed.com/Collections/Culture/MonkeyWrenches/DSCN4699S-133.shtml

Tuck Mfg. Co.
Brockton, Massachusetts, 1852-1915-
Tool Types: Bits, Chisels, Knives, Screwdrivers, Nail Sets, Punches, and Others
Identifying Marks: TUCK; TUCK MFG. CO.
Remarks: They used an 1852 establishment date in 1915, but the name may have been changed earlier at some point (Nelson 1999). Several Tuck screwdrivers and other hand tools make frequent appearances in New England tool kits and workshops (FFLTC).

Underhill Edge Tool Co.
Boston, Massachusetts and Nashua, New Hampshire, 1852-1890
Tool Types: Axes, Adzes, Chisels, Edge Tools, Hammers, Hatchets, Picks, and Shaves
Identifying Marks: UNDERHILL/EDGE TOOL CO. (sometimes on same line)
Remarks: George W. Underhill (July 19th, 1815-October 13th, 1882), John H. Gage and a few others formed this company initially as "Nashua Edge Tool Co." In 1879 they acquired the Amokeag Ax Co. and were bought out by the American Axe & Tool Co. in 1890. Though the plant closed, the brand was still used. The Nashua location was their center of manufacture while the Boston office operated strictly in sales and distribution. George W. Underhill acted as Superintendent until 1875 and a Director until his death (Nelson 1999). An Underhill Edge Co. ax was the murder weapon in the trial of Lizzie Borden. The Underhill clan of edge toolmakers can be documented as working as early as the seventh decade of the 18th century (Josiah, Chester, NH) in both southern NH and the Boston area. Underhill edge tools, ranging from steeled wrought iron to the finest cast steel timber framing tools, are frequently recovered from New England's boatyards and collections. Numerous references to their significance as one of a small group of the most important New England edge toolmakers, along with the Buck Bros., Timothy Witherby, and the Swan Co., are contained in this publication series. Nelson (1999) lists no less than 19 Underhill edge toolmakers working in the Merrimack River watershed area of New Hampshire and in Boston.
References: Klenman, Allen. (December 1998). Amoskeag Ax Company of Manchester, New Hampshire. *The Chronicle*. 51(4). pg. 129-130.
Underhill Edge Tool Co. (1859). *Wholesale Prices of Chopping Axes, Carpenter's, Cooper's, Butcher's, and Many Other Kinds of Mechanics' Tools, Manufactured by the Underhill Edge Tool Company*. Underhill Edge Tool Co., Nashua, NH.
Links: http://lizzieandrewborden.com/pdf%20files/TrialLBPearson.pdf -- A transcription of the trial of Lizzy Borden including a mention of the ax.
http://www.yesteryearstools.com/Yesteryears%20Tools/American%20Axe%20&%20Tool%20Co..html
http://www.davistownmuseum.org/bioUnderhill.html

Union Mfg. Co.
New Britain, Connecticut, 1880-1919-
Tool Types: Chisels, Drills, and Saws
Identifying Marks: Union Mfg Co./New Britain Ct/USA
Remarks: A major and prolific competitor of the Stanley Rule & Level Co., Union's wood and iron planes are frequently found in New England workshops.
References:
Union Tool Company. (1969). *Union Tool Company: Machinist and carpenter's tools: Combined 1969 catalog*. Orange, MA.
Union Mfg. Co. (no date). *Union iron and wood planes*. Reprinted in June 1981 by Ohio Tool Collectors Association.
Links: http://www.davistownmuseum.org/bioUnionMfg.html

Walter's Sons, William P.
Philadelphia, Pennsylvania, 1831-1899?
Tool Types: Miscellaneous Tools

Identifying Marks: W.P.WALTERS SONS PHILA; W.P.W. SONS 1233 MARKET ST. PHILA.

Remarks: This company was primarily a dealer but did make and mark some of its own tools. Their "Est. 1831" date cannot be confirmed, the earliest instance dating them at 1888. Both marked tools found were cast iron holding racks (Nelson 1999).

References:

Walter's, Wm. P. Sons. 1888. *Illustrated catalogue of wood workers' tools and foot power machinery.* Facsimile of the original. Lancaster, MA. The North Village Publishing Co.

Links: http://www.davistownmuseum.org/bioWilliamWallardsons.html

Walworth Mfg. Co.

Boston, Massachusetts, 1892-1911

Tool Types: Dies, Misc. Tools, and Wrenches

Identifying Marks: WALFWORTH MFG. CO./BOSTON, USA (plus patent holder names/dates, brand names, etc.); WALCO

Remarks: The original makers of the Stillson patent wrench, Walworth made a range of plumbing tools. While Stillson apparently patented this particular type of pipe wrench, it appears that the name was used generically by Bonney Vise & Tool Works, the Erie Tool Works, the Moore Drop Forging Co., the J.P. Danielson Co., and probably others. On dies, they used the brand names RUFF & TUFF and MILLER'S PATENT, though the latter was possibly not until after 1900. It appears they made a 1907 Parmelee patent wrench and may have been succeeded by Parmelee Co. (Nelson 1999). America's most prolific turn of the century (1900) adjustable non-monkey wrench manufacturer. A Walworth wrench is one component of a woven triptych by Alan Magee on display at the Davistown Museum (http://www.davistownmuseum.org/MAG%20Photos/Magee%20Photos/tih.jpg).

Links: http://www.davistownmuseum.org/bioWalworth.html

Watts, Joseph

Boston, Massachusetts, 1834-1849

Tool Types: Marking Gauges, Rules, and Squares

Identifying Marks: J.WATTS/BOSTON

Remarks: Though he worked in Charleston, Watts marked his tools Boston. One report of A. J. Watts, Boston, is thought to be a misrecording of the name (Nelson 1999). Phil Platt states "Joseph Watts' working dates were 1834 - 1849 (D.A.T.) He apparently worked at rule making, making gauges and squares in Charlestown, MA.; but, marked at least the rules 'BOSTON'. There are many 'Watts' family members in and around the city of Boston. Don and Anne Wing, Marion, MA (EAIA) are currently doing research on J. Watts and trying to connect him back to English rule makers. See: Milt Bacheller, "American Marking Gages" for an extensive write up on the Watts family."

References:

Bacheller, Milton H., Jr. (2000). *American marking gages, patented and manufactured.* Self-published, 185 South St., Plainville, MA 02762.

Massachusetts Charitable Mechanic Association, Boston. (1874) *Twelfth Exhibition.* Alfred Mudge & Son, Boston, MA.

Links: http://www.davistownmuseum.org/bioWatts.html

Wheeler, Madden & Bakewell
New York and Middletown, New York, 1853-1860
Tool Types: Saws, Trowels, and Other

Remarks: This company's facility was called the Monhagen Saw Works. In 1860, Elisha P. Wheeler, Edward M. Madden and Josiah Bakewell became "Wheeler, Madden & Clemson," using the brand names SEARS & CO., H. MILLSON & CO., and VERNON & CO. (possibly through acquisition) There is a questionable possibility they succeeded Cane, Weel & Co. (Nelson 1999).

References:
Wheeler, Madden & Bakewell. [1860] n.d. *Monhagen Saw Works, Middletown, Orange County, N.Y. Illustrated Price List.* South Burlington, VT: The "Press" Printing Establishment, Exchange Building.

Links: http://www.davistownmuseum.org/bioWheeler.html

White, L. & I. J.
Buffalo, New York, and Monroe, Michigan, 1837-1928
Tool Types: Adzes, Chisels, Coopers' Tools, Draw Knives, Plane Irons, Saws, and Wood Planes

Identifying Marks: L.&I.J.WHITE/BUFFALO;
L.&I.J.WHITE/1837/BUFFALO N.Y. (top and bottom lines curved into an oval)

Remarks: Leonard White (1810-1893) and Ichabod Jewett White (?-1880) moved from Monroe to Buffalo in 1844 (Nelson 1999). L. & I. J. White was one of America's most prolific manufacturers of a wide variety of coopers' tools, some of which frequently appear in New England tool collections (FFLTC).

References: The L. & I. J. White Company. *Catalogue of Coopers' Tools Including Turpentine Tools, Edge Tools and Machine Knives.* L. & I. J. White Co., Buffalo, NY.

Whitman & Barnes, Mfg. Co.
Akron, Ohio, and Philadelphia, Pennsylvania, 1848-1915
Tool Types: Bits, Chisels, Drills, Farm Tools, Files, Hammers, Hatchets, Knives, Screwdrivers, Taps, Wrenches, and Others

Identifying Marks: Various combinations of "W.&B." or "W.&B.Co." in a diamond outline; full name; WHITMAN BARNES; DIAMOND; HERCULES; ACME; ALWAYS READY; BULLDOG

Remarks: This company's widespread factories included locations in Syracuse, NY, West Pullman, IL and St. Catharines, OH. At some point their headquarters moved from Akron to Philadelphia, though the factory in Akron continued to produce tools. Patents include 27 February 1883, 1 July 1890, 19 May 1891, and 19 April 1898 (Nelson 1999). Whitman & Barnes got their start making some of the first quality blades and sickles for McCormick mowers.

References:
Page, Herb (Mr. Oldwrench). 2003. Reach for the wrench: Whitman & Barnes: Extraordinary wrench makers. *The Fine Tool Journal* 52(3): 15.

Links: http://www.akronhistory.org/kendig_agricultural.htm -- article on the agricultural boom of Akron, OH, including Whitman & Barnes' part in it.
http://www.davistownmuseum.org/bioWhitmanbarnes.html

Whittemore & Co., Amos
Bennington, New Hampshire, 1855-1860-
Tool Types: Cutlery, Edge Tools, Knives, and Leather Tools
Remarks: Amos Whittemore (1802-1881) worked with his brother Alfred as part of Baldwin & Whittemore from 1853 to 1855 and in this company, which was also known as A. & G. A. Whittemore, circa 1860. He may or may not have worked alone before George joined him in this company (Nelson 1999).

Wilkinson & Co., A. J.
Boston, Massachusetts, 1842-1993
Tool Types: Draw Knives, Drills, Grinders, Lathes, Machinists' Tools, Saw Tools, Saws, Vises, Wood Planes, and Other
Identifying Marks: Various straight and curved configurations of the full maker name, with or without "& Co.," city, state, "MAKERS," "2 WASH ST," or "A.J.W. & CO."
Remarks: This company marked and sold tools it did not manufacture in addition to its own lines. The "Est. 1842" date probably refers to Wilkinson's own start, though it's unclear when "& Co." was added. Patents they made include a draw knife with an 1895 patent, several micrometers patented between 1883 and 1886 by Merrick M. Barnes, the rights to which were later bought around 1890 by Brown & Sharpe (Nelson 1999). Considered one of the best manufacturers of adjustable draw knives in the early 20th century, examples of their craftsmanship are common occurrences in New England tool collections.

An adjustable draw knife, Davistown Museum IR Collection ID# 52304T4.

References:
Wilkinson, A. J. & Co. [1867] n.d. *Wilkinson, A. J. & Co.'s illustrated catalogue of hardware and tools*. Boston, MA: The Mid-West Tool Collectors Association.
Links: http://www.owwm.com/mfgindex/detail.aspx?id=898
http://www.davistownmuseum.org/bioWilkinson.html

Williams & Co., J. H.
New York and Buffalo, New York, 1882-1909-
Tool Types: Wrenches and Clamps
Identifying Marks: W in a diamond, J.H. WILLIAMS & CO./BROOKLYN N.Y., VULCAN, SUPERWRENCH, SUPERECTOR, AGRIPPA, SUPERSOCKET, SUPERJUSTABLE
Remarks: This company used the brands in the marks above. The Williams Co. was one of the most prolific early and mid-20th century manufacturers of box wrenches for auto mechanics (FFLTC).
References:
Williams & Co., J. H. (1937). *Superior Drop-forgings and Drop-forged Tools*. J. H. Williams & Co., New York City, New York.

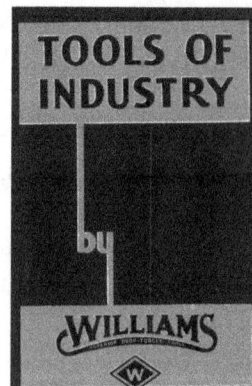

Winchester Arms Co.
New Haven, Connecticut, 1866-Present
Tool Types: Firearms, Knives, Sporting Gear, Planes, Chisels, and Other Tools
Identifying Marks: WINCHESTER and variations
Remarks: In 1848, Walter Hunt devised a self-propelling bullet, essentially a tiny rocket, the predecessor of modern cartridge propelled rounds. His business venture focused on this invention led to the formation of Volcanic Repeating Arms Co., incorporated in 1855 by forty different backers. The first rifle produced by the company was the 1866, a firearm very well received by the market. This company gave rise to several other arms companies, including Smith & Wesson and the eventual withdrawal of Horace Smith and Daniel Wesson that the company floundered, failed, and was eventually bought out by Oliver Fisher Winchester in 1866. Enjoying success with both military and private contracts, Winchester's company shrewdly planned for post-war production in the 1920's and expanded from firearms to cutlery, all manner of sporting goods and a variety of household tools such as hammers and screwdrivers (Williamson, 1952). Along with the Marbles Co., it is among the most collectable of all American 20th century toolmakers. Their tools are occasionally recovered by the Liberty Tool Co. and then quickly slip out the door of the tool store.

Winchester Arms Co. chisel, Davistown Museum IR Collection ID# 70209T4 and catalog cover.

References:
Williamson, H. F. 1952. *Winchester: The gun that won the West.* Washington, DC: A.S. Barnes and Co.
Winchester Repeating Arms Co. (1923). *Winchester: Trade mark: Pocket catalog of tools: 1923: For sale at the Winchester Store.* Winchester Repeating Arms Co., New Haven, CT.
Winchester Repeating Arms Co. (1926). *Winchester: 1926 - 27 product catalog: The Winchester store.* Winchester Repeating Arms Co., New Haven, CT. Reprinted in 1985 by R. L. Deckebach, 413 Walnut St., Royal Oak, MI.
Winchester Repeating Arms Co. (1931). *Winchester, world standard guns and ammunition, Winchester Repeating Arms Co., New Haven, Conn., U.S.A.* Reprinted by R. L. Deckebach, Royal Oak, Michigan in 1989.
Links: http://www.winchester.com/
http://www.winchestercollector.org/guns/w-history.shtml
http://www.davistownmuseum.org/bioWinchester.html

Witherby, Thomas H.
Millbury, Massachusetts and Winsted, Connecticut
Tool Types: Chisels and Draw Shaves
Identifying Marks: WITHERBY, sometimes with WARRANTED
Remarks: "Thomas H. Witherby made chisels and drawknives in Millbury, MA from 1849 to 1850.

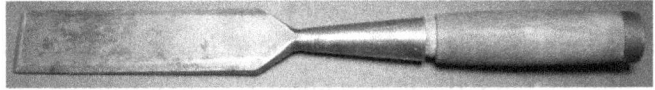

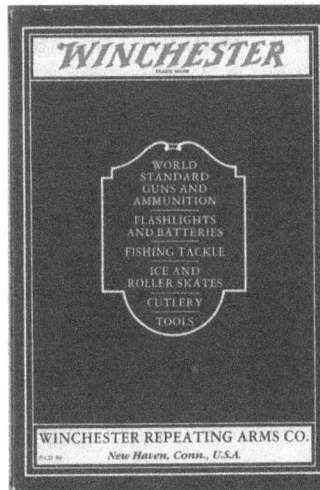

He used the mark T.H. WITHERBY, sometimes in a diamond shaped outline. It is assumed that he was succeeded by the Witherby Tool Co. of Millbury. The only recorded working date for this company is 1868. It is possible that the T. H. WITHERBY mark may actually belong to the company. Witherby is also known to have worked in Winsted, CT and the Witherby Tool Co. was also reported as located in Winsted, CT." (Nelson 1999, 871-3).

The Winsted Edge Tool Works of 1890 used the mark WINSTED EDGE TOOL WORKS / - WINSTED, CONN. U.S.A. - // T.H. WITHERBY. It is not clear whether Witherby adopted this name or if this company bought him out. They used the brand name RAZOR TEMPER. Thomas Witherby, often called Timothy, is, along with the Buck Bros., America's most famous edge toolmaker. Witherby edge tools are frequently recovered by the Liberty Tool Co. Numerous examples are in the Davistown Museum "Art of the Edge Tool" exhibition and illustrated in this publication series.

Links: http://www.geocities.com/sawnutz/witherby/index.htm
http://www.oldtoolsshop.com/z_pdf/1WinstedEdge/WinstedEdgeTools-ne.pdf
http://www.davistownmuseum.org/bioWitherby.html

Wood & Co., William T.
Arlington, Massachusetts, 1895-1905
Tool Types: Ice Tools
Identifying Marks: WM.T.WOOD & CO. (curved over a double fleur-de-lis)
Remarks: A surviving undated price list seems consistent with 1895 printing techniques, but is apparently a condensed version of a larger catalog. It deals mainly in

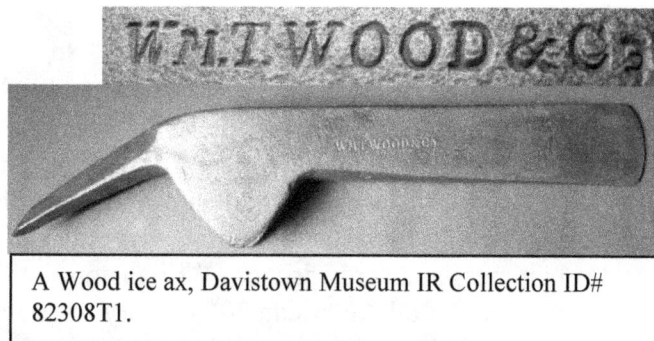

A Wood ice ax, Davistown Museum IR Collection ID# 82308T1.

ice-related tools, elevators, picks, shavers, saws, tongs, and more. DATM (Nelson 1999) lists William T. Wood & Co. in Arlington, MA, from 1845 - 1905. The book *Town of Arlington, Past and Present* by Charles Symme Parker (1907) notes that William T. Wood came there about 1841 and started working with Abner Wyman making and repairing ice tools. He purchased the business in 1845 and ran it by himself until partnering with his brother Cyrus in 1858. He died in 1871. His son, William E. Wood, took over, retaining the name of William T. Wood & Co. In 1905 the company was consolidated with Gifford Brothers of Hudson, NY, as Gifford-Wood Co. Dozens of Wood ice axes, which were widely used in the New England ice harvesting industry, have been recovered by the Liberty Tool Co. (FFLTC).

References:
Wood, Wm. T. & Co. (c. 1895). *Price list, Wm. T. Wood & Co. manufacturers of finest quality ice tools, Arlington, Mass.* The Early Trades & Crafts Society, Long Island, NY.
Links: http://www.davistownmuseum.org/bioWood.html

Yerkes & Plumb
Frankfort and Bridesburg, Pennsylvania, 1870-1887
Tool Types: Axes, Blacksmiths' Tools, Hammers, Hatchets, and Railroad Tools
Identifying Marks: YERKS & PLUMB; Y & P; brand names ARTISANS CHOICE, MECHANICS PRIDE, and ANCHOR.
Remarks: Fayette R. Plumb bought a half interest in an existing business of Jonathan Yerkes in 1870. They moved from Frankfort to Bridesburg in the early 1880s. Plumb bought Yerkes out in

1887 and changed the name to Fayette R. Plumb but continued to use the joint name as a brand name and Y. & P. tools were still being sold in 1900. The company was later absorbed into Philadelphia Tool Co. (Nelson 1999).

Yerkes Tool Co.
-1898
Tool Types: Tools
Remarks: Atha Tool Co. bought this company in 1898 (Nelson 1999).

Zenith Cutlery/Marshall Wells Hdwe. Co.
Duluth, Minnesota, 1884-1960
Tool Types: Adzes, Axes, Bevels, Braces, Edge Tools, Farm Tools, Knives, Planes, and Rules
Identifying Marks: M.W. CO.; M.W.H. CO.; MARSHALL WELLS HARDWARE CO.; ZENITH; MARSWELLS; HARTFORD; SUPERIOR; NORTHERN KING; DEFIANCE; VICTOR; FOUR-SQUARE; TWO-TONE; HANDYMAN (Nelson 1999)

Remarks: This company began as a hardware store and progressed to making and marking its own tools. Most marked tools are from after 1900. The catalog notes that their tools use Swedish pig iron and blister steel. They state their alloys include the use of vanadium, tungsten, chromium, and nickel.
References:
Marshall-Wells Hdwe. Co. [circa 1910] 1987. *Zenith tools and cutlery.* The special Publications Committee, Mid-West Tool Collectors Association.
Links: http://www.davistownmuseum.org/bioMarshallwells.html

Appendix D. Unidentified Toolmakers in the Davistown Museum Collection

The following is a list of tools in the Davistown Museum collection with maker's marks and signatures we have not yet been able to identify. None are listed in the Directory of American Toolmakers (Nelson 1999). Numerous additional unidentified makers' signatures from current un-cataloged tools will be added in the future. Any information on these marks would be appreciated. Email: curator@davistownmuseum.org.

MI: Historic Maritime I (1607-1676): The First Colonial Dominion

| 83102T1 | Moulding plane | "JB" twice upside down in 17th or 18th century script |

MII: Historic Maritime II (1720-1800): The Second Colonial Dominion & the Early Republic

43006T2	Hacksaw	"AOES Co" on the iron ferrule
12801T3	Fillister plane	"Ar. Ritchie" also marked "Stewart" in a smaller font, probably an owner's mark
121805T6	Socket chisel	"C KALER" with an obscured mark
TBE1001A	Framing square	"CUTTER"
42801T12	Beading plane	"H Goss"
101900T1	Plough plane	"H. R. WEBB" on the side of the plane
TBC1003	Hewing ax	"I H", "HARRISON", and "N∘.4"
50402T2	Skew plane	"J. C. Larrabee" with owner's initials "C.J.S." over stamped on the mark, partially obscuring it
81602T14	Framing square	"JBH"
080907T1	Mortising chisel	"KIMPTON" with a backwards N and a scalloped edge around the imprint, there is a first initial that might be "I" or "J"
TBH1002	Dado plane	"Marsh & Winn" and "J. Ho---"
81101T10	Screw auger	"Perkins 5" in 18th century script
TBF1003	Whetstone	"R S DAVIS"
TBD1003	Claw hammer	"TACONY 2"
913108T50	Drawshave	"VESEY" in a square in two places and "ES" in dots on the middle of the blade

913108T41	Drawshave	"Wm FISS" on the blade and "_E. LAUBER" on a band on the handle
040103T4	Bill hook	with a cartouche of a crown and the capital letter M
TBC1001	Socket chisel	with an 18 c. style touch mark "J.W." in a circle with triangles around the edge and dots in between the triangles

MIII: Historic Maritime III (1800-1840): Boomtown Years & the Dawn of the Industrial Revolution

121805T18	Tin snips	"_USESTAHL" and "___ STEEL" and "F W BRANT" with a sun stamp.
72801T8	Square file	"A Prior"
TCC3010	Wood chisel	"A. ARTHUR CAST STEEL"
914108T16	Back saw	"ABRIE"
913108T44	Gouge	"ASKHAM & MOSFORTH"
100605T1	Backsaw	"BARBER & GENN GERMAN STEEL"
TCH1003	Slitting cutter	"BARNEII" and marked "37"
TCH1003A	Cobbler's slitting cutter(?)	"BARNETT 37"
TCR3000	Saw set	"BORUEAU PARIS"
TCC2001	Drawknife	"BROWN & WALKER WARRANTED CAST STEEL"
33002T15	Snips	"Brown Germany Cast Steel"
032203T5	Adjustable bevel	"C G PINKHAM" possibly an owner's mark
31808PC2	Saw	"C. H. TUPPER" on handle and "SUPERIOR TEM__ WARRANTEE" on brass
913108T51	Drawshave	"CAST" and "STEEL" in a box and "I.POPE" in a box.
TCM1001	Cobblestone hammer	"COCKRHYMES & CO" on one side and "J.T. & CO" on the reverse side
51100T1	Block	"D ADAMS MAKER BOSTON"
111001T26	Double calipers	"E. A. Belcher"
TCX1001	Early ship's caulking tools (set)	"E. A. DEXTER"
914108T8	Hatchet	"E. COB"
72801T16	Block plane	"E. French", blade unmarked
TCR1020	Pliers	"Fletcher"
TCN1002	Claw hammer	"G LINDLEY"

914108T5	Dividers	"G. BUCK"
TCR1302	Hand vise	"G. KIPP"
121600T3	Punch	"G. Platte"
TCV1007	Awl	"GEO. LAUTE BOSTON."
TCC3005	Hatchet	"Gray's", with "0" above the touch mark
TCM1005A	Hammer	"H M CHRISTENSEN BROCKTON MASS"
TCR1005	Scraper	"H. M. INMAN"
TCU1001	Stone hammer	"H.C. Briggs"
TCE1002	Pod auger	"HARRESON"
102904T7	Auger	"HAYER T HAYER" and "8"
TCC2002	Gouge	"Holland & Turner, cast steel"
100400T16	Peen adz	"HOLLAND CAST STEEL" with 4 small suns and an oval with a keyhole inside it
TCR1003	Pliers	"HUBER TOOL WORKS 5 PHILADA" and on the reverse side of the handle marked "C. STEEL"
111002T2	Hewing ax	"I H. Harrison No 4"
12801T6	Hewing ax	"J HATCH CAST STEEL"
81200T14	Knife	"J Ward Riverside Mass"
TCC2004	Socket chisel	"J. BRIGGS" "CAST-STEEL" and "#" on the opposite side
TCL1001	Rasp	"J. DAY & CO."
TCC2006	Hewing ax	"J. Emory, cast steel"
TCQ1001	Framing square	"J. F. Brown"
TCU1003	Toothed stone chisel	"J. GERM" with a second illegible signature
111001T1	Hewing ax	"J. Hatch CAST STEEL"
51606T6	Drawknife	"J. MATLACK"
51100T3	Screwdriver	"J. W. Ferren"
041505T21	Drawknife	"J. Windly"
111001T2	Corner chisel	"J.CRAY" and "CAST.STEEL"
51100T5	Scissors	"Jonathan Crookes"

70701T10	Tweezers	"Joseph Lisaro Sheffield England" and "Jos. F. McCoy Co."
31602T10	Coopers' shave	"L Hardy CAST STEEL"
TCC3000	Hatchet	"L. OLSEN"
TCD1002	Low angle block plane	"L.O. Tappan" (probably the owner's signature)
101701T1	Drawknife	"LAVERY CAST STEEL"
TCE1003D4	Auger bit	"LG HALL 16"
81200T11	Cold chisel	"M Fognaty"
TCF1002A	Nail header	"P.S. CRONIN"
4105T3	Socket chisel	"R&HPORTER"
112400T2	Hammer	"R.A. FISH"
71401T12	Try square	"Ridgewell Middletown CONN"
101400T5	Coopers' broad ax	"Roxbury _____ EVRETT CAST STEEL
111900TX1	Coopers' jointer plane	"RYING" on the leg and an illegible signature on the blade
TCD1004	Spar plane	"S T. Livingston"
42602T5	Socket chisel	"S. W. DROWN CAST STEEL"
101701T3	Sheep shears	"Shear Steel W. Wilkinson"
TCU1002	Brick chisel	"SHEARER" in two different places, also marked "SCF"
51606T2	Hay cutter	"STINSON"
TCE1003G7	Auger bit	"T. DAVIS & CO No 6"
TCU1005	Toothed stone chisel	"T. GRANGER"
040904T5	Hewing ax	"T. ROGERS"
TCC2011A	Socket chisel	"TILTON & WHEELWRIGHT MANUFG. CO. WARRANTED CAST STEEL"
TCE1003F6	Auger bit	"TOWNE SNELL 5"
10700-T5	Gouge	"Tremont Co"
102100T17	Eyelet punch	"W F BINGHAM"
83102T8	Dividers	"W H Hale"
62202T1	Rabbet plane	"W. J. Foote", probably an owner

43006T7	Marking gauge	"W. R. Stone"
TCC2005	Shipwright's slick	"WARRANTED CAST STEEL" and "_. TINKHAM"
111001T13	Drawshave	"Wilson Lewiston" with an 8 point asterisk touchmark
TCJ3500	Howell (chiv)	"MORTON ARNOLD"

MIV: Historic Maritime IV (1840-1865): The Early Industrial Revolution

TCR1021A	Pliers?	"?. NISSEL"
100605T2	Hoop driver	"A. G. MORSE&CO" "BOSTON"
071704T5	Drawknife	"A.G.WOOD" and "CAST-STEEL"
TCM1006	Stone hammer	"AHEW"? (in a triangle)
11301T5	Coopers' broad ax	"Beardsley & Tyler"
914108T6	Spoon	"BERTOCCHI"
72801T4	Peen adz	"Boston Arnold"
102100T9	Pin vise	"C HAMACHER"
33002T16	Claw hammer	"C. BARNARD"
32502T5	Punch	"C. HARLTON CAST STEEL" and "6"
914108T10	Ax	"C. MAMM" "PHILAD" and "CAST STEEL"
TCX1003	Ship's caulking iron	"C.B. Timpson & Tucker"
TCT1006	Slick	"C.H.P." (probably the manufacturer's sign), also has other letters and touch marks
83102T4	Wire gauge	"CARANTIE" and marked 1 to 60
81200T13	Wood chisel	"Chas Mellor"
TCR1018	Block	"Clayville Iron Works NY"
040103T5	Marking gauge	"D. Cummings"
51100T6	Adjustable calipers	"E. F. Sibley"
51100T7	Vernier calipers	"E. F. Sibley"
914108T12	Cold chisel	"E. MILLER MS" and on the other side "S.W.T"
110404T1	Smooth plane	"E.R.KING" "MAKER" "E. BOSTON" on nose, ("CHARLES BUCK" "CAST STEEL" "WARRANTED" on blade, "MOULSON BROTHERS" "M B" "WARRANTED" "STEEL" on curling iron)
72801T15	Hand saw	"F Dowst Boston Warrented Cast Steel"

33002T4	Nail header	"F. E. Streeter"
913108T46A	Chisel	"G. H. TUCKER"
TCQ3500	Framing square	"H A WEST PATENTED WARRANTED ___?___ STEEL" AND "B HARMON"
7602T2	Coopers' broad ax	"H. A. W. KING" "LEWIS STNY" (?)
121805T5	Calipers	"H. O. Perry" and "H.O.P."
TCP1005A	Dividers	"H.A. PAGE CAST STEEL"
3405T6	Hand vise	"Heile and Quack"
913108T4	Float	"HWINSUGGLES"
6405T4	Rounding plane	"I. Spear"
14302T15	Gouge	"IH" and "J. Harrison Warranted"
914108T4	Chisel	"J. WADSWORTH" and "CAST STEEL"
041505T2	Caulking iron	"J.STOR"
32802T14	Gristmill stone hammer?	"JOHN HARTMAN Boston Mass"
7602T3	adz	"KING New York"
TCR1012	Clamp	"KNOTT BOSTON"
71401T9	Scraper	"L M Hildreth New Haven CONN PAT Applied For"
3405T2	Wire gauge	"LACENE Mfg. Co Manchester NH" also numerated 2 -12
070705T1	Tongue and groove plane	"LSHOREY" ("WILLIAM ASH & Co" "WARRANTED" and "CAST STEEL" on blade)
42405T6	Drawshave	"M & AM DARLING CAST STEEL WARRANTED"
72801T18	Saw set	"P ? Hopkins"
41801T6	Shears	"P H Hahn NY"
041505T37	Double calipers	"PAT APL'D FOR" "J.P. BARNES"
42607T4	Gouge	"PEUGEOT FRERES" with a man in the moon hallmark
121401T1	Rule	"Revere", marked with hand stamped numerals
42801T6	Dividers	"S H F Bingham Cast Steel"
71401T21	Gouge	"Schroder & Arete"
42405T5	Wood chisel	"Thamesville Co. Cast Steel"
62207T1	Chisel	"THAMESVILLECo" "CAST STEEL"

052107T1	Square	"TURNER & _____" and "GERMAN STEEL"
TCZ1006A	Open ended wrenches	"W. C. HASLAM" on two of them
100605T4	Drawknife	"W. FARNHAM"
TCP1004A	Dividers	"W.D.EVANS"
31701T1	Candlewick cutter	"W_BANNA_ PATD _ 25th ____"
913108T13A	Peavey or cant dog spike	"WILLARD", the first initial is obscured
TCZ1008	Open ended wrench	"YORK M. Co"
72801T3	Double bitted Ax	only "Oakland" is visible along with "S.S." who might be the owner.

IR: The Industrial Revolution (1865f.): Classic Period of American Machinist's Tools

30202T6	Bevel square	"Alworth Bevel Square Rule _ _ _ as made by Stark W_ _ _ _ss USA PATENTED Aug 7, 1888"
041505T33	T square	"C. EGGE"
102503T2	Inside calipers	"D. E. LYMAN"
121805T26	Calipers	"E R Wharton"
914108T2	Thickness gauge	"EINAR HANSON" "-TOOLS-" "WORCHESTER. MASS" and owner's mark "F. W. PAGE"
041505T31	Rule	"Fleming" in script and "U. S. A." and "Fleming Machine Co." "Worcester, Mass" and "No. 1334"
041505T38	Die	"G. S. PAGET" "CO." "-BOSTON-" and "WOOD."
090508T4A	Calipers	"GEO. PLUMPTON" with an X trademark
041505T35	Outside calipers	"H. A. ELLIS" and "J. P."
11301T3	Square	"J B Jopson"
041505T32	Double calipers	"J. HOOD"
72002T8	Wire gauge	"Lacene Mfg Co. Manchester NH"
TJG1001	Lathe tool holder	"MACHINISTS TOOL CO PROV. R.I. PATENTED MAY 26, 1868"
101701T7	Tap drill gauge	"Made by STERLING ELLIOTT NEWTON, MASS, USA"
62202T8	Inside calipers	"Murphy"
32405T1	Countersink	"PATENDED" "JAN 23, 1877." "D.J. ADAMS" "KITTERY, ME." and "R.L. MARKS"
31908T35	Framing square	"S. HAYES" "PATENT" "WARRANTED" and "STEEL"

TJS2201	Measuring device	"STALL & ATHERTON BROCKTON, MASS"
32802T10	Outside calipers	"W. E. TRUFANT Whitman Mass PAT Apr 18 03"
102800T7	Pencil compass	"WT. ATHERHEAD PATENT. REISSUED FEB 18 187?"

IR: The Industrial Revolution (1865f.): Other Factory Made Tools

31501T4	Peen adz	"A DRAUDAY" or "A DRAUBAY" "PAT JUNE 8, 1875" (?)
TTCI3000	Screw jack	"C E HOBBS CO BOSTON" and "3/4 X 2"
040103T6	Back saw	"C. H. BILL & SON WALTHAM MASS" "SPRING STEEL WARRANTED" and an eagle medallion
31602T13	Rasp	"Carver File Co USA"
11301T8	Tool holder	"Cooper & Phillips Patented July 3, 1866 4"
071704T3	Steam gauge	"Crosby Steam Gage and Valve Co. Boston U.S.A. & London, Eng."
102800T6	Pruning shears	"D. Bowers"
040103T14	Expansive bit	"DAVIS THE HARDWARE MAN"
121805T19	Tang chisel	"J. N. Cutler" and "Electric cast steel" and a pioneer on the reverse with "TM"
41302T2	Drawknife	"L. Palmer"
32502T48	Pliers-type tools (4)	"Lindstrom Sweden", "Halle IT Co", "T H Brown" and "P S Studeay"
10407T7	Tin snips	"MATEA"
83102T9A	Pliers	"PAZZANO PAT No. 19027 MADE IN USA"
914108T3	Auger bit	"PETEROR" is the best guess, it is badly obscured
TJG1001A	Carriage upholsterer's hammer	"R C CLAY 1874"
TCZ1004	Stillson wrench	"RED-HEAD" and "MADE IN U.S.A."
30202T12	Combination wrench	"Ryder's Combination Pat'D Nov 10 1896"
TTDA3000	Drawknife	"TINKHAM & CUMMINGS WARRANTED CAST STEEL" with a very unusual eagle and flag touchmark
040103T15	Drill gauge	"W & M Mfg Co. Worcester Mass"
914108T1	Tap wrench	"W. F. PAGE" and "NO. 1."
102904T18	Wrench	"W.--W. MFG. CO. INC." " WORCESTER, MASS." "PAT 7-9-20"
111001T28	Nail holder	"Williams Nail Holder & Guide Patent 1688446"

14302T17	Screwdriver	"Mullen Mfg. Co. Boston Mass Patented"
81602T18	Box scraper	on nut "Holmes PATENT May 6 1868"

IR: The Industrial Revolution (1865f.): Patented and Transitional Planes

101400T4	Smooth plane	"Jacob Reisser New York"
7602T1	Razee Plane	"Made by P Marshall" on the toe and "Buck Brothers Warranteed Cast Steel" on the blade

Tools of Historic Interest not in the Museum Collection

SCOM1001	Hand saw (8 point cross cut)	"Seth Wood 1XL Taunton Mass" in oval with "EXTRA" above the oval, "Spring Steel4 Warranted" below oval

Annotated Bibliographies

Introduction

Readers of this volume of the Davistown Museum *Hand Tools in History* Series please take note of our peculiar bibliographic format. Because of the huge number of citations within this series, we have made some attempt to organize them both by timeframe and by subject matter.

The Volume 6, *Steel- and Toolmaking Strategies and Techniques before 1870* bibliography *European Precedents and the Early Industrial Revolution*, covers subjects pertaining to ferrous metallurgy beginning in the Bronze Age, continuing through the Iron Age (Halstadt, La Téne), the florescence of Roman metallurgy at Noricum, followed by a general survey of citations pertaining to the development of early modern steelmaking techniques in continental Europe and in England. The reader may note this bibliography is limited to English language sources; the extensive foreign language citations pertaining to the study of Iron Age metallurgy in Europe are referenced (Wertime, etc.) but not included.

Volume 7, *Art of the Edge Tool: The Ferrous Metallurgy of New England Shipsmiths and Toolmakers 1607-1882*, has one bibliography with citations pertinent to the introductory essays on New England shipsmiths and edge toolmakers and New England's colonial and early American history.

Volume 8, *The Classic Period of American Toolmaking 1827-1930*, has the following bibliographic format:

I.   The Industrial Revolution in America
II.   U. S. and New England Toolmakers
III.   Tools of the Trades
IV.   Collector's Guides, Handbooks, and Dictionaries
V.   Tool Catalogs
VI.   Tool Journals and Auctions

In addition to these bibliographic citations, each company listing contains additional citations, catalogs, and links.

Volume 9, *Davistown Museum Exhibition: An Archaeology of Tools*, our catalog of tools in the museum collection does not contain any bibliography.

Volume 10, *Registry of Maine Toolmakers*, has a bibliography that is restricted to publications pertaining to registry information sources and Maine and Canadian maritime toolmakers and manufacturers.

Volume 11, *Handbook for Ironmongers*, has an extensive bibliography of sources cited in the handbook and a second special bibliography on metallurgy.

A few citations may appear in more than one bibliography, especially frequently cited texts often referenced in the publication series. If you are seeking a specific citation you may have to peruse the bibliographies in several of the volumes. Most citations pertaining to American toolmakers are in the more extensive bibliographies in this volume.

The Industrial Revolution in America

Also check the bibliographies on specific trades and metalworking and metallurgy.

No author. (1840). *British mechanic's and labourer's handbook, and true guide to the United States.* C. Knight and Co., London.

Abbot, Charles Greeley. (1932). *Great inventions.* In: Abbot, Charles Greeley, Ed. *Smithsonian Scientific Series: Volume 12.* Smithsonian Institution Series, Inc., NY, NY.

Adamson, Rolf. (1969). Swedish iron exports to the United States, 1783-1860. *Scandinavian Economic History Review.* 17.

Alexander, John Henry (1840). *Report on the manufacture of iron: Addressed to the governor of Maryland by J.H. Alexander, topographical engineer of the state.* printed by order of the Senate. William McNeir, Annapolis, MD.

Allen, Richard S. (1967). *Separation and inspiration, concerning the first industrial application of electricity in America.* Historical Publication No. 1. The Penfield Foundation, Crown Point, NY.

Allen, Richard S. (1983). The iron industry of northern New York. *Canadian Mining and Metallurgical Bulletin.* 76. pg. 85-89.

Allen, Ross F., Dawson, James C., Glenn, Morris F., Gordon, Robert B., Killick, David J., and Ward, Richard W. (1990). An archeological survey of bloomery forges in the Adirondacks. *IA, Journal of the Society for Industrial Archeology.* 16(1). pg. 3-20.

Allen, Zachariah. (1832). *The practical tourist.* 2 Vols. A.S. Beckwith, Providence, RI.

Anonymous. (May 8, 1879). Pittsburgh: Its blast furnaces, rolling mills and steelworks. *Iron Age.* 8. pg. 1-3.

Ardey, R.L. (July 18, 1901). The history of the steel plow. *Iron Age.* 68. pg. 26-27.

Asher, Robert. (1983). *Connecticut workers and technological change.* Center for Oral History, University of Connecticut, Storrs, CT.

Backert, Adolphus O., Ed. (1915). *The ABC of iron and steel.* Cleveland, OH.

Bagnall, William R. (1971). *The textile industries of the United States.* Vol. 1. Augustus Kelley, New York, NY.

Baldwin Locomotive Works. (1923). *History of the Baldwin Locomotive Works, 1831-1923.* Bingham, Philadelphia, PA.

Bale, M.P. (1913). *Woodworking machinery.* 3rd Ed.

Bardell, P.S. (1984). The origins of alloy steels. In: Smith, Norman, Ed. *History of Technology, 9th annual volume.* Mansell Publishing, London.

Barker, Elmer E. (1969). *The story of Crown Point Iron.* Historical Publication No. 3. Penfield Foundation, Ironville, NY.

Barraclough, K. C. (1984a). *Steelmaking before Bessemer: Blister steel, the birth of an industry.* Volume 1. The Metals Society, London.

Barraclough, K. C. (1984b). *Steelmaking before Bessemer: Crucible steel, the growth of technology.* Volume 2. The Metals Society, London.

Batchelder, Samuel. (1863). *Introduction and early progress of the cotton manufacture in the U.S.* Reprinted in 1969, Taylor, George Rogers, Ed. by Harper and Row, NY, NY.

Bathe, Greville and Bathe, Dorothy. (1935). *Oliver Evans: A chronicle of early American engineering.* Philadelphia, PA.

Bealer, Alex W. (1976). *The tools that built America.* Barre.

Bedini, Silvio. (1974). *Thinkers and tinkers: Early American men of science.* Scribner, NY, NY. Reprinted in 1989.

- "The best analysis of early American instrument manufacture; much on surveying instruments." (Tesgeract Catalog, Fall 1989, pg. 26).

Bell, Daniel. (1973). *The coming of post-industrial society.* NY, NY.

Bell, Isaac Lowthian. (1884). *Principles of the manufacture of iron and steel: With some notes on the economic conditions of their production.* G. Routledge, NY, NY.

Berthoff, Rowland T. (1953). *British immigrants in industrial America, 1790 - 1950.* Cambridge, MA.

Binder, Frederick M. (1974). *Coal age empire: Pennsylvania coal and its utilization to 1860.* Pennsylvania Historical and Museum Commission, Harrisburg, PA.

Bining, Arthur Cecil. (1933). *British regulation of the colonial iron industry.* University of Pennsylvania Press, Philadelphia, PA.

- The most important of all tracts on the colonial iron industry.
- Bining's bibliography contains a particularly comprehensive and useful listing of colonial era manuscripts and official printed sources as well as an excellent general bibliography

pertaining to 17th and 18th century publications on trade, travel, and American history. Bining's bibliography also includes newspaper and periodical citations and publications of historical societies. A reprint of this bibliography is available for perusal at the Davistown Museum library.

- See *Volume 7:Art of the Edge Tool: The Ferrous Metallurgy of New England Shipsmiths and Toolmakers: From the Construction of Maine's First Ship, the Pinnace Virginia (1607), to 1882* for a reprint of Bining's Appendix A, "a list of forges and furnaces within the Province of Massachusetts Bay", 1758.

Bining, Arthur Cecil. (1938). *Pennsylvania iron manufacture in the eighteenth century.* Pennsylvania Historical Commission, Harrisburg, PA.

Bishop, J. Leandar. (1866). *A history of American manufacturers from 1608-1866.* 3 vols. Edward Young and Co., Philadelphia, PA. Reprinted in 1967 by Johnson Reprint Corp. vol. 2 and 3.

- The definitive Victorian survey of American manufacturers.

Bixby, George F. (1911). The History of the iron ore industry on Lake Champlain. *New York Historical Association Proceedings.* 10. pg. 169-237.

Bober, Harry. (1981). *Jan van Vliet's book of crafts and trades: With a reappraisal of his etchings.* Early American Industries Association, Albany, NY.

Boileau, Etienne. (1268). *Le livre des metiers de Paris.* [The Book of Paris Trades.] Original is in the Bibliotheque Nationale de France.

Bolger, William C. (1980). *Smithville: The result of enterprise.* The Burlington County Cultural and Heritage Commission, Mount Holly, N.J.

Bolles, Albert S. (1878). *Industrial history of the United States.* Henry Bill Publishing Co., Norwich, CT.

Bourque, Bruce J. (1995). *Diversity and complexity in prehistoric maritime societies: A Gulf of Maine perspective.* Plenum Press, NY.

Brack, H. G. (2006). *Norumbega Reconsidered.* Pennywheel Press, Hulls Cove, ME.

Brack, H. G. (2008a). *Art of the edge tool: The ferrous metallurgy of New England shipsmiths and toolmakers.* Vol. 7. Pennywheel Press, Hulls Cove, ME.

Brack, H. G. (2008b). *Steel- and toolmaking strategies and techniques before 1870.* Vol. 6. Pennywheel Press, Hulls Cove, ME.

Brack, H. G. (2008c). *Handbook for ironmongers: A glossary of ferrous metallurgy terms.* Vol. 11. Pennywheel Press, Hulls Cove, ME.

Brack, H. G. (2008d). *Registry of Maine toolmakers*. Vol. 10. Pennywheel Press, Hulls Cove, ME.

Brain, Jeffrey Phipps. (2007). *Fort St. George: Archaeological investigation of the 1607-1608 Popham Colony*. Occasional Publications in Maine Archaeology. Number 12. The Maine State Museum, The Maine Historic Preservation Commission, and the Maine Archaeological Society, Augusta, ME.

Bridenbaugh, Carl. (1950). *The colonial craftsman*. The University of Chicago Press, Chicago, IL. Reprinted by Phoenix Books in 1966.

Brooke, David. (March 1986). The advent of the steel rail 1857-1914. *Journal of Transport History*. 3rd. s., 7.

Brown, Carrie. (1997). *Pedal power: The bicycle in industry and society*. American Precision Museum, Windsor, VT.

Brown, Carrie. (1999). *Carriage wheels to Cadillacs: Henry Leland and the quest for precision*. American Precision Museum, Windsor, VT.

Brown, Henry T. (1868). *Five hundred and seven mechanical movements*. Brown & Seward, NY, NY. Reprinted in 1995 by Astragal Press.

Brown, Sharon. (1988). The Cambria Iron Company of Johnstown, Pennsylvania. *Canal History and Technology Proceedings* 7. pg. 19-46.

Bucki, Cecelia, et al. (1980). *Metal, minds and machines: Waterbury at work*. Prepared by Cecelia Bucki and the staff of the Mattatuck Historical Society. Waterbury, CT.

Burlingame, Roger. (1953). *Machines that built America*. Harcourt, Brace and World.

Burn, D.L. (1931). The genesis of American engineering competition, 1850-1870. *Economic History*. 2. pg. 292-311.

Calder, Ritchie. (1968). *The evolution of the machine*. American Heritage Publishing Co., Inc. and The Smithsonian Institution, NY, NY.

Cameron, Edward Hugh. (1955). *The genius of Samuel Slater*. Technology Review, Cambridge, MA.

Campbell, H.I. (1936). *Metal castings*. John Wiley & Sons, Inc., NY, NY.

Carnegie, David, with Gladwyn, Sidney G. (1913). *Liquid steel: Its manufacture and cost*. Longmans, Green & Co.

Carnegie-Illinois Steel Corporation. (1938). *U·S·S carilloy steels*. United States Steel, Pittsburgh, PA.

Carroll, Charles F. (1975). The forest society of New England. In: Hindle, Brooke, Ed. *America's wooden age: Aspects of its early technology*. Sleepy Hollow Restorations, Tarrytown, NY.

Casterlin, W.S. (1895). *Forty years at cast steel and tool making*. Scranton, PA.

Chahoon, George. (1875). The making of iron in northern New York Catalan forges. *Iron Age*. 16(7). pg. 7.

Chamberlain, John. (1963). *The enterprising Americans: A business history of the United States*. Harper & Row, New York, NY.

Chamberlain, John. (1981). *Frontiers of change: Early undustrialism in America*. Oxford University Press, New York, NY.

Christensen, Erwin O. (1950). *The index of American design*. Macmillan Company, NY.

Clark, C.M. (1987). Trouble at t'Mill: Industrial archaeology in the 1980's. *Antiquity*. 61. pg. 169-179.

Clark, Victor S. (1929). *History of manufacturers in the United States, 1609-1928*. 3 Volumes. McGraw Hill, NY, NY. Reprinted in 1949.

- Another important listing of American manufacturers.

Clarke, Mary Stetson. (1968). *Pioneer iron works*. Chilton Book Company, NY, NY.

- A history of the first ironworks at Saugus, Massachusetts (Hammersmith).
- A good read and an excellent reference for a secondary school library or history course.
- See Hartley (1957) for the most important publication on the Saugus Ironworks.

Cochran, Thomas Childs. (1981). *Frontiers of change: Early industrialism in America*. Oxford University Press, NY, NY.

Colvin, Fred H. (1947). *Sixty years with men and machines*. Whittlesey House, New York, NY.

Colvin, William H. (1950). *Crucible steel of America: 50 years of specialty steelmaking in the USA*. Newcomen Society in America, New York, N.Y.

Conrad, James L. (1986). The making of a hero: Samuel Slater and the Arkwright frames. *Rhode Island History*. 45. pg. 3-13.

Conrad, James L. Jr. (1997). "Drive that branch": Samuel Slater, the power loom, and the writing of America's textile history. In: Cutcliffe, Stephen H. and Reynolds, Terry S., Eds. *Technology & American history: A historical anthology from Technology & Culture*. The University of Chicago Press, Chicago, IL.

Conzen, Michael P., Ed. (1990). *The making of the American landscape*. Unwin Hyman, Boston, MA.

Cooper, Carolyn C. (1987). The evolution of American patent management: The Blanchard lathe as a case study. *Prologue*. 19. pg. 245-259.

Cooper, Carolyn C. (1987). Thomas Blanchard's woodworking machines: Tracking 19th century technological diffusion. *IA: The Journal of the Society for Industrial Archeology*. 13(1). pg. 41-54.

Cooper, Carolyn C. (1988). A whole battalion of stockers': Thomas Blanchard's production line and hand labor at Springfield armory. *IA, Journal of the Society for Industrial Archaeology*. 14. pg. 37-57.

Cooper, Carolyn C. (1989). *Making shuttles by machine at Wilkinsonville, 1825-1984*. Report by the Museum of American Textile History for the Blackstone River Valley National Heritage Corridor. American Textile History Museum, Lowell, MA.

Cooper, Carolyn C. (1991). *Shaping invention: Thomas Blanchard's machinery and patent management in nineteenth-century America*. Columbia University Press, New York, NY.

Cooper, Carolyn C. (January 2003). Myth, rumor, and history: The Yankee whittling boy as hero and villain. *Technology and Culture*. 44. pg. 82-96.

Cooper, C.C., Gordon, R.B., and Merrick, H.V. (1982). Archaeological evidence of metallurgical innovation at the Eli Whitney armory. *IA: The Journal of the Society for Industrial Archaeology*. 8(1). pg. 1-12.

Cooper, Carolyn C., and Malone, Patrick, M. (March 17, 1990). *The mechanical woodworker in early nineteenth-century New England as a spin-off from textile industrialization*. Paper presented at Old Sturbridge Village Colloquium, Old Sturbridge Village, Sturbridge, MA.

Copeland, M.T. (1966). *The cotton manufacturing industry of the United States*. Kelley, New York, NY.

Copley, F.B. (1923). *Frederick W. Taylor, father of scientific management*. Vol. 2. NY.

Coxe, John Redman. (1812). *The emporium of arts & sciences*. Vol. 2. Joseph Delaplaine, Philadelphia, PA.

Cutcliffe, Stephen H. and Reynolds, Terry S., Eds. (1997). *Technology & American history: A historical anthology from Technology & Culture*. The University of Chicago Press, Chicago, IL.

- A number of excellent articles from this publication are also cited separately in this and other sections of this bibliography.

Daddow, Samuel H., and Bannan, Benjamin. (1866). *Coal, iron, and oil: or, The practical American miner: a plain and popular work on our mines and mineral resources, and text-book or guide to their economical development*. Benjamin Bannon, Pottsville, PA and, J.B. Lippincott, Philadelphia, PA.

Dalzell, Robert. (1987). *Enterprising elite: The Boston Associates and the world they made*. Harvard University Press, Cambridge, MA.

Dane, E. Surrey. (1973). *Peter Stubs and the Lancaster hand tool industry.* John Sherrat & Son, Altrincham.

- An influential early read (1975) for the buyer at the Liberty Tool Co.

Dayrup, Felicia. (1948). *Arms making in the Connecticut Valley: A regional study of the economic development of the small arms industry, 1798-1870*. Vol 33. Smith College Studies in History. Northampton, MA. Reprinted in 1970, George Shumway, York, PA.

Defebaugh, J.E. (1907). *History of the lumber industry of America.* 2 Vols. Chicago, IL.

DeGarmo, Paul E. (1979). *Materials and Processes in Manufacturing*. MacMillan Publishing Company, NY.

de Tousard, Louis. (1809). *American artillerist's companion, or elements of artillery.* C. and A. Conrad and Co., Philadelphia, PA.

Devanney, Joseph J. (Fall 1999). Henry Mercer and the Mercer Museum. *The Fine Tool Journal.* 49(2). pg. 5-7.

Dickinson, Henry W. (1913). *Robert Fulton, engineer and artist: His life and works*. Publisher unknown, London.

Diderot, Denis. ([1751-75] 1959). *A Diderot pictorial encyclopedia of trades and industry: Manufacturing and the technical arts in plates selected from "L'Encyclopédie, ou Dictionnaire Raisonné des Sciences, des Arts et des Métiers" of Denis Diderot: In two volumes*. Vol. 1 and 2. Dover Publications Inc., New York.

Disston, Jacob. (1950). *Henry Disston (1819-1878): Pioneer industrialist, inventor and good citizen.* Newcomen Society in America, New York, N.Y.

Dunwell, Steve. (1978). *The run of the mill: A pictorial narrative of the expansion, dominion, decline and enduring impact of the New England textile industry.* David R. Godine Publisher, Boston, MA.

du Pont, Henry Francis. (1964). *Winterthur portfolio: A journal of American material culture.* Winterthur Museum, Winterthur, DE.

Durfee, William Franklin. (1893/1894). The first systematic attempt at interchangeability in firearms. *Cassier's Magazine.* 5. pg. 469-477.

Egleston, Thomas. (1879-1880). The American bloomery process for making iron direct from the ore. *Transaction of the American Institute of Mining Engineers.* 8. pg. 515-550.

Erickson, C.J. (1957). *American industry and the European immigrant 1860-1885.* Cambridge, MA.

Evans, Oliver. (1795). *The young mill-wright and miller's guide.* Arno Press, NY, NY. Reprinted in 1990 by Oliver Evans Press, Wallingford, PA.

- This important reference was published in many editions.

Evans, Oliver. (1805). *The abortion of the young steam engineer's guide: Containing an investigation of the principles, construction and powers of steam engines. A description of a steam engine on new principles ... A description of a machine, and its principles, for making ice and cooling water in large quantities ... by the power of steam ... A description of four other patented inventions.* Printed for the author by Fry and Kammerer., Philadelphia, PA.

Faler, Paul. (1981) *Mechanics and manufacturers in the early Industrial Revolution: Lynn, Massachusetts, 1780-1860.* SUNY Press, Albany, NY.

Ferguson, Eugene S. (1980). *Oliver Evans: Inventive genius of the American industrial revolution.* Hagley Museum, Greenville, DE.

Fischer, David Hackett. (2008). *Champlain's Dream.* Simon & Schuster, New York, NY.

Fisher, Douglas Alan. (1949). *Steel making in America.* The United States Steel Corp., Bethlehem, PA.

Fisher, Douglas Alan. (1963). *The epic of steel.* Harper & Row, NY.

- One of the first modern writings on iron and steel production in the Europe and US. Now superseded by Gordon's *American Iron*, this text remains an old favorite of the Davistown Museum.
- The museum has reproduced an excerpt from this book on precursors of the blast furnace in the Volume 6 essay section of its website.

Fisher, Marvin. (1967). *Workshops in the wilderness: The European response to American industrialization, 1830 - 1860.* Oxford University Press, NY, NY.

Fitch, Charles H. (1883). Report on the manufactures of interchangeable mechanisms. In: *Bureau of the Census: Report on the manufactures of the United States at the tenth census.* Vol. 2. Census Office, Washington, DC.

Fitch, Charles H. (1884). The rise of a mechanical ideal. *Magazine of American History.* 11. pg. 516-527.

Flexner, James T. (1944). S*teamboats come true: American inventors in action.* Publisher unknown, NY.

Francis, James B. (1855). *Lowell hydraulic experiments.* Little, Brown & Co., Boston, MA.

Freedley, Edwin T., Ed. (1856). *A treatsie on the principal trades and manufactures of the United States.* E. Young & Co., Philadelphia, PA.

Freedley, Edwin T. (1867). *Philadelphia and its manufactures.* E. Young & Co., Philadelphia, PA.

French, Benjamin F. (1858). *History of the rise and progress of the iron trade of the United States from 1620 to 1857: With numerous statistical tables, relating to the manufacture, importation, exportation, and prices of iron for more than a century.* Wiley & Halsted, NY, NY.

Gardner, John. (March 1970). Cast steel. *CEAIA.* 23. pg. 6, 16.

Gibb, George S. (1950). *The Saco-Lowell shops: Textile machinery building in New England.* Harvard University Press, Cambridge, MA.

Giedion, Siegfried. (1948). *Mechanization takes command: A contribution to anonymous history.* Oxford University Press, NY, NY.

Gies, Joseph and Gies, Frances. (1976). *The ingenious Yankees.* Crowell, NY, NY.

Gilmer, Harrison. (1953). Birth of the American crucible steel steel industry. *Western Pennsylvania Historical Magazine.* 36.

Goodman, W.L. (September 1976). Tools and equipment of the early settlers in the New World. *The Chronicle.* In: Pollak, Emil and Pollak, Martyl, Eds. (1991). *Selections from The Chronicle: The fascinating world of early tools and trades.* The Astragal Press, Mendham, NJ.

Gordon, Robert B. (July 1982). The Metallurgical Museum of Yale College and nineteenth century ferrous metallurgy in New England. *Journal of Metals.* 34. pg. 26-33.

Gordon, Robert B. (1983). English iron for American arms: Laboratory evidence on the source of iron used at the Springfield Armory in 1860. *Journal of the Historical Metallurgical Society.* 17. pg. 91-98.

Gordon, Robert B. (1983). Material evidence of the development of metalworking technology at the Collins Axe factory. *Journal of the Society for Industrial Archeology.* 9(1). pg. 19-28.

Gordon, Robert B. (1983). Materials for manufacturing: The response of the Connecticut iron industry to technological change and limited resources. *Technology and Culture.* 24. pg. 602-634.

Gordon, Robert B. (1988). Material evidence of the manufacturing methods used in 'armory practice'. *IA, Journal of the Society for Industrial Archaeology.* 14. pg. 23-25.

Gordon, Robert B. (1988). Who turned the mechanical ideal into mechanical reality? *Technology and Culture.* 29. pg. 744-778.

Gordon, Robert B. (November 1988). Material evidence of the development of metalworking technology at the Collins Axe factory. *Journal of the Society for Industrial Archeology.* 14(2). pg. 19-28.

Gordon, Robert B. (1992). Industrial archeology of American iron and steel. *IA: The Journal of the Society for Industrial Archeology.* 18(1/2). pg. 5-18.

Gordon, Robert B. (January 1989). Simeon North, John Hall, and mechanized manufacturing. *Technology and Culture.* 30(1). pg. 179-188.

Gordon, Robert. (1996). *American iron, 1607 - 1900.* Johns Hopkins University Press, Baltimore, MD.

- The most comprehensive survey of the history of the iron industry in America up to the beginning of the 20th century.
- It contains an excellent bibliography pertaining to iron and steel production in America.
- A number of definitions in Volume 11, *Handbook for Ironmongers*, are taken from Gordon's glossary.

Gordon, Robert B. (2001). *A landscape transformed: The ironmaking district of Salisbury, Connecticut.* Oxford University, New York, NY.

Gordon, Robert B. and Killick, David J. (Summer 1992). The metallurgy of the American bloomery process. *Archeomaterials.* 6(2). pg. 141-167.

Gordon, Robert B. and Malone, Patrick M. (1994). *The texture of industry: An archaeological view of the industrialization of North America.* Oxford University Press, New York, NY.

Gordon, Robert B. and Tweedale, Geoffrey. (1990). Pioneering in steelmaking at the Collins Axe factory. *Journal of the Historical Metallurgy Society.* 24. pg. 1-11.

Graeff, A.D. (1949). *A history of steel casting.* Philadelphia, PA.

Greeley, Horace, Case, Leon, Howland, Edward, Gough, John B., Ripley, Philip, Perkins, F.B., Lyman, J.B., Brisbane, Albert, Hall, Rev. E.E. and others. (1872). *The great industries of the United States being an historical summary of the origins, growth, and perfection of the chief industrial arts of this country.* J. B. Burr & Hyde, Hartford, CT.

Green, Constance McLaughlin. (1939). *Holyoke, Massachusetts; a case history of the Industrial Revolution in America.* Yale University Press, New Haven, CT.

Greenwood, Richard. (1984). *A History of the Blackstone Canal, 1825-1849.* Report for the Rhode Island Historical Preservation Commission.

Grimshaw, Robert. (1882). *Saws: The history, development, action, classification of saws, etc.* E. Claxton & Co., Philadelphia, PA.

Gundrum, Paul Thomas. (1974). *The charcoal iron industry in 18th century America: An expression of regional economic variation.* University of Wisconsin Press, Madison, WI.

Habakkuk, H. John. (1962). *American and British technology in the nineteenth century.* Cambridge, England.

Habakkuk, H. John. (1967). *American and British technology in the nineteenth century: The search for labour saving inventions.* At the Press, Cambridge, England.

Hall, John W. D. (1885). Ancient iron works in Taunton. *Old Colony Historical Society Collections.* 3. pg. 131-162.

Harbord, Frank W. (1904). *The metallurgy of steel.* C. Griffin & Co.

Harte, Charles Rufus. (January 1948). The early American iron industry. *The Chronicle.* 3(14). pg. 119, 122-125.

Hartley, E.N. (1957). *Ironworks on the Saugus: The Lynn and Braintree ventures of the company of undertakers of the ironworks in New England.* University of Oklahoma Press, Norman, OK.

- "Within twenty years of the settlement of Boston in 1630, enterprising men had erected large integrated ironworks in the midst of a virgin wilderness... to convert bog iron into cast and wrought iron." (pg. 3).
- "These ironworks, however, were different in that they were large-scale factory enterprises involving joint financing, a complicated technology, specially imported workmen, and heavy

capital risk. They were big business and heavy industry. In the Bible Commonwealth they stood out as atypical, anachronistic and wonderful." (pg. 3).

- "At Braintree, scene of faltering beginnings, was a blast furnace of good size, a forge consisting of finery, chafery, and giant water-power-driven hammer, a charcoal house, and workmen's cottages. At Lynn, there was a complete ironworks, whose design and engineering were as bold as sophisticated. Here was a huge furnace, a forge comprising two fineries, a chafery, and a big hammer, an extensive water-power system, good storage facilities, workmen's accommodations, and a pier for the use of the small boats which plied the Saugus River laden with the ironworks products. Here was a rolling and slitting mill, the first in the New World, and set up when there were only about a dozen of which we have record in the British Isles and on the continent." (pg. 4-5).

- "Braintree Furnace achieved the first recorded and successful production of cast iron within the limits of what is now the United States." (pg. 10).

- "The indirect process of iron manufacture is more complicated than the direct. The latter calls only for the use of the bloomery, a hearth in which, in a single if discontinuous operation, wrought iron is made directly from the ore. In the indirect process, the blast furnace smelts the ore and turns out cast iron in the form of sows and pigs. By the Walloon method, these are then converted into wrought iron in a series of heating and hammering operations in the forge with its two types of hearth, the finery and the chafery, and its big water-power-driven hammer." (pg. 11-12).

- "It was the competition of iron from the Mother Country which was another of the factors responsible for the relatively early eclipse of New England's first ironworks." (pg. 16).

Harvey, David. (1986). *A progress report on the reconstruction of the American bloomery process.* Paper prepared for the Colonial Williamsburg Foundation, Williamsburg, VA.

Harvey, David. (1988). Reconstructing the American bloomery process. *The Colonial Williamsburg Historic Trades Annual.* 1. pg. 19-38.

Harvey, David. (1989). Replicating America's earliest bloomery process, part one. *The Journal of the Minerals, Metals and Materials Society.* 41(6). pg. 46-47.

Harvey, David. (1989). Replicating America's earliest bloomery process, part two. *The Journal of the Minerals, Metals and Materials Society.* 41(7). pg. 44-46.

Hatch, Charles E., Jr. and Gregory, Thurlow Gates. (July 1962). The first American blast furnace, 1619-1622. *The Virginia Magazine of History and Biography.* 70(3). pg. 259-296.

Heaton, Herbert. (October 17, 1951). The industrial immigrant in the United States, 1783 - 1812. *Proceedings of the American Philosophical Society.* 95(5). pg. 519 - 527.

Hey, David. (1997). The development of the English toolmaking industry during the seventeenth and eighteenth centuries. In: Gaynor, James M., Ed. *Eighteenth-century woodworking tools: Papers presented at a tool symposium: May 19-22, 1994.* Colonial Williamsburg Historic Trades. Volume III. The Colonial Williamsburg Foundation, Williamsburg, VA.

Hindle, Brooke. (1966). *Technology in early America*. Chapel Hill, NC.

Hindle, Brooke, Ed. (1975). *America's wooden age: Aspects of its early technology*. Sleepy Hollow Restorations, Tarrytown, NY.

Hindle, Brooke, Ed. (1981). *Material culture of the wooden age*. Sleepy Hollow Press, Tarrytown, NY.

Hindle, Brooke and Lubar, Steven. (1986). *Engines of change: The American Industrial Revolution, 1790 - 1860*. Smithsonian Institution Press, Washington, DC.

- "The classical Industrial Revolution did not involve all production. Many crafts did not change at all, and those that were mechanized and industrialized changed at very different rates. The four most dramatic changes occurred in (1) iron production, (2) the rise of the steam engine, (3) the mechanization of textile production, and (4) precision machine work." (pg. 15).
- Particularly good on the development of railroad transportation and its relationship with the steam engine.

Historic American Buildings Survey. (1971). *The New England textile mill survey*. Historic American Buildings Survey, Washington, D.C.

Hogan, William T. (1971). *Economic history of the iron and steel industry in the United States*. 5 vols. D.C. Heath & Co., Lexington, MA.

Hoke, Donald R. (1990). *Ingenious Yankees: The rise of the American system of manufacturers in the private sector*. Columbia University Press, NY, NY.

Holoway, M.O. and Squarcy, C.M. (1961). Colonial ironworkers. In: *A History of Iron and Steelmaking in the United States*. The American Institute of Mining, Metallurgical, and Petroleum Engineers, New York, NY.

Hopkins, George M. (1903). *Home mechanics for amateurs*. Scientific American Series, Munn & Co., Publishers, NY, NY.

Hounshell, David. (1978). *From the American system to mass production: The development of manufacturing technology in the United States 1850-1920*. Newark, DE.

Hounshell, David. (1984). *From the American system to mass production, 1800-1932: The development of manufacturing technology in the United States*. Johns Hopkins University Press, Baltimore, MD.

Howe, Henry M. (1890). *The metallurgy of steel*. Scientific Publishing Co., New York, NY.

Howell, Charles. (1975). Colonial watermills in the wooden age. In: Hindle, Brooke, Ed. *America's wooden age: Aspects of its early technology.* Sleepy Hollow Restorations, Tarrytown, NY.

Hummel, Charles F. (1965). *Winterthur portfolio, Volume II 1965.* Early American Industries Association, by permission of The Henry Francis du Pont Winterthur Museum. Reprinted in 1976.

Hummel, Charles F. (1968). *With hammer in hand: The Dominy craftsmen of East Hampton, New York.* University Press of Virginia, Charlottesville, VA.

Hunt, Freeman. (1858). *Lives of American merchants.* 2 volumes. Derby & Jackson, NY, NY.

Hunt, R.W. (1876-1877). A history of the Bessemer manufacture in America. *American Institute of Mining Engineers.* 5. pg. 207.

Hunter, Louis C. (1928-1929). Influence of the market upon technique in the iron industry in western Pennsylvania up to 1860. *Journal of Economic and Business History.* I.

Hunter, Louis C. (1949). *Steamboats on the western rivers: An economic and technological history.* Publisher unknown, Cambridge, MA.

Hunter, Louis C. (1975). Waterpower in the century of the steam engine. In: Hindle, Brooke, Ed. *America's wooden age: Aspects of its early technology.* Sleepy Hollow Restorations, Tarrytown, NY.

Hunter, Louis C. (1979). *A history of industrial power in the United States, 1780-1930. Volume one: Waterpower in the century of the steam engine.* Published for the Eleutherian Mills-Hagley Foundation by the University Press of Virginia, Charlottesville, VA.

- The very best summary of the role of water mills in the industrialization of America.
- Our annotations from this important text are so extensive we have placed them in a Davistown Museum online essay.

Hunter, Louis C. (1985). *A history of industrial power in the United States, 1780-1930. Volume two: Steam power.* Published for the Eleutherian Mills-Hagley Foundation by the University Press of Virginia, Charlottesville, VA.

- An excellent summary of the early role of the steam engine in water and rail transportation and the long delay in its use for most forms of industrial manufacturing.
- For additional annotations see the Davistown Museum online essay.

Hunter, Louis C. (1991). *A history of industrial power in the United States, 1780-1930. Volume three: The transmission of power.* The MIT Press, Cambridge, MA.

- A clear explanation of why steam turbines, replacing the earlier forms of reciprocating steam engines, played a key role in the development of electrical transmission of power.

- For additional annotations see the Davistown Museum online essay.

Hurt, R. Douglas. (1982). *American farm tools: From hand-power to steam-power.* Sunflower University press, Manhattan, KS.

Hutchins, John G. Brown. (1941). *The American maritime industries and public policy, 1789-1914; an economic history.* Harvard University Press, Cambridge, MA.

Hutton, Joseph. (March 31, 1849). Condition and future prospects of the staple trades of Sheffield. *Sheffield Times.*

Innes, Stephen. (1995). *Creating the Commonwealth: The Economic Culture of Puritan New England.* W.W. Norton, New York, NY.

The Iron Industry of Northern New York. (1983). *Canadian mining and metallurgical bulletin.*

Jardini, David. (1997). From iron to steel: The recasting of the Jones and Laughlins workforce between 1885 and 1896. In: Cutcliffe, Stephen H. and Reynolds, Terry S., Eds. *Technology & American history: A historical anthology from Technology & Culture.* The University of Chicago Press, Chicago, IL.

Jeans, J.S. (1880). *Steel, its history, manufacture, properties and uses.*

Jeremy, David. (1981). *Transatlantic Industrial Revolution: The diffusion of textile technology between Britain and America, 1790-1830's.* Harvard University Press, Cambridge, MA.

Jeremy, David. (1990). *Technology and power in the early American cotton industry.* American Philosophical Society, Philadelphia, PA.

Kauffman, H.J. (December 1969). Cast steel. *CEAIA.* 22. pg. 49-50.

Kaempffert, Waldemar B. Ed. (1924). *A popular history of American inventions.* 2 vols. C. Scribner's Sons, New York, NY.

Kebabian, Paul B. (1997). Eighteenth-century American toolmaking. In: Gaynor, James M., Ed. *Eighteenth-century woodworking tools: Papers presented at a tool symposium: May 19-22, 1994.* Colonial Williamsburg Historic Trades. Volume III. The Colonial Williamsburg Foundation, Williamsburg, VA.

Kebabian, Paul B. and Lipke, William, Eds. (1979). *Tools and technologies: America's wooden age.* Robert Hull Fleming Museum, University of Vermont, Burlington, VT.

Keown, Samuel. (1985). Tool steels and high-speed steels 1900-1950. *Historical Metallurgy.* 19(1).

Killebrew, Joseph Buckner. (1881). *Iron and coal in Tennessee.* Tavel and Howell, Nashville, TN.

King, Clarence David, (1948). *Seventy-five years of progress in iron and steel; manufacture of coke, pig iron, and steel ingots*. Published for the Seeley W. Mudd Fund, by the American Institute of Mining and Metallurgical Engineers, NY, NY.

Kirkland, Edward Chase. (date unknown). *Men, cities, and transportation: A study in New England history, 1820-1900*. 2 vols.

Kohn, Ferdinand. (March 27, 1869). *Iron and steel manufacture: A series of papers on the manufacture and properties of iron and steel: With reports on iron and steel in the Paris exhibition ... works in Great Britain and on the continent*. Published by William McKenzie, London, Edinburgh & Glasgow. UK.

Kranzberg, Melvin and Purcell, Carroll W. Jr., Eds. (1967). *Technology in western civilization*. 2 vols. Volume 1: *The emergence of modern industrial society, earliest times to 1900*. Oxford University Press, NY, NY.

Kulik, Gary, and Bonham, Julia C. (1978). *Rhode Island: An inventory of historic and engineering sites*. Government Printing Office, Washington, D.C.

Kulik, Gary, Parks, Roger and Penn, Theodore. (1982). *The New England mill village, 1790-1860*. MIT Press, Cambridge, MA.

Kurjack, Dennis C. (1954). *Hopewell Village: National historical site: Pennsylvania*. National Park Service Historical Handbook Series No. 8, Washington, DC. Reprinted in 1961.

Lake, E.F. (May 1912). Steels made in the electric furnace. *Cassier's Magazine*. 42. pg. 99-112.

Lathrop, William Gilbert. (1926). *The brass industry in the United States: A study of the origin and the development of the brass industry in the Naugatuck Valley and its subsequent extension over the Nation*. Revised Edition. W.G. Lathrop, Mount Carmel, CT.

Lenik, Edward J. (1970). The rediscovery of lower Longwood forge. *Bulletin of the Archaeological Society of New Jersey*. 26. pg. 12-21.

Lesley, J. Peter. (1859). *The iron manufacturer's guide to the furnaces, forges, and rolling mills of the United States*. Wiley, New York, NY.

Linklater, Andro. (2002). *Measuring America: How an untamed wilderness shaped the United States and fulfilled the promise of Democracy*. Walker Publishing Co.,Walker & Company, New York, NY.

Linney, Joseph R. (1934). *A History of the Chateaugay ore and iron company*. The Delaware and Hudson Railroad Company, Albany, NY.

Loewen, James W. (1995). *Lies my teacher told me: Everything your American history textbook got wrong*. The New Press, New York.

Loveday, Amos, J., Jr. (1983). *The rise and decline of the American cut nail industry.* Greenwood Press, Westport, CT.

Lyle, Capt. David Alexander and Porter, Samuel W. (1881). *Report on the manufacture and uses of files and rasps.* Ordnance Dept. Report of the Chief of Ordinance; Appendix 35. Washington, D.C.

Malone, Patrick M. (1983). *Canals and industry.* Lowell Museum, Lowell, MA.

Marx, Leo. (1964). *The machine in the garden; technology and the pastoral ideal in America.* Oxford University Press, NY, NY.

Mayr, Otto and Post, Robert C. (1981). *Yankee enterprise: The rise of the American system of manufactures.* Smithsonian Institution Press, Washington, DC.

McGaw, Judith A. Ed. (1994). *Early American technology: Making & doing things from the colonial era to 1850.* University of North Carolina Press, Chapel Hill, NC.

McHugh, J. (1980). *Alexander Holley and the makers of steel.* Baltimore, MD.

McMahon, A.M. and Morris, S.A. (1977). *Technology in industrial America.* Wilmington, DE.

McMahon, A.M. and Morris, S.A. (1850). *Journal of the Franklin Institute.* 20. pg. 41.

Mercer, Henry C. (1929). *Ancient carpenters' tools together with lumbermen's, joiners' and cabinet makers' tools in use in the eighteenth century.* Horizon Press. Fifth edition reprinted in 1975 by the Bucks County Historical Society.

- A most important and essential reference.
- See the Davistown Museum online essays for some photographs of axes from this text.

Mercer, Henry C. (1961). *The bible in iron: Pictured stoves and stoveplates of the Pennsylvania Germans.* The Bucks County Historical Society, Doylestown, PA.

Mirsky, Jeanette and Nevins, Allan. (1952). *The world of Eli Whitney.* Macmillan, New York, NY.

Moriarty, W.H. (1961). *A century of steel castings.* New York, NY.

Morison, Samuel Eliot. (1921). *The maritime history of Massachusetts 1783-1860.* Houghton Mifflin Company, Boston, MA.

Morrison, John H. (1958). *History of American steam navigation.* Stephen Daye Press, NY, NY.

Muir, Diana. (2000). *Reflections in Bullough's Pond: Economy and ecosystem in New England.* University Press of New England, Hanover, NH.

- "The Industrial Revolution began in New England because there were more young men than there was arable land, forcing some of these surplus men to invent new kinds of machinery." (pg. 250).
- Muir defines the Industrial Revolution as a "transition from the use of this year's sunlight to the use of the products of fossil sunlight (coal and petroleum.)" (pg. 248).
- "By the 1780s, there was scarcely a village in New England that could not boast a gristmill, a sawmill, a cooper's shop, a tannery, a fulling mill to finish homespun woolen cloth with mechanical beaters, and a smithy capable of fashioning hardware for kitchen and farm. Where there was clay, there was likely to be a pottery; where the fuller was enterprising, there might be a dyehouse beside his mill." (pg. 74 - 75).
- "In 1810, the year of the shoe peg, every town in southern New England was extensively engaged in manufacturing for distant markets. Across the region, artisans plaited straw bonnets, felted fur and wool for hats, poured brass into molds to make buttons, filed cow horn into combs, and tempered wrought iron to shape augers. With so many hands performing repetitive tasks for so many hours, it seems obvious that someone would think of a way to do a job more quickly, more cheaply, or more easily." (pg. 92).
- "Elisha Root reconceptualized the making of axes [at Samuel Collins's ax factory]. He eliminated the labor of flattening wrought iron, folding it around a steel pin, and forging the two sides together under a trip-hammer, by arranging a series of dies and rollers that could 'die forge' -- or apply pressure to a mold, forming a solid piece of yellow-hot wrought iron into the shape of an ax, with a eye already punched to receive the handle. Root also automated the process of tempering axes with a machine that regulated oven heat and moved axheads through the oven on a rotating wheel. Another machine reduced the amount of hand labor required to give axes a sharp edge by 'shaving' the hardened steel until it need only receive its final finish by hand on a grindstone." (pg. 131).

Mulholland, James A. (1981). *A history of metals in colonial America.* University of Alabama Press, Birmingham, AL.

Murdock, Bartlett. (1937). *Blast furnaces of Carver, Plymouth County.* Privately printed, Poughkeepsie, NY.

National Park Service. (1979). *Ironmaking.* GPO: 1991-283-602/20158. U.S. Department of the Interior, America's National Parks, 470 Maryland Drive, Suite 2, Ft. Washington, PA.

- This is a folded pamphlet that may also be used as a poster. It has an excellent diagram of a blast furnace, is suitable for classroom use, and may be seen on display at the Davistown Museum. Available online from eParks (http://www.eparks.com/eparks/).

National Park Service. (1983). *Hopewell Furnace: A guide to Hopewell Furnace National Historical Site: Pennsylvania.* Handbook 124. U. S. Department of the Interior, Washington, DC.

Needham, J. (1958). *The development of iron and steel technology in China.* 2 vols. Newcomen Society, London.

Nelson, T. Holland. (1922-1923). Comparison of American and English methods of producing high-grade crucible steel. *TASST.* 3. pg. 279-298.

Nichols, Nick. (September 2006). In search of the three-lipped auger. *The Chronicle.* 59(3). pg. 109-10.

North, S.N.D. and North, Ralph H. (1913). *Simeon North; First official pistol makers of the United States; A memoir.* Rumford Press, Concord, NH.

Osborn, Henry S. (1869). *The metallurgy of iron and steel, theoretical and practical.* H.C. Baird, Philadelphia, PA.

Page, Herb. (2004). *The brothers Coes and their legacy of wrenches.* Sunset Mercantile Enterprises, Davenport, IA.

Parker, Margaret T. (1940). *Lowell: A study of industrial development.* Publisher unknown, NY.

Parsons, John E. Ed. (1949). *Saml. Colt's own record: Samuel Colt's own record of transactions with Captain Walker and Eli Whitney, Jr. in 1847.* Connecticut Historical Society, Conn. Printers.

Paskoff, Paul F. (1983). *Industrial evolution: Organization, structure, and growth of the Pennsylvania iron industry, 1750 - 1860.* The Johns Hopkins University Press, Baltimore, MD.

Pearse, John Barnard. (1876). *A concise history of the iron manufacture of the American colonies up to the revolution, and of Pennsylvania until the present time.* Allen, Lane & Scott, Philadelphia, PA.

Penn, Theodore Z. (1981). The development of the leather belt main drive. *IA: Journal of the Society for Industrial Archeology.* 7(1). pg. 1-14.

Percy, John. (1864). *Metallurgy: The art of extracting metals from their ores, and adapting them to various purposes of manufacture.* John Murray, London, UK.

Perry, E.G. (1903). *A trip around Buzzards Bay Shores and Vineyard Sound.* C. S. Binner Corp, Boston, MA.

- On page 197 is a reference to the Pocasset Iron Works in Pocasset.

Pollak, Emil and Pollak, Martyl, Eds. (1991). *Selections from The Chronicle: The fascinating world of early tools and trades.* Astragal Press, Mendham, NJ.

Pollak, Emil and Pollak, Martyl. (1994). *A guide to the makers of American wooden planes, third edition.* Astragal Press, Mendham, NJ.

Pollak, Emil and Pollak, Martyl. (2001). *A guide to the makers of American wooden planes, fourth edition*. Revised by Thomas L. Elliott. Astragal Press, Mendham, NJ.

- The definitive reference on American planemakers.

Pool, J. Lawrence and Pool, Angeline. Eds. (1982). *America's valley forges and valley furnaces*. J.L. Pool, West Cornwall, CT.

Pressnell, L.S. Ed. (1960). *Studies in the Industrial Revolution*.

Pring, J.N. (1921). *The electric furnace*.

Prude, Jonathan. (1983). *The coming industrial order: Town and factory life in rural Massachusetts, 1810-1860*. Cambridge University Press, Cambridge, MA.

Pursell, Carroll W. (1969). *Early stationary steam engines in America; a study in the migration of a technology*. Smithsonian Institution Press, Washington, DC.

Pursell, Carroll W. Jr., Ed. (1981). *Technology in America: A history of individuals and ideas*. The MIT Press, Cambridge, MA.

Quimby, Ian M. G. and Earl, Polly Anne, Eds. (1974). *Technological innovation and the decorative arts*. 19th Winterthur Conference, 1973. Published for the Henry Francis du Pont Winterthur Museum by the University Press of Virginia, Charlottesville, VA.

Ransom, J.H. (1966). *Vanishing ironworks of the Ramapos*. Rutgers University Press, New Brunswick, NJ.

Rebus, Inc. (1984). *New Orleans decorative ironwork*. The Knapp Press, Los Angeles, CA.

Reck, Franklin Mering. (1952). *Sand in their shoes: The story of American steel foundries*. American Steel Foundries.

Jane, Rees and Rees, Mark. (1997). *Christopher Gabriel and the tool trade in 18th century London*. The Astragal Press, Mendham, NJ.

Renwick, James. (1830). *Treatise on the steam engine*. Publisher unknown, NY.

Reynolds, Terry S. (1983). *Stronger than a hundred men: A history of the vertical water wheel*. The John Hopkins University Press, Baltimore, MD.

Rhode Island Department of Environmental Management. (1987). *Working water - A guide to the historic landscape of the Blackstone River valley*. Rhode Island Department of Environmental Management, Providence, RI.

Rhode Island Parks Association. (1997). *Working water - A guide to the historic landscape of the Blackstone River valley.* Rhode Island Department of Environmental Management and the Rhode Island Parks Association, Providence, RI.

Rivard, Paul E. (1974). *Samuel Slater: A short interpretive essay on Samuel Slater's role in the birth of the American textile industry.* Slater Mill Historic Site, Pawtucket, RI.

Rivard, Paul E. (1988). *Samuel Slater, father of American manufactures.* Slater Mill Historic Site, Pawtucket, RI.

Roberts, G.A., Hamaker, J.C. and Johnson, A.R. (1962). *Tool steels.* Cleveland, OH.

Roberts, Kenneth. (1970). *The contributions of Joseph Ives to Connecticut clock technology, 1810-1862.* American Clock and Watch Museum, Bristol, CT.

Roberts, Kenneth. (1976). *Tools for the trades and crafts: An eighteenth century pattern book.* Originally published by R. Timmins & Sons, Birmingham, England. K. Roberts Pub. Co., Fitzwilliam, NH.

Roberts, K.D. and J. (1971). *Planemakers and other edge tool enterprises in New York state in the nineteenth century.* Cooperstown, NY.

Robson, C., Ed. (1875). *The manufactories and manufacturers of Pennsylvania in the nineteenth century.* Philadelphia, PA.

Roden-Hauser, W., Schoenawa, J. and Vom Baur, C.H. (1913). *Electric furnaces in the iron and steel industry.* 2nd Ed. NY.

Roger, Bob. (September 2006). The three-pod auger. *The Chronicle.* 59(3). pg. 111-12.

Rolando, Victor R. (1975). *Taconic ironworks and slagheaps.* Self-published, Box 281, Nassau, NY.

- This includes the locations of blast furnaces in all the New England states.

Rolando, Victor R. (1992). *200 years of soot and sweat: The history and archeology of Vermont's iron, charcoal, and lime industries.* Vermont Archaeological Society, Burlington, VT.

Roper, Stephen. (1874). *A catechism of the high pressure or non-condensing steam engines.* Publisher unknown, Philadelphia, PA.

Rorabaugh, W. J. (1986). *The craft apprentice: From Franklin to the machine age in America.* Oxford University Press, NY, NY.

Rosenberg, Nathan. (1963). Technological change in the machine tool industry, 1840-1910. *Journal of Economic History.* 23. pg. 414-443.

Rosenberg, Nathan, Ed. (1969). *The American system of manufactures.* Chicago, IL.

Rosenberg, Nathan, Ed. (1969). *The American system of manufactures: The report of the committee on the Machinery of the United States, 1855, and the special reports of George Wallis and Joseph Whitworth, 1854.* Edinburgh University Press, Edinburgh. UK.

Rosenberg, Nathan. (1972). *Technology and American economic growth.* NY, NY.

Rosenberg, Nathan. (1975). America's rise to woodworking leadership. In: Hindle, Brooke, Ed. *America's wooden age: Aspects of its early technology.* Sleepy Hollow Restorations, Tarrytown, NY.

Rosovsky, Henry, Ed. (1966). *Industrialization in two systems.* John Wiley & Sons, NY.

Rowe, Adam Ward. (1977). *Connecticut's cannon: The Salisbury Furnace in the American Revolution.* The American Revolution Bicentennial Commission of Connecticut, CT.

Rowe, F.H. (1938). *History of the iron and steel industry in Scioto County, Ohio.* Columbus, OH.

Russell, Carl P. (1967). *Firearms, traps, and tools of the mountain men.* Alfred A. Knopf, New York, NY.

Rutsch, Edward S. and Morrell, Brian H. (1992). An industrial archeological survey of the Long Pond Ironworks, West Milford Township, Passaic County, New Jersey. *Journal of the Society for Industrial Archeology.* 18(112). pg. 40-60.

Saul, Samuel B. Ed. (1970). *Technological change: The United States and Britain in the nineteenth century.* Methuen, London, UK.

Sawyer, John E. (1954). The social basis of the American system of manufacturing. *Journal of Economic History.* 14. pg. 361-279.

Sayward, Elliot M. (1971). *Of plates and purlins: Grandpa builds a barn: Being a compendium of information in the words of those who saw and took part in the building of timber-framed barns.* Early Trades and Crafts Society & Friends of the Nassau County Museum, Old Bethpage Village Restoration, NY.

Sayward, Elliot M. (1972). *The cooper and his work: Definitions, operations, materials, tools.*

Sayward, Elliot M., Ed. (1989-92). *The Chronicle of the early American Industries Association.* Early American Industries Association, Inc., Albany, NY.

- Elliot Sayward edited *The Chronicle* for several years and also authored articles in the following issues.
- 43(1). pg. 9, 43(2). pg. 46, 44(2). pg. 41, 48(1). pg. 28, 48(3). pg. 57, 59(3). pg. 109.

- Chips and sawdust. 43(4). pg. 100; 44(2). pg. 53.
- Dutch treat: The van Vliet Etchings of crafts and trades, part I. 46(1). pg. 7–11.
- The pith of the matter: A response from your editor. 41(1). pg. 9–10.
- Letter. 45(3). pg. 94.
- Two letters. (July/August 2004; September 2004).
- Elliot Sayward also published *Plane Talk*, a quarterly journal, for ten years up until 1991.

Sayward, Elliot M. (1996). The Oyster Bay connection: The first American newspaper advertisement offering items for sale. *The Freeholder*. www.oysterbayhistory.org/Freeems.htm.

Schallenberg, Richard H. and Ault, David A. (1997). Raw materials supply and technological change in the American charcoal iron industry. In: Cutcliffe, Stephen H. and Reynolds, Terry S., Eds. *Technology & American history: A historical anthology from Technology & Culture*. The University of Chicago Press, Chicago, IL.

Schenck, Helen R. and Knox, Reed. (1985). Wrought iron manufacture at Valley Forge. *MASCA Journal*. 3. pg. 132-141.

Schenck, Helen R. and Knox, Reed. (1986). Valley Forge: The making of iron in the eighteenth century. *Archaeology*. 39. pg. 27-33.

Shlakman, Vera. (1935). *Economic history of a factory town; a study of Chicopee, Massachusetts*. The Department of History of Smith College, Northampton, MA.

Siracusa, Carl. (1979). *A mechanical people: Perceptions of the industrial order in Massachusetts: 1815 - 1880*. Wesleyan University Press, Middletown, CT.

Sisco, F.T. (1924). *The manufacture of electric steel*. New York, NY.

Sisson, William. (1992). A revolution in steel: Mass production in Pennsylvania, 1867 - 1901. *Journal of the Society for Industrial Archeology*. 18(1/2). pg. 79-93.

Sloane, Eric. (1964). *A museum of early American tools*. Funk & Wagnalls, Inc. Reprinted in 1973 by Ballantine Books, NY, NY.

Sloane, Eric. (1965). *Diary of an early American boy Noah Blake 1805*. Ballantine Books, New York, NY.

Sloane, Eric. (1986). *Eric Sloane's sketches of America past*. Promontory Press, NY, NY. Originally published in three volumes (1962, 1964, 1965) as *Diary of an early American boy, A museum of early American tools* and *A reverence for wood*.

Smith, Cyril S. (1961). The interaction of science and practice in the history of metallurgy. *Technology and Culture*. 2.

Smith, Cyril S. (1964). The discovery of Carbon in steel. *Technology and Culture.* 5.

Smith, Elmer L. (1973). *Early tools and equipment.* Applied Arts Publishers, Lebanon, PA.

Smith, H.R. Bradley. (1966). *Blacksmith's and farriers' tools at Shelburne Museum: A history of their development from forge to factory.* Museum Pamphlet Series, Number 7. The Shelburne Museum, Inc., Shelburne, VT.

- The most important primary source of information on blacksmithing including a concise explanation of forging and tempering steel for use in hand tools. See especially pages 156-163 for an explanation of how axes were made and the differences between the American ax and the European trade ax.
- This book also contains one of the best explanations of how the production of steel evolved from the first blacksmith shops in colonial America including descriptions of the first steel poured by the Bessemer process in 1864, the development of the open hearth steel process in 1868, the electric arc furnace in 1911, the low frequency and high frequency electrical furnaces of 1913 and 1930, and the basic oxygen process introduced in 1955.
- "...wrought iron has qualities which make it superbly fitted for hand forging, which are not equaled by mild steel. Its ductility and softness, and the way it flows under the hammer make it prized by smiths who must forge it with hand hammers." (pg. 46).
- "The smith used a metal pattern ... to form the eye of an axe ...by splitting the metal lengthwise and wrapped it around the swage. A steel bit was then welded into the open end and the entire axe drawn to the proper shape. The bit and *poll* (opposite end to the bit) is made of steel. The poll is weighted with hardened steel to give balance and to protect the iron body when the axe is used to drive wedges into wood." (pg. 156).
- "Today there is not a trip hammer to be found in the entire plant. They went out in 1930 when the axe was completely forged of high carbon steel." (pg. 160).
- "Welding is the fusion of two pieces of metal by heating and hammering. ...Fluxes, such as borax and ammonium-chloride, and clean sand, are put sparingly on the metal just before the welding heat is reached. The flux lowers the melting point of the scale [oxide of iron], allowing the scale to run off, and at the same time it serves as a prophylactic, sealing out the air which would cause more oxide of iron to form. After the flux is applied, the metal is brought to a welding heat and welded on the anvil with the hammer." (pg. 47).
- Clear, well organized, concise, enjoyable reading and an indispensable reference for anyone interested in the history of tools.
- This is one of the references used to construct the chronology of tool manufacturing in the Davistown Museum's *Hand Tools in History* publication series.
- An early influence on the missions and tool picking strategies of the Liberty Tool Co.

Smith, Merritt Roe, Ed. (1985). *Military enterprise and technological change.* MIT Press, Cambridge, MA.

Smithurst, Peter. (Spring 2001). The second Industrial Revolution: New England and the British connection. *Tools & Technology.* 19(1). pg. 1 - 4.

Spring, Laverne W. (1917). *Non-technical chats on iron and steel and their application to modern industry*. Frederick A. Stokes Company, NY, NY. Reprinted in 1992 by Lindsay Publications Inc., Bradley, IL.

St. George, Robert Blair. (1979). *The wrought covenant: Source material for the study of craftsmen and community in southeastern New England: 1620-1700*. Brockton Art Center, Fuller Memorial, Brockton, MA.

Stanley, Autumn. (1995). *Mothers and daughters of invention: Notes for a revised history of technology*. Rutgers University Press, New Brunswick, NJ.

- The most comprehensive listing of women inventors and their inventions.

Stansfield, A. (1907). *The electric furnace.* New York, NY.

Stapleton, Darwin H. (1987). *The transfer of early industrial technologies to America.* American Philosophical Society, Philadelphia, PA.

Steinberg, Theodore. (1991). *Nature incorporated: Industrialization and the waters of New England.* Cambridge University Press, Cambridge, UK.

Steeds, S. (1969). *A history of machine tools: 1700 - 1910.* Clarendon Press, Oxford, England.

Stephens, David. (1973). *Forgotten trades of Nova Scotia.* Petheric Press Ltd., Canada.

Stone, Orra. (1930). *History of Massachusetts industries: Their inception, growth and success.* 3 vols. S. Clark, Boston, MA.

- With respect to the vigorous iron industries of the Massachusetts Bay Colony, and Massachusetts during the early Republic, this publication is totally useless. It has no individual listings of foundries, forges, edge toolmakers, shipsmiths, blast furnaces, and makes no mention of such famous iron foundries as that at Bridgewater, where the iron sheathing for the Monitor was made, nor of the many blast furnaces of Taunton, Carver, and elsewhere, which supplied shot and cannon in both the Revolutionary War and the War of 1812.

Strandh, Sigvard. (1989). *The history of the machine.* Dorset Press, NY, NY.

Strassmann, W. Paul. (1956). *Risk and technological innovation: American manufacturing during the nineteenth century.* Cornell University Press, Ithaca, NY.

Swank, James M. (1884). *History of the manufacture of iron in all ages, and particularly in the United States for three hundred years, from 1585 to 1885.* Published by the author, Philadelphia, PA.

Swank, James M. (1892). *History of the manufacture of iron in all ages and particularly in the United States from colonial times to 1891*. The American Iron and Steel Association, Philadelphia, PA.

Swank, James M. (1897). The manufacture of iron in New England. In: Davis, William, Ed. *The New England states*. D.H. Hurd and Co., Boston, MA.

Taber, M. Van H. (1959). *A history of the cutlery industry in the Connecticut Valley.* Northampton, MA.

Tann, Jennifer. (1970). *The development of the factory.* Cornmarket Press, London, England.

Taylor, George Rogers. (1951). *The transportation revolution.* Holt, Reinhart,Winston, NY, NY.

Temin, Peter. (1964). *Iron and steel in nineteenth-century America.* MIT Press, Cambridge, MA.

Temin, Peter. (2000). *Engines of enterprise: An economic history of New England.* Harvard University Press, MA.

Thallner, O. (1902). *Tool steel.* Philadelphia, PA.

Thompson, Elroy S. (1928). *The history of Plymouth, Norfolk and Barnstable counties.* NY.

Thompson, Michael D. (1976). *The iron industry of western Maryland.* Morgantown, W.VA.

Thurston, Robert Henry. (1884). *A history of the growth of the steam-engine.* D. Appleton and Co., NY, NY.

Tryon, Rolla M. (1917). *Household manufactures in the United States, 1640 - 1860. A study in industrial history.* University of Chicago Press, Chicago, IL.

Tucker, Barbara. (1984). *Samuel Slater and the origins of the American textile industry, 1790-1860.* Cornell University Press, Ithaca, NY.

Tunis, Edwin. (1999). *Colonial craftsmen: And the beginnings of American industry.* The Johns Hopkins Press, Baltimore, MD.

- The contents are excellent for classroom preparation.

Turnbull, Archibald D. (1928). *John Stevens: An American record.* Publisher unknown, NY.

Turner, Ella M. (1930). *James Rumsey: Pioneer in steam navigation.* Publisher unknown, Scottsdale, PA.

Tweedale, Geoffrey. (1983). *Sheffield steel and America: Aspects of the Atlantic migration of special steelmaking technology, 1850-1930.* F. Cass & Co., London, England.

Tweedale, Geoffrey. (1983). *The Sheffield steel industry and its allied trades and the American market, 1850-1930.* publisher unknown.

Tweedale, Geoffrey. (1987). *Sheffield steel and America: A century of commercial and technological interdependence, 1830-1930.* Cambridge University Press, NY, NY.

Tweedale, Geoffrey. (1995). *Steel city: Entrepreneurship, strategy, and technology in Sheffield 1743-1993.* Oxford, UK.

Tylecote, Ronald F., Austin, J.N. and Wraith, A.E. (1971). Mechanics of the bloomery process in shaft furnaces. *Journal of the Iron and Steel Institute.* 210. pg. 342-363.

United States Bureau of the Census. (1954). *Facts about the census of manufactures, 1810-1954: The only complete and authentic body of statistics on all manufacturing industries in the United States, being taken for 1954.* U.S. Dept. of Commerce, Bureau of the Census, Washington, DC.

United States Treasury Dept. (1892). *Alexander Hamilton's famous report on manufactures, made to Congress, December 5, 1791, in his capacity as Secretary of the Treasury.* Home Market Club, Boston, MA.

Uselding, P. (October 1974). Elisha K. Root, forging and the "American system". *Technology and Culture.* 15. pg. 561-562.

Vander Voort, George F. (1984). *Metallographic principles and practice.* McGraw-Hill, New York, NY.

Vander Voort, George F., Ed. (1986). *Applied metallography.* Nan Nostrand Reinhold, New York, NY.

Wagoner, H.D. (1966). *The US machine tool industry from 1900 to 1950.* Cambridge, MA.

Walton, Perry. (1912). *The story of textiles: A bird's-eye view of the history of the beginning and the growth of the industry by which mankind is clothed.* John S. Lawrence, Boston, MA.

Ware, Caroline F. (1931). *The early New England cotton manufacture in the United States.* Houghton Mifflin, Boston. Reprinted in 1966 by Russell & Russell, NY, NY.

Warren, Kenneth. (1973). *The American steel industry, 1850-1970: A geographical interpretation.* Clarendon Press, Oxford. UK.

Warren, William L. (July 1992). The Litchfield Iron Works in Bantam, Connecticut 1729 - 1825. *Tools & Technology.* Special Edition.

Wayman, M.L., King, J. CH. and Cuddock, P.T. (c. 1990). *Aspects of early North American metallurgy*. British Museum Occasional Paper 79, London, England.

- The subject of this publication is restricted to Pacific coastal Native American use of iron and steel knives by indigenous coastal traders. These steel and iron tools originating in Siberia and were brought across the Bering Sea as trade items. Athapascans among other Native American groups utilized these knives prior to contact with European settlers.

Welsh, Peter C. (1965). United States patents 1790-1870. New uses for old ideas. *Contributions from the Museum of History and Technology, Paper 48. United States National Museum, Bulletin 241*. Washington, DC, pg. 109-152.

Welsh, Peter C. (1965). *United States patents: 1790 to 1870, new uses for old tools*. Smithsonian Institution, Washington, DC.

Welters, Linda. (June 2001). Early power loom fabrics in Rhode Island quilts. *The Chronicle*. 54(2). pg. 67-74.

Westcott, Thompson. (1857). *The life of John Fitch, the inventor of the steamboat*. Publisher unknown, Philadelphia, PA.

White, George Savage. (1836). *Memoir of Samuel Slater: The father of American manufactures; connected with a history of the rise and progress of the cotton manufacture in England and America; with remarks on the moral influence of manufactories in the United States*. Printed at no. 46, Carpenter Street, Philadelphia, PA.

White, John H. (1968). *A history of the American locomotive: Its development: 1830-1880*. Johns Hopkins University Press, Baltimore, MD.

White, John H. (1978). *The American railroad passenger car*. Baltimore, MD.

Whitehill, Walter Muir. (1965). *The arts in early American history*. Chapel Hill, NC.

Wilkinson, Norman B. (1997). Brandywine borrowings from European technology. In: Cutcliffe, Stephen H. and Reynolds, Terry S., Eds. *Technology & American history: A historical anthology from Technology & Culture*. The University of Chicago Press, Chicago, IL.

Williamson, Scott Graham. (1960). *The American craftsman*. Crown Publishers, NY.

Wilson, Mitchell. (1960). *American science and invention: A pictorial history: The fabulous story of how American dreamers, wizards, and inspired tinkerers converted a wilderness into the wonder of the world*. Bonanza Books, NY, NY.

Winsor, Justin, Ed. (1881). *The memorial history of Boston including Suffolk County, Massachusetts. 1630-1880*. 4 vols. James R. Osgood and Company, Boston, MA.

Worssan, W. S. (1892). *The history of the bandsaw*. Manchester.

Zimiles, Martha and Murray. (1973). *Early American mills*. Clarkson Potter, NY, NY.

U. S. and New England Toolmakers

Including Toolmakers of the Maritime Provinces and Important Continental Toolmakers

Also check the Maine toolmakers in the *Registry of Maine Toolmakers*, the tool catalogs bibliography, and citations for the tool manufacturers. A selection of European makers may be found in the trades bibliography.

No author. (April 1872). The Morse twist-drill. *The Manufacturer and Builder.* 4(4). pg. 78-79.

- See this article online at: http://cdl.library.cornell.edu/cgi-bin/moa/moa-cgi?notisid=ABS1821-0004-150.

No author. (December 1893). The Morse Twist Drill Co. and its products. *The Manufacturer and Builder.* 25(12). pg. 269.

- See this article online at: http://cdl.library.cornell.edu/cgi-bin/moa/moa-cgi?notisid=ABS1821-0025-712. It includes a nice illustration of the drills.

No author. (Fall 1982). Lufkin's History. *Cooper Canada Newsletter.* 4(4). Reprinted by Ken Roberts Publishing Co.

No author. (June 21, 1964). 100 years of leadership in metal cutting. *The New Bedford Sunday Standard-Times.* New Bedford, MA.

- The Morse Twist Drill & Machine Co.'s 100th anniversary in 1964.

Aber, R. James. (November 19, 1978). *Some notes on Gage planes.* Crafts of NJ meeting.

- A Xerox of a short bibliography of John Porcius Gage is attached to the back of one of the copies of these notes in our library.

Anonymous. (December 1967). Auger and auger bit makers. *The Chronicle.* 20(supplement). pg. 1-4.

Avila, Richard T. (1999). *Cesar Chelor and the world he lived in.* Smithsonian Institution, Anacostia Museum, Washington, DC.

- Chelor was a freed slave and prominent planemaker.

Baker, Phil. (2006). The Henry Disston fullback. *Fine Tool Journal.* 56(1). pg. 13.

Baker, Phil. (2006). Roe's patent backsaw. *Fine Tool Journal.* 56(2). pg. 29.

Baker, Phil. (Fall 2008). Phil's saws: Henry Disston and the ergonomic backsaw. *Fine Tool Journal*. 58(2). pg. 10-2.

Baldwin Tool Co. (1857). *Illustrated supplement to ... [tools] manufactured at the Arrowmammett Works, Middletown, Conn.* Charles H. Pelton, Middletown, CT. Reprinted in 1976 by K.D. Roberts Publishing Co., Fitzwilliam, N.H.

Barnard, Henry. (1866). *Armsmear: The home, the arm, and the armory of Samuel Colt: A memorial.* Reprinted in 1976. Private printing, Alvord, NY.

Bates, Alan. (1986). *Thomas Napier, the Scottish connection.* The Early American Industries Association, Inc.

Batory, Dana M. (n.d.). *Vintage woodworking machinery: An illustrated guide to four manufacturers.* The Astragal Press, Mendham, NJ.

- Oliver, Defiance, Fay & Egan and Yates-America.

Batory, Dana. (Fall 2004). Early woodworking machine manufacturers: The terrific twins: Greenlee Brothers & Co. *The Fine Tool Journal*. 54(2). pg. 23-24.

Batory, Dana. (2006). Early wood-working machine manufacturers. *Fine Tool Journal*. 56(1). pg. 22.

Beatty, Charles I. (June 2007). Untangling the Beattys -- a hundred years of edge-tool makers, Part I. *The Chronicle*. 60(2). pg. 49-67.

Beatty, Charles I. (September 2007). Untangling the Beattys -- a hundred years of edge-tool makers, Part II. *The Chronicle*. 60(3). pg. 99-111.

Blaisdell, Katharine. (1982). *Over the river and through the years: Book four: Mills and mines.* Courier Printing Company, NH.

Blanchard, Clarence. (Spring 1997). The number one: Cute and useful. *The Fine Tool Journal*. 46(4). pg. 8-10.

Blanchard, Clarence. (Summer 2004). Leonard Bailey's adjustable try square, patented May 9, 1871. *Fine Tool Journal*. 54(1). pg. 6.

Blanchard, Clarence. (2006). F.M. Bailey's patents. *Fine Tool Journal*.55(3). pg. 19.

Blanchard, Clarence. (2006). Stanley leveling stands. *Fine Tool Journal*. 56(2). pg. 22.

Blanchard, Clarence. (Fall 2007). Stanley plane truth & exceptions to the rule: Birdsill Holly and the Stanley Rule & Level Co. *The Fine Tool Journal*. 57(2). pg. 20-23.

258

Blumenberg, B. (1989). Newbury/Newburyport Massachusetts: A unique cluster of American Planemakers and Cabinetmakers. *The Gristmill.* 54(2,3). pg. 10-13.

Blumenberg, Ben. (September 1995). An early, rare mark of Francis Nicholson. *The Chronicle.* 48(3). pg. 63-66.

- An extraordinary and possibly unique early mark of Francis Nicholson has been found on a plow plane recently acquired by Robert Wheeler of Pepperell, Massachusetts, an 18th-century plow plane collector and tool historian.

Bopp, Carl. (November 19, 1978). *The Gage Tool Company and the Gage plane.* CRAFTS of NJ meeting transcription by R. James Aber.

Boyd, Milton. (June 1995). Rufus Bliss: Premier maker of wooden clamps. *The Chronicle.* 48(2). pg. 62-65.

Brewer, Alden R. and Moore, Charles H. (September 18, 1917). *Keyhole saw patent application.* Serial number 1,240,173. United States Patent Office, Washington, DC.

Broehl, Wayne G. (1984). *John Deere's company: A history of Deere & Company and its times.* Doubleday, New York, NY.

Brown & Sharpe Mfg. Co. (no date). *The micrometer's story: 1867 - 1902.* Brown & Sharpe, Providence, RI. Reprinted by Martin J. Donnelly Antique Tools, Bath, NY.

Browne, George Waldo. (1915). *The Amoskeag Manufacturing Company of Manchester, New Hampshire.* Amoskeag Manufacturing Co., Manchester, NH.

Brundage, Larry. (March 1993). Harvey W. Peace's fan club. *The Chronicle.* 46(1). pg. 19.

Burchard, S. W. (December 1978). Butcher blades in America. *The Chronicle.* 31(4). pg. 61.

- William Butcher, Sheffield maker of plane blades.

Burchard, Seth W. and Walker, Philip. (1985). The earliest planemaking industry? Some notes for the Dutch candidature. *Tool and Trades History Society Journal.* 3. pg. 96f.

Burdick, James M. (ca. 1930). *History of the Bailey Plane business from 1869.* Internal history of the Stanley Rule & Level Company, New Britain, CT.

Butterworth, Dale and Blanchard, Clarence. (2006). Valentine Fabian and his rules. *Fine Tool Journal.* 55 (3). pg. 11.

Chalk, Charles. (February 2003). A history of the Russell Cutlery Company. *Muzzle Blasts.* 64(6). pg. 4-7.

Clark, Neil McCullough. (1937). *John Deere. He gave to the world the steel plow.* Desaulniers & Co., Moline, IL.

Collins, Samuel W. (1868). *The Collins Company historical memoranda, 1826-1871.* MS 72190. Connecticut Historical Society, Hartford, CT.

Cope, K. (December 1992). The Coes Wrench Company. *The Gristmill.* 69. pg. 16.

Cope, Ken. (March 1993). Harvey W. Peace, sawmaker. *The Chronicle.* 46(1). pg. 18-9.

Copeland, Jennie F. (1936). *Every day but Sunday: The romantic age of New England Industry.* Stephen Daye Press, Brattleboro, VT.

- Chapter 9, Machine Shops, includes information on Samuel B. Shcenck, Mansfield Machine Company, Fletcher Manufacturing Company, John Birkenhead, Simon W. Card, John J. Grant, Card Company, David E. Harding and S. W. Card Manufacturing Company.

Crucible Steel Company of America. (n.d.). *A brief history of Park Works.* Crucible Steel Company of America, Pittsburgh, PA.

Crucible Steel Company of America. (1951). *50 years of fine steelmaking.* Crucible Steel Company of America, New York, NY.

DeAvila, Richard. (date unknown). Levi Tinkham. *Plane Talk.* VII. pg. 2-14.

DeAvila, Richard. (December 1989). The Jethro Jones -- Cesar Chelor connection. *The Chronicle.* 42(4). pg. 87-88.

DeAvila, Richard. (December 1993). Ceasor Chelor and the world he lived in: Part II. Continued from 46, 2 (June 1993). *The Chronicle.* 46(4). pg. 91-97.

Devanney, Joe. (Spring 2003). Keen Kutter. *Fine Tool Journal.* 52(4). pg. 10-12.

Devitt, Jack. (2000). *Ohio toolmakers and their tools.* Tavenner Pub Co., Anderson, SC.

DeVoe, Shirley Spaulding. (1968). *The tinsmiths of Connecticut.* Wesleyan University Press, Middletown, CT.

- Available as a loan from the E.A.I.A. Library.

Disston, Henry & Sons, Inc. (no date). *Disston saw, tool and file manual: How to choose and use tools.* Philadelphia, PA.

Disston, Henry & Sons. (1920). *The Disston history: History of the works.* Vol. 2. Disston Saw Co., Philadelphia, PA.

Disston, Henry & Sons. (1922). *Saw, tool and file book*. Philadelphia, PA.

Disston, Henry & Sons, Inc. (April 1942). *Disston saw, tool and file manual*. Philadelphia, PA.

- Published in the 102nd year of the company (1840 - 1942). Rose Antique Tools has some excerpts from the 1941 version of this manual on its website.

Drew, Emily Fuller. (1937). *A century of C. Drew & Company*. Transcript of speech given before a regular meeting of the Jones River Village Club, Inc., on the evening of December the eleventh, 1937. Original located at the Kingston Public Library, Kingston, MA.

- The Museum has set up an online information file on C. Drew & Company. Excerpts from this speech have been included in this file.
- A most important information source on a key New England Toolmaker.

Elliott, Carrie. (2005). *Life on the river: The flow of Kingston's industries*. Town of Kingston and Jones River Village Historical Society, Kingston, MA.

Englund, David V. (2006). Chelor's deluxe version planes. *The Chronicle*. 59 (1). pg. 22.

Farnham, Alexander. (1984). *Early tools of New Jersey and the men who made them*. Kingwood Studio Publications, Stockton, NJ.

Farnham, Alexander. (1992). *Search for early New Jersey toolmakers*. Kingwood Studio Publications, Stockton, NJ.

Fitz Water Wheel Co. (1928). *Fitz Steel Overshoot Water Wheel*. Bulletin No. 70. Fitz Water Wheel Co., Hanover, PA.

Freeman Supply Company. (no date). *Freeman Supply Company conversion tables for making patterns for shell molding of aluminum and cast iron*. Freeman Supply Company, Toledo, OH.

Gage, John P. (April 15, 1913). *Plane patent application*. Serial # 1,059,137. United States Patent Office, Washington, DC.

Gaier, Dan. (Winter 2007). Reach for the wrench: Daniel Chapman Stillson: Inventor of the pipe wrench. *The Fine Tool Journal*. 56(3). pg. 18-21.

Gaier, Dan. (Summer 2007). Reach for the wrench: John C. Speirs - bicycle wrench maker. *The Fine Tool Journal*. 57(1). pg. 20-23.

- A survey of the many bicycle wrenches made by Speirs for other bike-makers.

Gaier, Dan. (Winter 2009). Reach for the wrench: The Mossberg Wrench Company. *The Fine Tool Journal*. 58(3). pg. 22-4.

Garvin, James L. and Garvin, Donna-Belle. (1985). *Instruments of change: New Hampshire hand tools and their makers 1800 - 1900.* New Hampshire Historical Society, Canaan, NH.

- Available for perusal at the Davistown Museum and the Concord, NH, library of the State Historical Society.
- An important reference on New Hampshire toolmakers.

Gibb, G.S. (1943). *The whitesmiths of Taunton.* Cambridge, MA.

Gordon, Robert B. (1991). Machine archeology: The John Gage planer. *Journal of the Society for Industrial Archeology.* 17(2). pg. 3-14.

Graham, John H. & Co. (1970). *A century of hardware, John H. Graham & Co. Inc. 1870 – 1970.* John H. Graham & Co., Inc., Oradell, NJ.

Hall, Elton W. (June 2001). Braddock D. Hathaway a New Bedford toolmaker. *The Chronicle.* 54(2). pg. 60-66.

Halsey, Doris. (1972). *Byram of Sag Harbor.* Unpublished, prepared for the April 18, 1972 meeting of the Early Trades and Crafts Society.

Hammond, John Winthrop. (1941). *Men and volts: The story of General Electric.* J.B. Lippincott Co., Philadelphia, PA & New York, NY.

Hatch, Alden P. (1956). *Remington arms in American histroy.* Rinehart & Co., New York, NY.

Heckel, David E. (no date). *The Stanley "forty-five" combination plane.* David E Heckel, 1800 McComb St., Charleston, IL 61920.

Heckel, David E. (Fall 1996). Winchester tools "as good as the gun". *The Fine Tool Journal.* 46(2). pg. 6-8.

Hoopes, Penrose R. (1930). *Connecticut clockmakers of the eighteenth century.* Dodd, Mead and Company, New York, NY. Reprinted in 1974 by Dover Publications Inc., New York, NY.

Hummel, Charles F. (1968). *With hammer in hand; the Dominy craftsmen of East Hampton, New York.* Published for Henry Francis du Pont Winterthur Museum, University Press of Virginia, Charlottesville, VA.

- A classic reference on an important New York family of craftsmen.
- See our commentary on this text in our publication *The Florescence of American Toolmakers 1713 - 1930.*

Huntslar, Donald A. (2002). *Ohio gunsmiths and allied tradesmen 1750 - 1950.* 5 vols. Association of Ohio Longrifle Collectors, 4443 Johnstown Rd., Gahanna, OH 43230.

Hurd, S. Webster. (April 1933). John Porcius Gage. *The Vineland Historical Magazine*. XVIII(2). pg. 257.

- The in stock Xerox of this article is attached to the back of one copy of Aber's *Some Notes on Gage Planes*.

Ingraham, Ted. (March 2001). Francis, John and Cesar: A different view of their planes. *The Chronicle*. 54(1). pg. 1-8.

- Cesar Chelor, John Nicholson, Deacon Francis Nicholson, known as the "Wrentham Trio".

Jacob, Walter W. (September 1998). The Stanley Rule & Level Company: Its historic beginning. *The Chronicle*. 51(3). pg. 80-84.

- See annotations in the online Davistown Museum information file on Stanley Tool.

Jacob, Walter W. (December 1998). The Stanley Rule & Level Company: Charles L. Mead and the acquisition of E. A. Stearns. *The Chronicle*. 51(4). pg. 120-122.

Jacob, Walter W. (June 2000). Brace up for a bit of Stanley history part III: 1917 to 1958. *The Chronicle*. 53(2). pg. 62-65.

- See annotations in the online Davistown Museum information file on Stanley Tool.

Jacob, Walter W. (March 2001). Stanley tapes measure the world part II. *The Chronicle*. 54(1). pg. 31.

Jacob, Walter W. (June 2001). Stanley tapes measure the world part III. *The Chronicle*. 54(2). pg. 75-80.

Jacob, Walter W. (September 2001). Stanley tapes measure the world part IV. *The Chronicle*. 54(3). pg. 121-125.

Jacob, Walter W. (March 2002). The turn of the screw: The history of Stanley screwdrivers. *The Chronicle*. 55(1). pg. 31-34.

Jacob, Walter W. (2006). Stanley's adjustable cabinetmakers' rabbet planes. *The Chronicle*. 59(1). pg. 25.

Jacob, Walter W. (June 2006). The advertising signs of the Stanley Rule & Level Co. – script logo period (1920-1920). *The Chronicle*. 59(2). pg. 70-74.

Jacob, Walter W. (September 2006). The history of Stanley four-square household tools part I. *The Chronicle*. 59(3). pg. 114-19.

Jacob, Walter W. (March 2008). Stanley hand drills -- Part III: Early twenties steel-frame drills. *The Chronicle*. 61(1). pg. 35-37.

Jacob, Walter W. (December 2008). Stanley hand drills part VI: Defiance brand hand drills. *The Chronicle*. 61(4). pg. 166-70.

Jacob, Walter W. (June 2009). Stanley hollow-handle tool sets. *The Chronicle*. 62(2). pg. 77-81.

Johnson, Laurence A. (December 1958). The hasher. *The Chronicle*. 11(4). pg. 57-59, 64.

- This article contains important information on Laroy S. Starrett.

Johnson, Laurence A. (June 1959). The Barlow knife. *The Chronicle*. 12(2). pg. 17-21.

Jones, Robert S. (May 6, 2002). *Wooden plane iron makers marks*. Self published. 3042 Cochise Cir. SE, Rio Rancho, NM.

Kaufman, Michael D. (September 1979). Some nineteenth century Massachusetts level makers. *The Chronicle*. 32(3). pg. 37-41.

Kebabian, John S. (September 1970). Early American factories: The Disston factory, Philadelphia, Pennsylvania. *The Chronicle*. 23(3). pg. 39-40.

Kebabian, John S. (March 1972). Early American factories: Buck Brothers, Millbury, Mass. *The Chronicle*. 25(1). pg. 10-11.

Kebabian, John S. (September 1972). Early American factories: The Douglas Axe Manufacturing Company: East Douglas, Mass. *The Chronicle*. 25(3). pg. 43-47.

Kebabian, John S. (September 1978). An American axe. *The Chronicle*. 31(3). pg. 45.

- Captain Joseph Allen, Gloucester, Massachusetts.

Kebabian, John S. (September 1984). Isaiah J. Robinson: The St. Johnsbury Tool Company. *The Chronicle*. 37(3). pg. 47-48.

Keller, Charles M., and Benson, Paul. (1989). Ignatius Streibich, blacksmith. *The Chronicle*. 42. pg. 47-49.

Keller, David N. (1983). *Cooper Industries: 1833 - 1983*. Ohio University Press, Athens, OH.

Klenman, Allen. (September, 1984). The smart axe! *The Chronicle*. 37(3). pg. 45-46.

Klenman, Allen. (1984?). *Axes made, found or sold in Canada*. Booklet, self-published.

Klenman, Allen. (1990). *Axe makers of North America*. Whistle Punk Books, Currie's Forestgraphics Ltd., Victoria, B.C.

Klenman, Allen. (March 1998). Josiah Fowler of New Brunswick. *The Chronicle*. 51(1). pg. 25.

- See a copy of this article and it's illustrations in the online Davistown Museum information file on Josiah Fowler.

Klenman, Allen. (December 1998). Amoskeag Ax Company of Manchester, New Hampshire. *The Chronicle*. 51(4). pg. 129-130.

- See the annotations for this article in the online Davistown Museum information file on axes.

Kosmerl, Frank. (March 1995). Rochester, New York: A 19th-century edge tool center: Part 2. *The Chronicle*. 48(1). pg. 7-12, 24, 27.

- Thomas Morgan, Henry W. Stager, Stager & Guild, Stager & Barton, Charles Guild, Samuel Silsby & Company, Henry Gilman, Isaac Marsh, Calvin Morgan, Luman Squires, Globe Building, Henry Bush, T. & B. Bush, Bush & Butler, Bush & Bryan.
- Other than Barton, Rochester marks are rather uncommon in coastal New England. Perhaps their market was to the west and south and ship's carpenter's tools were not a specialty.

Kosmerl, Frank. (June 1996). Rochester, New York: A 19th-century edge-tool center: Part III. *The Chronicle*. 49(2). pg. 35-42.

- Leonard Kennedy, Jr., Benton, Evans & Co., E. & J. Evans, D. O. Crane, Samuel G. Crane, D. M. Shepard, H. Haight & Company, Silsby & Rand, P. Locklin, B. W. Schuyler, Schuyler & Sanborn, W. W. Bryan, Stevens & Co., William & J. Lovecraft, F. Weed, William Walker, Bernard Hughes, Hughes & Buell, M. Gregg, Gregg & Hamilton, Worden, Cole & Gior.

Lamond, Thomas C. (Spring 1996). The innovations of a shave maker: Leonard Bailey ... 1858 - 1869. *The Fine Tool Journal*. 45(4). pg. 12.

Lamond, Thomas C. (Fall 1996). The Bailey Tool Co. -- The Stanley Rule & Level Co. -- Edw. Preston & Sons, Ltd.: A circumstantial connection? or... whatever happened to the Defiance spokeshave line? *The Fine Tool Journal*. 46(2). pg. 11-12.

- "After the Bailey Wringing Machine Co. sold their plane division to the Stanley Rule & Level Co. (January 26, 1880) it is believed that the remainder of the Defiance plane inventory was sold off and no additional planes completely matching the Defiance design were subsequently manufactured. Curiously, none of the Bailey Tool Company spokeshaves were included in the supplement, or updated edition, of the 1879 *Stanley Rule & Level Catalog* but Stanley did offer a variety of the Defiance planes in that updated catalog." (pg. 11).

- "There were eleven different shave models offered by the Bailey Tool Company by the late 1870s." (pg. 11).

Lamond, Thomas C. (Spring 1997). Unconfirmed spokeshaves: Patented by Edward Preston (1908 - 1910). *The Fine Tool Journal*. 46(4). pg. 11.

Lamond, Thomas C. (Fall 1997). Sargent spokeshaves: A brief company history and an overview of their spokeshaves. *The Fine Tool Journal*. 47(2). pg. 10-12.

Lamond, Thomas C. (Fall 2002). Geo. D. Mosher and the Birmingham Plane Co. *The Fine Tool Journal*. 52(2). pg. 13-16.

- Located in Birmingham, Connecticut.
- Two exquisite Birmingham planes are on exhibition at the Davistown Museum.

Lamond, Tom. (Spring 2008). Bailey scraper-spokeshaves. *The Fine Tool Journal*. 57(4). pg. 13-15.

Lamond, Tom. (Winter 2009). Joshua Davies spokeshave. *The Fine Tool Journal* 58(3). pg. 26-7.

Larsen, Ray. (September 2001). "A satisfied customer is the best advertisement" C. Drew & Co.'s caulking irons. *The Chronicle*. 54(3). pg. 102-109.

- See the annotations of this citation in the online Davistown Museum C. Drew & Co. information file.

Lasswell, Pat. (Fall 2001). Jo Fuller bench planes. *Sign of the Jointer*. 3(3). pg. 124-131.

Leach, Patrick. (2002). Stanley Planes: *The superior works: Patrick's blood and gore*.

- One of the most comprehensive websites on Stanley planes (http://www.supertool.com/StanleyBG/stan0.htm).

Lytle, Thomas G. (1984). *Harpoons and other whalecraft*. The Old Dartmouth Historical Society Whaling Museum, New Bedford, MA.

- The single most important reference on New Bedford whalecraft.
- Excerpts from this text are reproduced in an Appendix of the *Registry of Maine Toolmakers*.

Mayer, John. (1994). The mills and machinery of the Amoskeag Manufacturing Company of Manchester, New Hampshire. *Journal of the Society for Industrial Archeology*. 20(1/2). pg. 69-79.

Merriam, Robert L., et al. (1976). *The history of the John Russell Cutlery Company, 1833-1936*. Bete Press, Greenfield, MA.

Morgan, Paul W. (June 1985). The Henry Disston family enterprise: Part one of a three-part article. *The Chronicle*. 38(2). pg. 17-20.

Morison, Samuel E. (1950). *The rope makers of Plymouth*. Publisher unknown, Boston, MA.

Navin, T.R. (1950). *The Whitin machine works since 1831: A textile machinery company in an industrial village*. Harvard Studies in Business History 15. Harvard University Press, Cambridge, MA.

Nicholson, William T. (1878). *A treatise on files and rasps descriptive and illustrated: For the use of master mechanics, dealers, & c ...: With a description of the process of manufacture, and a few hints on the care and use of the file*. Nicholson File Co., Providence, RI.

Page, Herb (Mr. Oldwrench). (Summer 2001). Reach for the wrench: E. F. Dixie, early manufacturer of "Hewet" patent wrenches. *The Fine Tool Journal*. 51(1). pg. 23-24.

- Herb has many other informative articles on wrench makers that are not included below.

Page, Herb (Mr. Oldwrench). (Fall 2001). Reach for the wrench: Coes key model. *The Fine Tool Journal*. 51(2). pg. 6-8.

- A photo from this article is reprinted in the online Davistown Museum Loring and Aury Gates Coes information file.

Page, Herb (Mr. Oldwrench). (Spring 2002). Reach for the wrench: Dodge & Wellington, Worcester wrench makers. *The Fine Tool Journal*. 51(4). pg. 14-15.

Page, Herb (Mr. Oldwrench). (Summer 2002). Reach for the wrench: The evolution of baby Coes wrenches. *The Fine Tool Journal*. 52(1). pg. 15-17.

Page, Herb (Mr. Oldwrench). (Winter 2003). Reach for the wrench: Whitman & Barnes: Extraordinary wrench makers. *The Fine Tool Journal*. 52(3). pg. 15.

Page, Herb (Mr. Oldwrench). (Spring 2003). Reach for the wrench: Hawkeye wrenches. *Fine Tool Journal*. 52(4). pg. 18-20.

Page, Herb. (2004). *The brothers Coes and their legacy of wrenches*. Sunset Mercantile Enterprises, Davenport, IA.

Page, Herb. (Spring 2008). Reach for the wrench: Bemis & Call wrenches. *Fine Tool Journal*. 57(4). pg. 10-12.

Page, Herb. (Fall 2008). Reach for the wrench: The wrenches of Peck, Stow & Wilcox Co. *Fine Tool Journal*. 58(2). pg. 20-3.

Park, David L., Jr. (September 1983). Augustus Rothery -- file cutter. *The Chronicle*. 36(3). pg. 49-52.

Parks, Michelle. (Thursday, September 21, 2006). Plane dealing: Arkansan elevates woodworking to an art form and is named a Living Treasure. *Arkansas Democrat-Gazette*. Style section.

- Larry Williams of Eureka Springs, AK.
- One of the only active American wood planemakers circa 2006.

Pernis, Paul Van. (Summer 1998). Leonard Bailey's first planes. *The Fine Tool Journal*. 48(1). pg. 5-7.

Pierce, Cecil. (Fall 1989). Fifty years a planemaker. *Plane Talk*. 13(3). pg. 196-197.

- Letter from planemaker Cecil Pierce about a smooth plane he made.

Poffenbaugh, John. (1981). *Union iron and wood planes*. Reprinted by the Ohio Tool Collectors Association, Galion, OH.

Price, Jim. (Winter 2003). Asa Weeks' patented expansive bit. *The Fine Tool Journal*. 52(3). pg. 18.

Prine, Charles W. Jr. (2000). *Planemakers of western Pennsylvania and environs*. Astragal Press, Mendham, NJ.

Prosser, Treat T. and Gillette, George W. (February 11, 1873). *Improvement in tools for facing bung-holes: Patent application*. Serial number 135,843. United States Patent Office, Washington, DC.

Rathbone, P. T. (1999). *The history of old time farm implement companies and the wrenches they issued*. P.T. Rathbone, Marsing, ID.

- "This 520 page book pictures over 3000 wrenches from over 750 companies. It has the history of over 500 companies and includes old advertising pictures from them. Included is a 176 page soft cover supplement that lists over 3300 part numbers matched to the company that issued the wrench as well as a price guide." (Amazon.com).

Reeder, Layla. (date unknown). Development of modern bench planes: Millers Falls - the "Buck Rogers" planes. *Tool Talk*.

Rees, Mark. (March 1995). Four planes by Thomas Granford. *The Chronicle*. 48(1). pg. 13-16.

- Granford is one of London's first documented planemakers.

Roberts, Kenneth D. (1975). *Wooden planes in 19th century America*. Ken Roberts Publishing Co., Fitzwilliam, NH.

- See annotations in the online Davistown Museum collector's guides bibliography.

Roberts, Kenneth D. (1978/1983). *Wooden planes in nineteenth century America, volume II: The Union factory, Pine Meadow, Connecticut, 1826-1929.* 2nd Ed. K.D. Roberts Publishing Co., Fitzwilliam, N.H.

Roberts, Kenneth D. and Roberts, Jane W. (1971). *Planemakers and other edge tool enterprises in New York state in the nineteenth century.* New York State Historical Association and Early American Industries Association, self-published.

Robinson, Charles. (date unknown). *Vermont cabinetmakers and chairmakers before 1855.* Publisher unknown.

Rodengen, J. L. (date unknown). *The legend of Stanley: 150 years of the Stanley Works.* Publisher unknown.

Rosebrook, Donald. (1999). *American levels and their makers, volume I, New England.* Astragal Press, Mendham, NJ.

Sayer, William L. Ed. (1889). *New Bedford: Massachusetts: Its history, industries, institutions, and attractions.* Published by order of the Board of Trade.

- The Morse Twist Drill & Machine Company of New Bedford is described on pg. 253 - 256. An illustration from this text can be seen in the online Davistown Museum information file on Morse.

Schiffer, Margaret Berwind. (1966). *Furniture and its makers of Chester County, Pennsylvania.* University of Pennsylvania Press, West Chester, PA.

Schoenky, Jim. (Spring 2003). The old boring tool notebook: McClellan's History of Patents. *Fine Tool Journal.* 52(4). pg. 14-17.

Schwarz, Christopher. (2006). Veritas router plane. *Fine Tool Journal.* 55(3). pg. 22.

Schwer, Wilbert. (September 1993). The Sandusky Tool Company story. *The Chronicle.* 46(3). pg. 57-68.

Sellins, Alvin. (1975). *The Stanley plane: A history and descriptive inventory.* The Early American Industries Association, South Burlington, VT.

- Prior to Walter, this was the only guide to Stanley tools.
- A most important reference on the Stanley Company.

Semel, Daniel M. (1978). *Thomas Grant, ironmonger.* Early American Industries Association & Frauncis Tavern Museum, NY.

Semowich, Charles J. (1984). *American furniture craftsmen working prior to 1920: An annotated bibliography*. Greenwood Press, Westport, CT.

Silver Lake News. (January 29, 1970). *Famed 'Cat's Paw' tool firm reorganizes: C. Drew sold for $99,000*. From the Helen D. Foster - Papers - Clippings, Kingston Public Library, Kingston, MA.

Simonds Manufacturing Co. (1907). *75 years of business progress and industrial advance, 1832-1907*. Cambridge, MA.

Smart, Charles E. (1962-67). *The makers of surveying instruments in America since 1700*. 2 vols. in one. Regal Art Press, Troy, NY.

- Available as a loan from the E.A.I.A. Library.

Smith, Philip Chadwick Foster. (June 2000). Notes on a visit to the Buff & Buff surveying instrument factory, Jamaica Plain, Massachusetts, Monday, 3 June 1968. *The Chronicle*. 53(2). pg. 76, 80.

- See annotations in the online Davistown Museum Buff & Buff information file.

Smith, Roger K. (1981-1992). *Patented transitional & metallic planes in America 1827 - 1927*. 2 vols. North Village Publishing Co., Lancaster, MA.

- See annotations in the online Davistown Museum collector's guides bibliography.

Smith, Roger K. (June 1997). Notes on New England edge tool makers, I: Edge tool makers, Hinsdale, NH, 1840-1900. *The Gristmill*. 87.

- The article lists the following toolmakers:
  - Pliny Merrill 1840-1858
  - Merrill & Wilder 1858-1866
  - Wilder & Thompson 1866-1868
  - Wilder & Hopkins 1870-1873(?)
  - G.S. Wilder 1873-1883
  - C.E. Jennings 1883-1885
  - Jennings & Griffin 1885-1900+(?)
- This article is online at http://www.mwtca.org/OTC/ar000021.htm.

Smith, Roger K. (Summer 1997). Caleb Jewett Kimball: Edge tool maker: Bennington, NH. *The Fine Tool Journal*. 47(1). pg. 5-7.

- A photograph of Kimball's shop from this article is in the online Davistown Museum information file on C. J. Kimball.

Smith, Roger K. (June 2009). The Stanley companies pictorial promotions circa 1860-1960. *The Chronicle* . 62(2). pg. 47-52.

Stanley, Philip E. (1985). *A concordance of major American rule makers*. Iron Horse Antiques, Inc., RD #2, Poultney, VT.

The Stanley Rule and Level Company. (1973). *Combination planes: Historical development, patents and uses*. Ken Roberts Publishing Co., Fitzwilliam, NH.

Stanley Tools. (1937). Grey iron castings. *Tool Talks*. Stanley Tools, New Britain, CT. pg. 15-17.

- This article is reproduced with copies of the illustrations in the online Davistown Museum tool information file on grey iron castings.

Stanley Tools. (1937). Stanley Tool history. *Tool Talks*. Stanley Tools, New Britain, CT. pg. 17.

- This short article has been reprinted in the online Davistown Museum tool information file on Stanley Tool.

Stanley Works. (1946). *Facts about tools: A message from Stanley*. Stanley Tools, New Britain, CT.

Tantillo, Joe. (2003). *Historic past enduring land: A tribute to the machine shops & countryside of Springfield, VT*. Self-published.

- This small book includes information on Jones & Lamson Machine Company, Fellows Gear Shaper Company, Parks & Woolson Machine Company, Bryant Grinder Company and Lovejoy Tool Company.

Taran, Pete. (Summer 1998). Disston, the good, the better, the best. *The Fine Tool Journal*. 48(1). pg. 13-15.

Taylor, Frank Hamilton. (1920). *History of the Alan Wood iron and steel company, 1792-1920*. Private Circulation, Philadelphia, PA.

Tidey, M.B. (Winter 2005). Early wood-working machine manufacturers. *Fine Tool Journal*. 54(3). pg. 24-26.

Turner, Kenneth E. (September 1996). The Starrett wheels that led to more wheels. *The Chronicle*. 49(3). pg. 69-71.

United States Patent Office. (Fall 1990). Augustus Stanley, of New Britain, Connecticut: Letters Patent No. 97,455, dated November 30, 1869; antedated November 20, 1869: Improvement in bit-stock. *Stanley Tool Collector News*. 1(1). pg. 8.

Walter, John. (c. 1993). Reproduction of a Stanley Tools newsletter: *The Iron Age: Thursday, November 3, 1898: The making of the cast iron carpenters' plane.*

Waltham Watch Company. (1940). *Waltham: Watch and clock material catalog.* Waltham Watch Company, Waltham, MA.

Waltham Watch Company. (1948). *Waltham: Watch and clock material catalog.* Waltham Watch Company, Waltham, MA.

Wanamaker, George. (August 1987). *Siegley block, jack, joiner, and smooth planes 1893 - 1920's.* Second Edition. Self-published. 515 S. Madison, Macomb, IL 61455.

Warren, William L. (1978). *Isaac Fitch of Lebanon, Connecticut: Master joiner: 1734 - 1791.* The Antiquarian & Landmarks Society, Inc., of Connecticut, Hartford, CT.

Welcker, Peter, Welcker, Merrill L. and Welcker, Anne P. (September 2000). Generations of augermakers in Kingston, Massachusetts. *The Chronicle.* 53(3). pg. 120-122.

- For more information on the C. Drew & Co. mentioned in this article, see the online Davistown Museum C. Drew information file page.

Wells, John G. and Van Pernis, Paul. (June 2005). Bailey, Woods & Co.: Who was the mysterious Mr. Woods? *The Gristmill.* 9. pg. 8-9.

West, Karl H. (September 1996). Ames Shovel & Tool Company. *The Chronicle.* 49(3). pg. 86-93.

Westley, Robert. (June 1996). Drury-Marsh: St. John, New Brunswick. *The Chronicle.* 49(2). pg. 43-49.

Wheeler, Robert. (March 1993). Nathaniel Potter: Could he be our earliest planemaker? A plane talk special feature. *The Chronicle.* 46(1). pg. 3-4.

- See our annotations plus excerpts and illustrations from this article in older editions of the *Registry of Maine Toolmakers.*

Whelan, John M. (September 1996). Plane chatter: The first American planemaker? *The Chronicle.* 49(3). pg. 85-86.

- "The acceptance of Robert Wooding as the first professional English planemaker rests on the number of surviving planes that bear his name. The smaller number of Thomas Granford planes found to date are enough to qualify him as England's first, under the broader definition. And recently, several other English candidates have arisen based on a few planes and some historical data." (pg. 86).

White, Frank G. (September 1979). A checklist of tinners' tool manufacturers. *The Chronicle*. 32(3). pg. 42-45.

White, Marcus and Humason, H.B. (1953). *New Britain, the center of hardware manufacture ... containing a brief review of its miraculous industrial development.*

- P & F Corbin, Russell & Erwin Mfg. Co. and the Stanley Works, etc.

White, Stephen W. (September 1977). A checklist of woodworking tool manufacturers in New Brunswick and Nova Scotia. *The Chronicle*. 30(3). pg. 47-51.

- So many of their tools appear in New England tool chests that they are included in this section.
- Josiah Fowler commenced toolmaking shortly after he was discharged as a Civil War soldier.
- Woodworking tool manufacturers (pg. 49-51):

New Brunswick

J. R. Googin Hardware	-1889-	hardware dealer
Andrews, John S.	-1889-	small dealer
Broad, E. & H.	-1862-64-	
Burpee, I. F.		rolling mills
Spiller Brothers	1867-75-	
Wood James	-1817-	
Mann Axe & Tool Co. Ltd	-1915-30-	adzes and axes
Maritime Edge Tool Co.	-1908-	axes
Veazy, John	-1889-	
Biggs, Robert B.	-1908-	axes
Moir, Robert		

Nova Scotia

Blenkhorn, James	1842-46	
Blenkhorn and Sons	1846-1962	edge tools
Conolly	-1877-	axes
Cameron, William J.	-1908-	axes
Campbell & Fowler	-1865-77-	carriage springs
Campbell, William	-1881-91-	edge tools
Campbell Brothers	-1891-1920-	axes
Campbell, G. Wilfred & Son	1920-21	axes and edge tools
Campbell & Fowler Ltd.	1921-22	axes and edge tools
Campbell & Campbell	1922-23	axes and edge tools
Edwards, John C.	1863-64-	

Josiah Fowler Co. Ltd.	-1881-1920	axes and edge tools
Ridgeway, Joseph	-1862-64-	files and rasps
The Somerset Axe and Tool Co. Ltd.	-1943-	adzes, axes, cold chisels
Spiller, Hanford B.	-1862-	
Spiller, Samuel	1820-67	axes

Wills, Geoffrey. (1974). *Craftsmen and cabinet-makers of classic English furniture*. St. Martin's Press, NY, NY.

Wing, Anne and Wing, Donald. (June 1983). The Nicholson family - Joiners and tool makers. *The Chronicle*. 36(2). pg. 41-43.

Wing, Anne and Wing, Donald. (Autumn/Winter 1984). *Chronology of 18th century planemakers in southeastern New England*. The Mechanick's Workbench, P.O. Box 420, Marion, MA.

Wing, Anne and Wing, Donald. (Summer 1991). Jonathan Ballou: On the west side of the great bridge, Providence, RI, 1723-1770. *Plane Talk*. V.XV(2).

Wing, Anne and Wing, Donald. (June 2002). A rural rulemaker: Captain Anthony Gifford. *The Chronicle*. 55(2). pg. 46-51.

- This article contains important information about all the early New England and New York City rulemakers including William Belcher, Lemuel Hedge, William Rook and Lawrence Watts.
- "By the early 1820s... William and Thomas Belcher, who had emigrated from Sheffield, were establishing in New York what was to become one of the larger American rulemaking firms of the period." (pg. 46).
- "Undoubtedly the single most important influence on rulemaking in this country was American inventor Lemuel Hedge of Windsor, Vermont. ...Hedge patented his rule-graduating machine in 1827, beginning the machine stamping of wood rules in this country; the British did not use machines for marking rules until decades later." (pg. 46).
- "By the late 1840s the Boston rule trade was beginning to diminish, and Gifford must have been aware of the growing success of the Connecticut firms of Stephens, Chapin, and the Middletown shops, as well as E.A. Stearns in Brattleboro, Vermont, and Belcher in New York. How could he hope to compete with them from the little coastal town of Westport? And yet, the Gifford factory appears to have been reasonably successful for a number of years." (pg. 48).
- "The very fact that Mortimer Hedge had joined the army is an indication that the Westport rule business was extinct by 1862. The small Boston rule factories were mostly finished as well, while Stanley, Chapin, and Stephens in Connecticut were burgeoning. The era of the small rule shop had ended." (pg. 51).

Wing, Donald and Wing, Anne. (September 1995). Simeon Doggett: Loyalist planemaker. *The Chronicle*. 48(3). pg. 35-44.

- This article also contains information on Levi Tinkham and other Middleboro, MA, planemakers.

Wright, Frazer. (1986). *Stanley 1936-1986.* Sheffield.

Wyllie, Robin H. (June 1984). Edge tools of the lower Saint Croix. *The Chronicle.* 37(2). pg. 21-24.

Wyllie, Robin H. (September 1985). New Brunswick revisited. *The Chronicle.* 38(3). pg. 50-51.

- Information about New Brunswick, Canada, edge tool and ax makers.

Tools of the Trades

Agriculture

Cobleigh, Rolfe. (1909). *Handy farm devices and how to make them.* Orange Judd Company, Canada. Ninth reprinting in 1996 by Lyons Press, NY, NY.

Hopf, Carroll. (September 1976). Additional thoughts about grafting tools. *The Chronicle.* 29(3). pg. 53-4.

Keller, Charles M. (June 1988). Scythe anvils and scythe spellings. *The Chronicle.* 41(2). pg. 33.

Lewis, M.J.T. (July 1994). The origins of the wheelbarrow. *Technology and Culture: The International Quarterly of the Society for the History of Technology.* 35(3). pg. 453-475.

Markwell, Dawes. (June 1972). Handy oxen. *The Chronicle.* 25(2). pg. 17-21.

Rathbone, Pembroke Thom. (1999). *The history of old time farm implement companies and the wrenches they issued including buggy, silo, cream separator, windmill, and gas engine companies.* R-Lucky Star Ranch, Rt. 1, Box 734, Marsing, Idaho 83639.

Roger, Bob. (March 2003). Hand-held green corn shredders. *The Chronicle.* 56(1). pg. 16-24.

Roger, Bob. (June 2004). A compendium of hay knives. *The Chronicle.* 57(2). pg. 54-61.

Roger, Bob. (June 2004). Addendum I: U.S. patents for hay knives. *The Chronicle.* 57(2). pg. 61-69.

Roger, Bob. (June 2004). Addendum II: Chronology of hay-knife articles in The Chronicle. *The Chronicle.* 57(2). pg. 69-71.

Russell, Howard S. (1982). *A long, deep furrow: Three centuries of farming in New England.* University Press of New England, Hanover, NH.

Tresemer, David. (1982). *The scythe book: Mowing hay, cutting weeds, and harvesting small grains with hand tools.* By Hand & Foot, Ltd., Brattleboro, VT.

Watkins, Malcolm. (July 1947). The ox yoke. *The Chronicle.* 3(12). pg. 103-104, 106.

Architects, Architecture, and Design

Cummings, Abbott Lowell. (1984). *Architecture in early New England.* Old Sturbridge Village Booklet Series, Old Sturbridge, Inc., Sturbridge, CT.

Félibien, A. (1767). *Principes de l'architecture, etc.* Paris.

Isham, Norman Morrison. (1968). *A glossary of colonial architectural terms*. American Life Foundation, Watkins Glen, NY.

Karp, Ben. (1966). *Wood motifs in American domestic architechture/phantasy in wood*. A.S. Barnes and Co, Inc., Cranbury, NJ. Revised in 1981 and published as *Ornamental carpentry on nineteenth century American houses* by Dover Publications, Inc., NY, NY.

Kimball, Fiske. (1966). *Domestic architecture of the American colonies and of the early republic*. Dover, NY, NY.

Lewandoski, Jan, et al. (2006). *Historic American roof trusses*. Timber Framer's Guild, Becket, MA.

McAlester, Virginia and McAlester, Lee. (1984). *A field guide to American houses*. Alfred A. Knopf, NY, NY.

Mercer, Henry C., Dr. (1923). *The dating of old houses*. Vol. V. Reprinted from a collection of papers read before the Bucks County Historical Society.

Morrison, Hugh. (1952). *Early American architecture: From the first colonial settlements to the national period*. Dover Publications, Inc., NY, NY.

Pierson, William H., Jr. (1978). *American buildings and their architects: Technology and the picturesque*. Doubleday, Garden City, NY.

Pratt, Richard. (1949). *A treasury of American homes*. Whittlesey House, NY, NY.

Rogers, Meyric R. (1947). *American interior design*. W. W. Norton, NY, NY.

Seymour, Howard. (October 1985). The steel pen and the modern line of beauty. *Technology and Culture*. 26.

Waterman, Thomas T. (1950). *The dwellings of colonial America*. University of North Carolina Press, Chapel Hill, NC.

Woodward, George E. and Thompson, Edward G. (1869). *Woodward's national architect; containing 1000 original designs, plans and details, to working scale, for the practical construction of dwelling houses for the country, suburb and village. With full and complete sets of specifications and an estimate of the cost of each design*. Geo. E Woodward, NY, NY. Reprinted in 1988 as *A Victorian housebuilder's guide: "Woodward's national architect" of 1869* by General Publishing Company, Ltd., Toronto, Canada.

Armorers

Creswell, K.A.C. (1956) *The bibliography of arms and armor in Islam*. The Royal Asiatic Society. London.

Demmin, Auguste. (1911). *An illustrated history of arms and armour*. G. Bell & Sons, Ltd., London.

Grafton, Carol Belanger Ed. (1995). *Arms and armor: A pictorial archive from nineteenth-century sources*. Dover Publications, Inc., New York, NY.

Hime, Henry W.L. (1915). *The origin of artillery*. Longmans, Green & Co, New York, NY.

Lynch, Kenneth. *The armourer and his tools*.

- The Davistown Museum has Lynch's preliminary guide to this book, which was never published, as well as a number of tools from his collection.
- The Davistown Museum has recently republished Lynch's catalog of armourer's tools. A copy is available for visitor perusal in the main hall of the museum. Copies may be ordered from the museum; please allow three weeks ($12.00 post paid.)
- See the blacksmith section below for Lynch's 1975 tool catalog.

Miller, Jeffrey. (August 2000). The armourer's corner: English lock and dog locks. *The Pine Tree Shilling*. 2(3). pg. 5, 12.

Robertson, Gredrick. (1921). *The evolution of naval armament*. Constable & Co., London.

Schubert, Hans. (1942). The first cast iron cannon made in England. *Journal of the Iron and Steel Institute*. XLVI(2). pg. 132.

Williams, A.R. (1977). Roman arms and armour. *J. Arch Sci*. 4. pg. 77-87.

Artisans and Craftsmen (arts and crafts)

Blandford, Percy W. (1974). *Country craft tools*. Newton Abbot.

Caulfeild, Sophia and Saward, Blanche. (1885). *The dictionary of needlework: An encyclopedia of artistic, plain, and fancy needlework illustrated with more than 800 wood engravings*. Second edition. London. Reprinted in 1989 as a facsimile of the original second edition by Blaketon Hall Ltd., Exeter, UK.

Desiant, A. (1927). *Ideas in stenciling*. Charles Griffin & Co. Ltd., London, UK. Reprinted in 1990 by Wellfleet Press, Secaucus, NJ.

Eaton, Allen H. (1954). *Handicrafts of New England*. Bonanza Books, NY, NY.

Frost, Rodney. (2000). *The quiet pleasures of crafting by hand*. Sterling Pub. Co., Inc., NY, NY.

Meyer, Carolyn. (1975). *People who make things: How American craftsmen live and work*. Atheneum, NY, NY.

Rock, Howard B. (1979). *Artisans of the new republic: The tradesmen of New York City in the age of Jefferson*. New York University Press, NY, NY.

Romaine, Lawrence B. (March 1939). Basket making. *The Chronicle*. 2(8). pg. 57-58.

Schnirring, Melissa. (date unknown). Life in early America. *Colonial Crafts*. volume unknown, pg. 489 - 497.

Smitten, Stanley. (1976). *Chair caning and furniture restoring*. Unpublished, prepared for the April 1976 meeting of the Early Trades and Crafts Society.

Sprague, William B. (April 1939). The brush maker. *The Chronicle*. 2(9). pg. 65, 67-68.

Sprague, William B. (April 1942). The comb maker. *The Chronicle*. 2(20). pg. 169, 171-172.

Stockham, Peter, Ed. (1976). *The little book of early American crafts & trades*. Dover Publications, NY, NY.

- This book is a reprint of Jacob Johnson's *The book of trades, or library of the useful arts, in three parts. Illustrated with copper plates*. Part I. London. 1800.

Stockham, Peter, Ed. (1992). *Old-time crafts and trades*. Dover Publications, NY, NY.

- This book is a reprint of Jacob Johnson's *The book of trades, or library of the useful arts*. Part II. London. 1807.

Stockham, Peter, Ed. (1992). *Early nineteenth-century crafts and trades*. Dover Publications, NY, NY.

- This book is a reprint of Jacob Johnson's *The book of trades, or library of the useful arts*. Part III. London. 1807.

Wamsley, James S. (1982). *The crafts of Williamsburg*. Colonial Williamsburg Foundation, Williamsburg, VA.

Yates, Raymond F. and Yates, Marguerite W. (1954). *Early American crafts & hobbies: A treasury of skills, avocations, handicrafts, and forgotten pastimes and pursuits from the golden age of the American home*. Funk & Wagnalls, NY, NY.

Axes

Berg, Elizabet, ed. (1997). *The axe book*. Gransfors Bruks AB, Sweden.

Cook, Dudley. (2005). *The ax book: The lore and science of the woodcutter*. Hood, Alan C. & Company, Inc., Chambersburg, PA.

Foley, Vernard and Moyer, Richard H. (June 1977). The American ax - was it better? *The Chronicle.* 30.

Gamble, James D. (1986). *Broad axes.* Tanro Co., Los Altos, CA.

Gordon, Robert B. (1988). Material evidence of the manufacturing methods used in "armory practice". *Journal of the Society for Industrial Archeology.* 14(1). pg. 23-35.

Grismer, Jerome T. and Kendrick, Clyde H. (1985). *Embossed American axes: A photographic guide.* Jerome T Grismer and Clyde H. Kendrick, publishers, Columbia, Missouri.

- This text has no mention of the famous axe factories of Oakland, Maine.

Hartzler, Daniel and Knowles, James. (1995). *Indian tomahawks & frontiersmen belt axes.* Privately printed.

Hallman, Richard. (1988). *Handtools for trail work.* U. S. Department of Agriculture, Forest Service, Missoula Technology and Development Center, Missoula, MT.

Heavrin, Charles A. (1998). *The axe and man: The history of man's early technology as exemplified by his axe.* The Astragal Press, Mendham, NJ.

Hodgkinson, Ralph. (1965). *Tools of the woodworker: axes, adzes and hatchets.* History Tech. Leaflet 28. American Association for State and Local History.

Kauffman, Henry J. (April 1954). Some notes on axes. *The Chronicle.* 7(2). pg. 18-20.

- "The introduction of cast-steel coincided with the factory production of axes in America. The earlier three-piece axe of the village blacksmith was discarded for a one-piece axe of cast-steel produced by the incipient machine age. Quickly the eye of the axe was punched from the stock and the edge was thinned on a water motivated trip hammer which worked long and tirelessly. The cutting edge could be heat treated and made hard, while the balance of the axe remained sufficiently soft to absorb the shock of impact." (pg. 20).

Kauffman, Henry J. (1972). *American axes: A survey of their development and their makers.* S. Greene Press, Brattleboro, VT.

Lamond, Tom. (2006). 200 years of axe design evolution. *Fine Tool Journal.* 55(3). pg. 7.

McLaren, Peter. (1929). *Axe manual of Peter McLaren.* Fayette R. Plumb, Inc., Philadelphia, PA.

Peterson, Harold. (1965). *American Indian tomahawks.* Museum of the American Indian, Heye Foundation, NY.

Russell, Carl. (March 1966). Double-bitted axe in America. *The Chronicle.* 19(1). pg. 8-9.

Weisgerber, Bernie and Vachowski, Brian. (1999). *An ax to grind: A practical ax manual*. Technical Report 9923-2823-MTDC. U. S. Department of Agriculture, Forest Service, Missoula Technology and Development Center, Missoula, MT.

Bicycles

Johnson, Laurence. (December 1960). Bone-shakers. *The Chronicle*. 13(4). pg. 41-43.

Blacksmiths

(also see farriers and the metallurgy bibliography in Volume 11, *Handbook for Ironmongers*)

Andrews, Jack. (1994). *New edge of the anvil: A descriptive book for the blacksmith*. Skipjack Press, Ocean River, MD.

Bacon, John Lord. (1904). *Forge-practice (elementary)*. Original copyright by John L. Bacon. Second edition, enlarged, published by John Wiley & Sons, Inc., New York, NY, 1914. Reprinted in 1986 by Lindsay Publications, Inc., Bradley, IL 60915.

Bealer, Alex W. (1976). *The art of blacksmithing*. Castle Books, Edison, NJ.

- The classic text on this subject.
- Frequently cited in Vol. 11, *Handbook for Ironmongers*.

Carlberg, Per. (July 1958). Early industrial production of Bessemer steel at Edsken. *Journal of the Iron and Steel Institute*. 189. pg. 201-204.

Chapman, Gene. (2004). *Hot shop (small potatoes) blacksmith suff.* Oak and Iron Publishing, Kingston, WA.

Clay, W. (1858). On the manufacture of puddled or wrought steel with an account of some of the uses to which it has been applied. *Journal of the Society of Arts*. VI. pg. 140-148.

Council for Small Industries in Rural Areas. (1952). *The blacksmith's craft: An introduction to smithing for apprentices and craftsmen*. Council for Small Industries in Rural Areas, London. Reprinted in 1975.

Council for Small Industries in Rural Areas. (1953). *Wrought ironwork: A manual of instruction for craftsmen*. Council for Small Industries in Rural Areas, London. Reprinted in 1968. Reprinted in 1987 by Macmillan Publishing Company, NY, NY.

Council for Small Industries in Rural Areas. (1962). *Decorative ironwork*. Council for Small Industries in Rural Areas, London. Reprinted in 1973. Reprinted in 1987 by Macmillan Publishing Company.

Didsbury, J. (December 1964). Tool study -- blacksmith's hammers. *The Chronicle*. 17(4). pg. 46-48.

Didsbury, J. (March 1965). Tool study -- blacksmith's hammers (continued). *The Chronicle*. 18(1). pg. 12-14.

Didsbury, J. (June 1965). Tool study -- blacksmith's (continued). *The Chronicle*. 18(2). pg. 28-30.

Didsbury, J. (September 1965). Blacksmith's cutting tools. *The Chronicle*. 18(3). pg. 46-48.

Didsbury, J. (December 1965). Blacksmith's tools (continued). *The Chronicle*. 18(4). pg. 61-62, 64.

Didsbury, J. (March 1966). Blacksmith tools (continued). *The Chronicle*. 19(1). pg. 14-15.

Didsbury, J. (June 1966). Blacksmith's tools shop accessories. *The Chronicle*. 19(2). pg. 25.

Didsbury, J. (September 1966). Blacksmith tools: 18th century drilling machines. *The Chronicle*. 19(3). pg. 43-44.

Didsbury, J. (December 1966). Blacksmith tools (continued). *The Chronicle*. 19(4). pg. 60-62.

Fisher, Leonard E. (2000). *The blacksmiths*. Benchmark Books, NY.

Gardner, J. Starkie. (1892). *Victoria & Albert Museum: Ironwork: Part I. From the earliest times to the end of the mediaeval period*. Printed under the authority of the Board of Education, London. Photolitho impression with supplementary bibliography compiled by Marian Campbell in 1978.

Gardner, J. Starkie. (1896). *Victoria & Albert Museum: Ironwork: Part II. Continental ironwork of the Renaissance and later periods*. Printed under the authority of the Board of Education, London. Photolitho impression with supplementary bibliography compiled by Marian Campbell in 1978.

Gardner, J. Starkie. (1922). *Victoria & Albert Museum: Ironwork: Part III. The artistic working of iron in Great Britain from the earlist times*. Printed under the authority of his Majesty's Stationary Office, London. Photolitho impression with supplementary bibliography compiled by Marian Campbell in 1978.

Gentry, George. (1950). *Hardening and tempering engineers' tools*. Model Aeronautical Press Ltd., Hertsfordshire, England. Revised in 1966 by Edgar T. Westbury.

Hall, R.G. (June 1976). Datable characteristics of anvils. *CEAIA*. 29. pg. 21-24.

Harcourt, Robert Henry. (1917). *Elementary forge practice*. Stanford University Press, Stanford, CA. Reprinted in 1995 by Lindsay Publications, Bradley, IL.

Holmstrom, John Gustaf and Holford, Henry. (1982). *American blacksmithing and twentieth century toolsmith and steelworker*. Greenwich House, NY, NY.

International Correspondence Schools. (1906). *Machine molding: Foundry appliances; Malleable casting; Brass founding; Blacksmith-shop equipment; Iron forging; Tool dressing; Hardening & tempering; Treatment of low-carbon steel; Hammer work; Machine forging; Special forging operations*. International Textbook Co., Scranton, PA. Reprinted in 1983 by Lindsay Publications, Bradley, IL.

Kelleher, Tom. (September 2002). Nuts and bolts. *The Chronicle*. 55(3). pg. 112-115.

- This article covers early 19th century blacksmith production of nuts and bolts utilizing iron screw dies, screw plates and threading dies.
- The article contains an illustration of a Stubs screw plate. "Peter Stubs manufactured tools in Warrington, Lancashire, England, from about 1826 until the late-nineteenth century. Like modern threading dies, solid screw plates were not adjustable and therefore made identical threads by cutting away metal from the bolt blank." (pg. 113).

Kalman, Bobbie. (2001). *The blacksmith*. Crabtree Publishing Company, NY, NY.

- See annotations in the Davistown Museum online Children's bibliography.

Larsen, Ray. (September 1978). Modern hand forging. *The Chronicle*. 31(3). pg. 49-53.

Larsen, Ray. (December 1984). The art of the draw forger. *The Chronicle*. 37(4). pg. 57-60.

Larsen, Ray. (March 1985). The art of the draw forger - part II. *The Chronicle*. 38(1). pg. 5-9.

- See the annotations of this citation in the Davistown Museum online C. Drew & Co. information file.

Leslie, Candace and Hopkins-Hughs, Diane. (2000). *From forge & anvil*. Astragal Press, Mendham, NJ.

Light, John D. (1984). The archeological investigation of blacksmith shops. *IA: The Journal of the Society for Industrial Archeology*. 10(1). pg. 55-68.

Lyman, Paul H. (May 1945). The ancient anvil. *The Chronicle*. 3(4). pg. 37.

Lynch, Kenneth. (c. 1975). Catalog: *The Kenneth Lynch Tool Collection*. 78 Danbury Rd., Wilton, CT 06897.

- This is one of the most comprehensive pictoral surveys of blacksmiths, silversmiths and armourers hammers and tools in the literature.

- The introduction to this catalog is excerpted in the Davistown Museum online Lynch biography.
- This catalog is not available from any source, but there is one copy in the Museum's library.

McDaniel, Randy. (2004). *A blacksmithing primer: A course in basic & intermediate blacksmithing.* 2nd Edition. Dragonfly Enterprises.

McDaniel, Randy. (2004). *A blacksmithing primer: A course in basic & intermediate blacksmithing.* 2nd Edition. Hobar Publications.

McRaven, Charles. (1981). *Country blacksmithing: A complete, step-by-step guide to working with iron: Everything you need to know to • make your own forge • rework scrap iron and steel • create anything from decorative nails and hinges to broadaxes and farming tools.* Harper & Row, Publishers, NY, NY.

Meilach, Dona and Seiden, Don. (1966). *Direct metal sculpture: Creative techniques and appreciation.* Crown Publishers, Inc., NY, NY.

Miller, Jeffrey. (November 1999). Tradesman's arts & mystery ~ ironmaking in colonial America. *The Pine Tree Shilling.* 1(4). pg. 15-16, 18.

Naujoks, Waldemar, B.S., M.E. and Fabel, Doanld C.,B.S., M.S. (1953). *Forging Handbook.* The American Society for Metals, Cleveland, OH.

Plummer, Don. *Colonial wrought iron: The Sorber collection.* Astragal Press, Mendham, NJ.

Postman, Richard. (1998). *Anvils in America.* Self-published, 10 Fischer Ct., Berrian Springs, MI.

Richardson, M.T., Ed. (1978). *Practical blacksmithing: The original classic in one volume.* Weathervane Books, NY.

- A classic but somewhat redundant reference.
- This is a reprint of an important late 19th century guide to blacksmithing.

Robins, F.W. (1953). *The smith.* Rider and Co., London.

Schiffer, Herbert, Schiffer, Peter and Schiffer, Nancy. (1979). *Antique iron: Survey of American and English forms fifteenth through nineteenth centuries.* Shiffer Publishing Ltd., Box E., Eaton, Pennsylvania.

Schwarzkopf, Ernst. (2000). *Plain and ornamental forging.* Astragal Press, Mendham, NJ.

Simmons, Marc and Turley, Frank. (2007). *Southwestern Colonial Ironwork.* Sunstone Press, Santa Fe, NM.

Smith, H.R. Bradley. (1966). *Blacksmith's and farriers' tools at Shelburne Museum: A history of their development from forge to factory*. Museum Pamphlet Series, Number 7. The Shelburne Museum, Inc., Shelburne, VT.

- See the annotations in the Industrial Revolution bibliography.

Sonn, Albert H. (1922). *Early American wrought iron*. Scribners, NY, NY. Reprinted in 1989, 3 volumes in 1, Bonanza Books, NY, NY.

Streeter, Donald. (1995). *Professional smithing*. The Astragal Press, Mendham, NJ.

Tucker, Ted. (1980). *Practical projects for the blacksmith*. Rodale Press, Emmaus, PA.

Uselding, Paul. (October 1974). Elisha K. Root, forging, and the 'American system'. *Technology and Culture*. 15(4). pg. 543-568.

Watson, Aldren A. (2000). *The blacksmith: Ironworker and farrier*. W. W. Norton & Company, NY, NY.

Weygers, Alexander G. (1973). *The making of tools*. Van Nostrand Reinhold Company, NY, NY.

Weygers, Alexander G. (1974). *The modern blacksmith*. Van Nostrand Reinhold Company, NY, NY.

Weygers, Alexander G. (1997). *The complete modern blacksmith*. Ten Speed Press, Berkeley, CA.

Bookbinding

Samford, C. Clement. (January 1954). Bookbinding in colonial America. *The Chronicle*. 7(1). pg. 4-9.

Wallace, William. (1975). *Some notes on bookbinding*. Unpublished, prepared for the May 1975 meeting of the Early Trades and Crafts Society.

Watson, Aldren A. (1996). *Hand bookbinding: A manual of instruction*. Dover Publications, NY, NY.

Candlemaking

Johnson, Leonard. (1971). *Some light on candle making*. Unpublished, prepared for the Dec. 4, 1971 meeting of the Early Trades and Crafts Society.

Carriagemaking

Berkebile, Don H. (1977). *American carriages, sleighs, sulkies, and carts: 168 Illustrations from Victorian sources.* Dover Publications, Inc. New York, NY.

Berkebile, Don H. (1978). *Carriage terminology: An historical dictionary.* Reprinted by The Astragal Press, Mendham, NJ.

Berkebile, Don H. (1989). *Horse-drawn commercial vehicles: 255 Illustrations of nineteenth-century stagecoaches, delivery wagons, fire engines, etc.* Dover Publications, Inc., New York, NY.

Locher, Floyd J. (December 1970). Coach maker's and cooper's tools. *The Chronicle.* 23(4). pg. 54-55.

Richardson, M.T., Ed. (1892). *Practical carriage building.* 2 Vols. M. T. Richardson Co., Publishers. Reprinted in 1981 in one volume by the Early American Industries Association, Scarsdale, NY.

Clockmakers and Watchmakers

The American Waltham Watch Factory. (November 17, 1888). The American Waltham watch factory. *Electrical Review.* pg. 2.

Bailey, Chris H. (1975). *Two hundred years of American clocks & watches.* Prentice-Hall, Englewood Cliffs, N.J.

Bruton, Eric. (1968). *The Longcase clock.* Frederick A. Praeger, Publisher, New York, NY.

Goodrich, Ward. (1952). *The watchmaker's lathe: Its use and abuse.* North American Watch Tool & Supply Co., Chicago, IL.

Ralph, William. (1972). *A time for clocks, being a discourse on timekeeping and related matters.* Unpublished, prepared for the March 21, 1972 meeting of the Early Trades and Crafts Society.

Roberts, Kenneth. (1973). *Eli Terry and the Connecticut shelf clock.* Ken Roberts Publishing Co., Bristol, CT.

Scientific American. (March 4, 1899). The building of a watch. *Scientific American.* pg. 132.

Tripplin, Julien and Rigg, Edward. (1880). *The watchmaker's hand-book intended as a workshop companion for those engaged in watchmaking and the allied mechanical arts.* Translated from the French of Claudius Saunier. Reprinted in 1948. The Technical Press Ltd., London.

Ullyett, Kenneth. (1950). *In quest of clocks.* Spring Books, New York, NY.

Wyke, John. (1978). *A catalogue of tools for watch and clock makers.* University Press of Virginia, Charlottesville, VA.

- "The John Wyke tool catalogue, the earliest known English printed source illustrating the extensive range of tools available to watch and clock makers, first appeared during the third quarter of the eighteenth century. French publications such as Antoine Thiout's *Traite de l'horlogerie, mechanique, et practique* (1741) and *Horlogerie* from Diderot and D'Alembert's *Encyclopedie ou Dictionaire raisonne des sciences des arts, et des metiers* (1765) illustrated many similar tools, but these books were horological textbooks and not for trading purposes. The Wyke catalogue was a pictorial listing of tools for order or sale at the manufactory and warehouses in Wyke's Court, Dale Street, Liverpool. The firm of John Wyke and Thomas Green supplied a wide range of tools needed by 'Clock and Watchmakers, Jewellers, Braziers, and other Mechanics.' The catalogue was issued under Wyke's name, and several of the plates had been engraved some five to fifteen years before he formed his partnership with Green in or soon after 1770." (pg. 1).
- "Sometime after the appearance of the Wyke publication, other suppliers of tools to the clock and watch trade issued illustrated catalogues, the important ones being those from Ford, Whitmore and Brunton of Birmingham, which appeared about 1785, and from Peter Stubs of Warrington, which came out soon after 1800 and was reprinted with few alterations at various dates in the nineteenth century. In both cases the tools illustrated closely resembled those in John Wyke's catalogue, and certain plates are virtually identical. This similarity is not surprising, for all three firms sold to the growing number of British watch and clock makers the very wide range of files and other tools produced by specialist outworkers in the cottage industry of southwest Lancashire." (pg. 1).
- "The origins of watch and watch tool making in southwest Lancashire are obscure, but from documentary evidence it is clear that the trade was well established in that corner of northwest England long before the end of the seventeenth century." (pg. 1).
- Some of the illustrations and descriptions of tools from this text are posted in the Davistown Museum online Peter S. Stubs information file.

Zea, Philip and Cheney, Robert. (date unknown). *Clockmaking in New England*. Old Sturbridge Village, Sturbridge, MA.

Cobblers

Mitchell, Robert. (December 1960). Some interesting shoemakers tools. *The Chronicle*. 13(4). pg. 46-47.

Townsend, Raymond. (May 1956). 18th century shoemaker's tools. *The Chronicle*. 9(2). pg. 18-19.

Coopers

Anonymous. (December 1991). A trussing of staves -- a miscellany of pieces. *The Chronicle*. 44(34). pg. 106-108.

Anonymous. (September 1965). The cooper's chamfer knife. *The Chronicle*. 18(3). pg. 40-41.

Anonymous. (1969). *The cooper and his work.* Unpublished, In honor of John S Kebabian, prepared for the May 20, 1969 meeting of the Early Trades and Crafts Society.

Anonymous. (June 1972). Modern cooper's tools. *The Chronicle.* 25(2). pg. 28.

Bailey, Gillian W. (September 1950). Cooperage. *The Chronicle.* 3(24). 217, 222-223.

Cooper, Miner J. (March 1962). Early American Industries tool study: Hoop-setting gauges. *The Chronicle.* 15(1). pg. 11-12.

Cope, Kenneth L. (2003). *American cooperage machinery and tools.* Astragal Press, Mendham, NJ.

Coyne, Franklin E. (1947). *The development of the cooperage industry in the United States, 1620-40.* Lumber Buyers Publishing Company, Chicago, IL.

Haines, Ethel. (January and April 1946). The handmade barrel hoop. *The Chronicle.* 3(6,7). pg. 60.

Hankerson, F. P. (1951). *The wooden barrel manual: Revised, 1951.* The Associated Cooperage Industries of America, Inc., 2100 Gardiner Lane, Suite 100E, Louisville, KY. Reprinted in 1983.

Kebabian, John S. (September 1972). Richard Cromwell as cooper, 1659. *The Chronicle.* 25(3). pg. 33-35.

Keller, David N. (1983). *Cooper Industries: 1833 - 1983.* Ohio University Press, Athens, OH.

Kilby, Kenneth. (1971). *The cooper and his trade.* Linden Publishing Co., Inc., Fresno, CA.

- "The three main branches were wet, dry and white. The wet cooper made casks with a bulge to hold liquids. The dry cooper made casks with a bulge to hold a wide variety of dry commodities. The white cooper made straight-sided, splayed vessels. But within these branches one could differentiate between the type and the quality of the work done. The beer, wine and spirit side of wet coopering demanded much greater precision than the oil, tar and pitch side where casks of a poorer quality were used. Internal pressure resulting from fermentation was a factor to be considered by the former, as well as the greater value of their contents, but amongst these the brewers' coopers might well have claimed that their casks needed to be stronger in order to undertake repeated journeys, and to be smooth inside for easy sterilization as opposed to the wine and spirit casks which were deliberately blistered and left rough. Dry coopering differed with regard to every commodity for which the casks were used. Herring coopers, and others who in earlier years made casks for preserving fish, were called dry-tight coopers, since although their work came under the category dry, it still had to be capable of holding brine. Butter and soap cask-making was also regarded as dry-tight work. At the other end of the scale there were the apple and tobacco barrel coopers whose frail, dry-slack casks were made from the cheapest timber just sufficiently bound to last one journey. Most hardware was shipped in strong dry casks." (pg. 42-43).

Louisville Cooperage Company/Chess and Wymond Company. (n.d.). *Cooper's craft*. Reprinted unrevised in 1983 by The Associated Cooperage Industries of American, Inc., 2100 Gardiner Lane, Suite 100E, Louisville, KY.

Martin, Richard A. (September 1992). The manufacture of wooden pails by manual labor. *The Chronicle. 45(3)*. pg. 67-69.

Miller, Jeffrey. (August 2000). Tradesman's arts & mystery: An introduction to coopering, the art of making barrels. *The Pine Tree Shilling.* 2(3). pg. 1-2.

Roberts, Kenneth D. (December 1968). Some cooper's tools: Part I. *The Chronicle.* 21(4). pg. 55-58.

Roberts, Kenneth D. (March 1969). Some cooper's tools: Part II. *The Chronicle.* 22(1). pg. 6-8.

Roberts, Kenneth D. (September 1969). Some cooper's tools: Part III. *The Chronicle.* 22(3). pg. 41-42.

Shagena, Jack L. (2006). *An illustrated history of the barrel in America*. Self published by Jack L. Shagena, Bel Air, MD.

Sprague, William B. (June 1938). The cooper. *The Chronicle.* 2(5). pg. 33, 35-36, 38.

Sprague, William B. (May 1956). The cooper. *The Chronicle.* 9(2). pg. 16-17, 19-20, 24.

Walsh, James D. (March 1998). Pipe: A lost part of Americana. *The Chronicle.* 51(1). pg. 13-14.

Wolcott, Stephen C. (January 1935). A cooper's shop of 1800. *The Chronicle.* 1(9). pg. 1-2, 7.

- The description of the contents of this shop provides an idea of what the cooper's shops in Davistown Plantation were like.

Crooked knives

Blodgett, Wentworth P. (December 1964). The crooked knife. *The Chronicle*. In: Pollak, Emil and Pollak, Martyl, Eds. (1991). *Selections from The Chronicle: The fascinating world of early tools and trades*. The Astragal Press, Mendham, NJ.

Brainard, Newton C. (September 1962). The crooked knife. *The Chronicle*. In: Pollak, Emil and Pollak, Martyl, Eds. (1991). *Selections from The Chronicle: The fascinating world of early tools and trades*. The Astragal Press, Mendham, NJ.

Sayward, Elliot. (Summer 2001). The crooked knife. *The Freeholder*. Oyster Bay Historical Society.

Starr, Nicholas. (September 2008). The crooked knife - a New World phenomenon: A brief overview with proposals for descriptive nomenclature. *The Chronicle.* 61(3). pg. 99-105.

Witthof, John. (March 1963). Notes on the American Indian crooked knife. *The Chronicle.* In: Pollak, Emil and Pollak, Martyl, Eds. (1991). *Selections from The Chronicle: The fascinating world of early tools and trades.* The Astragal Press, Mendham, NJ.

Cutlers

(also see blacksmiths, metallurgy, and European Precedents and the Early Industrial Revolution)

Abels, R. (1967). *Classic Bowie knives.* Robert Abels, Inc. New York, NY.

Himsworth, Joseph Beeston (1953). *The story of cutlery: From flint to stainless steel.* Benn, London, UK.

Grayson, Ruth and Hawley, Ken. (1995). *Knifemaking in Sheffield and the Hawley Collection.* Exhibition Catalog, The Hawley Collection, University of Sheffield, Sheffield, UK.

Leader, R.E. (1905). *History of the Cutlers' company.* 2 vols. Sheffield, UK.

Lloyd, G.I.H. (1913). *The cutlery trades.* London.

Pankiewicz, Philip. (1986). *New England cutlery.* Hollytree Publications, Gilman, CT.

Petersen, Harold Leslie. (1958). *American knives: The first history and collectors' guide.* Scribner, New York, NY.

Platts, Harvey (1978). *The Knifemakers who went west.* Longs Peak Press, Longmont, CO.

Taber, Martha Van Hoesen. (1955/9). *A history of the cutlery industry in the Connecticut Valley.* Smith College Studies in History, Smith College, Northampton, MA.

Unwin, Joan and Hawley, Ken. (1999). *Sheffield industries: Cutlery, silver and edge tools.* Images of England Series, Stroud, Tempus.

Unwin, Joan and Hawley, Ken. (2003). *A cut above the rest - the heritage of Sheffield's blade manufacture.* Exhibition Catalog, The Hawley Collection, University of Sheffield, Sheffield, UK.

Washer, Richard. (1974). *The Sheffield Bowie knife and pocket knife makers, 1825-1925.* Zeros Printers Ltd., Nottingham, UK.

Zalesky, Mark D. (May 2005). Bowies on a budget? *Knifeworld.* 31(5).

Electro-plating

Warburton, L. (1950). *Electro-plating for the amateur.* Model and Allied Publications Limited, Hertfordshire, England.

Engineers and Engineering

American Society of Tool and Manufacturing Engineers. (1959). *Tool engineers handbook: A reference book on all phases of planning construct design tooling and operations in the manufacturing industries*. 2nd edition. ASME, NY.

Appleton D. & Co. (1852.) *Appleton's dictionary of machines, mechanics, engine-work, and engineering*. D. Appleton & Co., NY, NY.

Aubuisson de Voisins, Jean François d'. (1852).*A treatise on hydraulics: For the use of engineers*. Little, Brown, and Co., Boston, MA.

Billings, William R. (1889). *Some details of waterworks construction*. The Engineering and Building Record Press, NY, NY.

Clapp, Howard Wm. and Clark, Donald Sherman. (1944). *Engineering materials and processes*. International Textbook Press, Scranton, PA.

Keuffel & Esser Company. (1976). Calendar: *Early American engineers: The early craftsman, his tools, his creations and his design*. K&E Co., Long Island City, NY.

Kirby, Richard Shelton, Withington, Sidney, Darling, Arthur Burr and Kilgour, Frederick Gridley. (1956). *Engineering in history*. Dover Publications, Inc., NY, NY.

Moore, Stanley H. (1908). *Mechanical engineering and machine shop practice*. McGraw-Hill Book Company, NY, NY.

North American Model Engineering Exhibition. (2002). Program: *13th annual North American Model Engineering Exposition*. North American Model Engineering Society, 36506 Sherwood, Livonia, MI, 48154.

Roper, Stephen. (1880). *Roper's questions and answers for engineers*. Edward Meeks, Philadelphia, PA.

White, Alfred H. (1939). *Engineering materials*. McGraw-Hill Book Co., Inc., NY, NY.

Engraving

Rees, F. H. (1909). *The art of engraving: A text-book and practical treatise on the engraver's art, with special reference to letter and monogram engraving*. The Keystone Publishing Co., Philadelphia, PA.

Farriers
Also see blacksmiths.

Lungwitz, A. and Adams, Charles F. (1981). *The complete guide to blacksmithing horseshoeing, carriage and wagon building and painting.* Bonanza, NY.

Lungwitz and Adams, J. (1884). *A textbook of horse shoeing.* Reprinted in 1966 by Oregon State University Press, Convallis, OR.

Straffin, Dean. (March 1989). Farrier's tools - a rewarding study. *The Chronicle.* 42(1). pg. 8-9.

Files

Fremont, Charles. (1920). *Files and filing.* Translated by Taylor, Geo. and dedicated to the Sheffield Files Trades Technical Society. Isaac Pitman & Sons, London, UK.

Harris, J. R. (1985). First thoughts on files. *Tools & Trades.* 3. pg. 27-35.

Nicholson File Company. (1878). *A treatise on files and rasps: Descriptive and illustrated: For the use of master mechanics, dealers, &c. in which the kinds of files in most common use, and the newest and most approved special tools connected therewith, are described -- giving some of their principal uses. With a description of the process of manufacture, and a few hints on the use and care of the file.* Reprinted in 1983 by the Early American Industries Association.

Nicholson File Company. (1945). *Nicholson files and rasps and F. Swiss pattern files.* Nicholson File Company, Providence, RI.

Nicolson File Company. (1956). *File filosophy and how to get the most out of files (-being a brief account of the history, manufacture, variety and uses of files in general.)* Twentieth Edition. Nicolson File Company, Providence, R.I.

- The definitive guide to files and their use.
- Also see the US & New England Toolmakers bibliography for many more Nicolson publications.

Sawers, John W. (March 1987). Recycled files. *The Chronicle.* 40(1). pg. 1-2.

Flax Dressing
(also see Milliners and Textiles)

Sprague, William B. (May 1936). Flax dressing by hand. *The Chronicle.* 1(17). pg. 1, 3.

Sprague, William B. (July 1936). Flax dressing by hand. *The Chronicle.* 1(18). pg. 4, 6.

Van Wagenen, Jared. (March 1947). Flax and the loom. *The Chronicle.* 3(11). pg. 93, 95, 97-99, 101-102 .

Foundry Operation

(also see metallurgy, blacksmiths and Vol. 11, *Handbook for Ironmongers*)

Alloy Casting Institute Division. (1973). *High alloy data sheets: Corrosion series*. Steel Founders' Society of America, Cast Metals Federation Building, Rocky River, OH.

Alloy Casting Institute Division. (1973). *High alloy data sheets: Heat series*. Steel Founders' Society of America, Cast Metals Federation Building, Rocky River, OH.

Batory, Dana Martin. (Summer 2004). Cast in iron. *Fine Tool Journal*. 54(1). pg. 8-13.

- "Carbon is the most important element in cast iron. Its greatest influence is its effect on the melting point. Pure iron melts at 2735 degrees, a temperature very difficult to reach. The presence of just 3.55 carbon in pig iron reduces the melting point to 2075 degrees. Therefore, cast iron is easily and cheaply melted and can be produced more economically than any other form of iron." (pg. 8).
- "H. B. Smith, founder of the H. B. Smith Machine Co. (1847) of Smithville, New Jersey, pioneered the use of cast iron in woodworking machinery. Unlike his competitors, he used all iron construction in his major machines from the very beginning. Even so, some minor Smith machines were built with wooden frames to meet the needs of smaller shops." (pg. 11).
- "William C. Bolger pointed out its importance in *Smithville: The Result of Enterprise* (1980): 'The statement that 'It is all iron' is significant, both in terms of the man and his future career as a manufacturer of woodworking machines. His reputation in the machine business was due as much to his use of iron as to the designs he patented.'" (pg. 11).

Hutton, R.S. (December 7, 1906). The electric furnace and its applications to the metallurgy of iron and steel. *Engineering*. 7. pg. 779-781.

Miller, J.K. (1936). *Forging dies*. International Textbook Company, Scranton, PA.

Sopcak, James E. (1968). *Handbook of lost wax or investment casting*. Gembooks, Mentone, CA.

- "A how-to-do manual that shows you how to make the equipment you will need as well as how to use it to make patterns, molds and castings for jewelry and small metal parts." (pg. 1).

Simpson, Bruce. (1948). *Development of the metals casting industry*. American Foundrymens Association, Chicago.

Stimpson, William C., Gray, Burton L. and Grennan, John. (1930). *Foundry work: A practical handbook on standard foundry practice, including hand and machine molding with typical problems, casting operations, melting and pouring equipment, metallurgy of cast metals, etc.* American Technical Society, Chicago.

Wedel, Ernst von. (1960). The history of die forming. *Metal treatment and drop forging*. 27. pg. 401-408.

West, Thomas D. (1883). *American foundry practice: treating of loam, dry sand and green sand moulding: And containing a practical treatise upon the management of cupolas and the melting of iron*. J. Wiley & Sons, NY, NY.

West, Thomas Dyson. (1891). *West's moulders' text-book: Being part II. of American foundry practice*. J. Wiley, NY, NY.

Woodworth, Joseph V. (1903). *Dies, their construction and use for the modern working of sheet metals: A treatise on the design, construction and use of dies, punches, tools, fixtures and devices, together with the manner in which they should be used in the power press*. N.W. Henley & Co., New York, NY.

Woodworth, Joseph V. (1907). *Punches, dies and tools for manufacturing in presses ... / by Joseph V. Woodworth ... a companion and reference volume to the author's elementary work entitled "Dies, their construction and use for the modern working of sheet metals"*. N.W. Henley & Co., New York, NY.

Glassblowing

Heddle, G. M. (1961). *A manual on etching and engraving glass*. Alec Tiranti Ltd., London.

Oppenheim, A. Leo et al. (1971) *Glass and glass making in ancient Mesopotamia*. Corning Museum of Glass, Corning, NY.

Starbuck, David R. (1986). The New England Glassworks. *New Hampshire Archaeologist*. 27.

Wilson, Kenneth M. (1969). *Glass in New England*. Old Sturbridge Village Publications, Sturbridge, MA.

Grinding

Behr-Manning. (1944). *How to sharpen*. Behr-Manning a division of Norton Abrasives, Troy, NY.

Norton Abrasives. (no date). *How to use truing and dressing tools for better grinding*. Norton Company, Worcester, MA.

Norton Abrasives. (1943). *The abc of internal grinding: A handbook for operators of internal grinding machines*. Norton Company, Worcester, MA.

Norton Abrasives. (1951). *The abc of o. d. grinding: Cylindrical centerless*. Norton Company, Worcester, MA.

Norton Abrasives. (1952). *The abc of surface grinding*. Norton Company, Worcester, MA.

Norton Abrasives. (1956). *A handbook on tool room grinding including the sharpening of high-speed steel, nonferrous cast alloy and cemented carbide tools*. Norton Company, Worcester, MA.

Norton Abrasives. (1956). *Mounted wheels: Principles of safe and efficient operation with tables of maximum operating speeds*. Norton Company, Worcester, MA.

Norton Abrasives. (1957). *A handbook on abrasives and grinding wheels*. Norton Company, Worcester, MA.

Tone, Frank J. (no date). *Abrasives in the service of industry*. The Carborundum Company, Niagara Falls, NY.

Woodbury, Robert S. (1959). *History of the grinding machine; a historical study in tools and precision production*. Technology Press, Massachusetts Institute of Technology, Cambridge, MA.

Graining

Wall, William E. (1905). *Graining: Ancient and modern*. Norwood Press, Norwood, MA.

Gunsmiths

Annonymous. (1985). Researching early Maine craftsmen: John H. Hall and the gunsmith's trade. *Maine Historical Society Quarterly*. 24. pg. 410-15.

Bonhams & Butterfields. 2006. *The William H. Guthman Collection part I: Arms and militaria: Thursday October 12, 2006 the Frank Jones Center Portsmouth, New Hampshire*. Bonhams & Butterfields, San Francisco, CA.

Brown, M. L. (1980). *Firearms in colonial America: The impact on history and technology, 1492-1792*. Smithsonian Institution Press, Washington, D.C.

Clephan, Robert Coltman, F.S.A. (1906). *An outline of the history and development of hand firearms, from the earliest period to about the end of the fifteenth century*. Walter Scott Publishing Co.,LTD. New York, NY.

Cooper, Carolyn C. (1988). A whole battalion of stockers': Thomas Blanchard's production line and hand labor at Springfield Armory. *IA: The Journal of the Society for Industrial Archeology*. 14(1). pg. 37-57.

Cooper, Carolyn C., Gordon, R.B. and Merrick, H.V. (1982). Archeological evidence of metallurgical innovation at the Eli Whitney Armory. *IA: The Journal of the Society for Industrial Archeology*. 8(1). pg. 1-12.

Deyrup, Felicia Johnson. (1948). *Arms makers of the Connecticut River Valley*. Smith College Studies in History. Vol. 33. Smith College, Northampton, MA.

Drepperd, Carl W. (March 1947). The American rifle and firearms industry. *The Chronicle*. 3(11). pg. 93-94, 96, 102.

Dunlap, Roy F. (1963). *Gunsmithing: A manual of firearms design, construction, alteration and remodeling. For amateur and professional gunsmiths and users of modern firearms.* Stackpole Books, Harrisburg, PA.

Efoulkes, Charles J. (1938). *The gunfounders of England.* University Press, London.

Edwards, William Bennett. (1962). *Civil War guns; the complete story of Federal and Confederate small arms: design, manufacture, identification, procurement, issue, employment, effectiveness, and postwar disposal.* Stackpole Co., Harrisburg, PA.

Gilbert, Keith Reginald. (1963). The Ames recessing machine: A survivor of the original Enfield Rifle machinery. *Technology and Culture*. 4. pg. 207-211.

Greener, W.W. (1910). *The gun and its development.* Reprinted by Bonanza Books, New York, NY.

- This classic on gunsmithing was first published in 1881; the ninth edition published in 1910 has been reprinted by Bonanza Books.
- A particularly comprehensive overview about the history of firearms and the many German, Italian, French and English arms smiths.
- Contains interesting documentation of what was called in the 19th century "Whitworth steel." Joseph Whitworth adapted Huntsman's cast steel to gunsmithing, manufacturing in the mid-19th century a brand of steel known as "wheat sheaf"; Greener notes this steel was the best high carbon "fluid compressed steel" available to gunsmiths at the beginning of the age of the steel. Guns prior to 1850 had been made of either wrought iron or the pattern welded twisted iron and steel barrels popular with English sportsmen and known as Damascus steel. Weldless cast steel guns were much more reliable than the pattern welded sporting guns of the English upper classes, which often developed gray mars from the flecks of iron oxide accidentally left by the pattern welding of the damascened gun barrel.

Guthman, William. (1975). *U.S. Army weapons, 1784-1791.* The American Society of Arms Collectors.

Hatch, A.P. (1956). *Remington arms in american history.* New York, NY.

Hubbard, Howard G. (November 1937). Gun flints. *The Chronicle*. 2(2). pg. 11.

Irwin, John Rice. (1980). *Guns and gunmaking tools of Southern Appalachia: The story of the Kentucky rifle.* Atglen, PA. Reprinted in 1983 by Schiffer Publishing, Exton, PA.

Kauffman, Henry J. (1924). *The Pennsylvania-Kentucky rifle.* Dillin Press, Minneapolis, MN.

Kindig, Joe. (1983). *Thoughts on the Kentucky rifle in its golden age*. G. Shumway, York, PA.

Nonte, George C. Jr. (1978). *Pistolsmithing*. Stackpole Books, Harrisburg, PA.

Raber, Michael S. (1988). Conservative innovators, military small arms, and industrial history at Springfield Armory, 1794-1968. *IA: The Journal of the Society for Industrial Archeology*. 14(1). pg. 1-21.

Raber, Michael S., and Raber Associates. (1989). *Conservative innovators and military small arms: An industrial history at Springfield Armory, 1794-1968*. Unpublished report for the National Park Service. Springfield Armory National Historic Site, South Glastonbury, CT.

Russell, Carl P. (1967). *Firearms, traps and tools of the mountain men*. Knopf, New York, NY.

Smith, Merritt Roe. (1973). John H. Hall, Simeon North, and the milling machine: The nature of innovation among antebellum arms makers. *Technology and Culture*. 14. pg. 573-591.

Smith, Merritt Roe. (1977). *Harpers Ferry Armory and the new technology: The challenge of change*. Ithaca, NY.

Steele, Brett D. (April 1994). Muskets and pendulums: Benjamin Robins, Loenhrad Euler, and the ballistics revolution. *Technology and Culture: The International Quarterly of the Society for the History of Technology*. 35(2). pg. 348-382.

Swinney, H.J. (March 1999). Gun iron and mild steel. *Muzzle Blasts*. 60(7). pg. 79-82. Reprinted from *Bulletin Number 78* of the American Society of Arms Collectors.

van Patten, R.E. (February 2003). A brief history of swivel guns - circa 1480-1837. *Muzzle Blasts*. 64(6). pg. 65-66.

Williamson, H.F. (1952). *Winchester the gun that won the west*. Washington, DC.

Hatters

Ogden, Oliver J. (March 1990). Hatters and hat making at the Harmony Society, Economy, Pennsylvania 1826-1875. 43(1). pg. 3-5.

Homemakers

(Also see the Women and Technology bibliography)

Anonymous. (June 1965). Women's world. *The Chronicle*. 18(2). pg. 23-24.

- Bread rasp, whisks and beaters, pans and bowls.

Anonymous. (September 1965). Women's world. *The Chronicle*. 18(3). pg. 44-45.

- Mincing knives.

Anonymous. (December 1965). Women's world. *The Chronicle*. 18(4). pg. 55-56.

- Lamp chimney accessories, stove pipe shelf, sad-iron heater or long pan.

Anonymous. (June 1966). Women's world. *The Chronicle*. 19(2). pg. 32.

- Wire kitchen utensils.

Anonymous. (June 1966). Women's world. *The Chronicle*. 19(3). pg. 45-48.

- Box mangles (wringer).

Anonymous. (December 1966). Women's world. *The Chronicle*. 19(4). pg. 62-63.

- Hair curling and crimping irons.

Anonymous. (March 1970). The churn. *The Chronicle*. 23(1). pg. 9-10.

Bacon, J. Earle. (August 1942). The busy Yankee girl. *The Chronicle*. 2(21). pg. 185-186.

Cooper, Miner J. (March 1963). Laundry irons. *The Chronicle*. 16(1). pg. 6-7.

Durell, Edward. (September 1962). More irons. *The Chronicle*. 15(3). pg. 28-30.

Gould, Mary Earle. (September 1960). Larding and larding needles. *The Chronicle*. 13(3). pg. 33.

Gould, Mary Earle. (December 1961). Mortars and pestles and their many uses. *The Chronicle*. 14(4). pg. 41-43.

Hynson, Garret. (April 1942). Maine butter molds. *The Chronicle*. 2(20). pg. 176.

Johnson, Laurence A. (June 1957). The Indian broom. *The Chronicle*. 10(2). pg. 13-14, 24.

Johnson, Laurence A. (June 1957). Fruit lifters. *The Chronicle*. 10(3). pg. 28-30.

Keillor, James A. (December 1971). Bread-making tools. *The Chronicle*. 24(4). pg. 57.

Lessey, Ruth. (December 1961). Hearth ovens. *The Chronicle*. 14(4). pg. 44.

McConnel, Bridget. (1999). *The story of antique needlework tools*. Schiffer Publishing, Ltd.

Packham, Jim. (September 1999). Shears and scissors. *The Chronicle*. 52(3). pg. 100-107.

- An excellent survey of the wide variety of scissor-type implements.

Whiting, Gertrude. (1928). *Old-time tools & toys of needlework*. Reprinted in 1971 by Dover Press.

Woloson, Leah. (March 1966). Women's world. *The Chronicle*. 19(1). pg. 13-14.

- Pastry jagger, graters, and apple parers.

Woodhull, Charlotte. (December 1961). Irons. *The Chronicle*. 14(4). pg. 37-40, 44.

Ice Harvesting

Cummings, Richard O. (1949). *The American ice harvests: A study in historical technology, 1800-1918*. University of California Press, Berkeley, CA.

Hall, Henry. (1974). *The ice industry of the United States, with a brief sketch of its history*. Early American Industries Association, from U.S. Dept. of the Interior, Census Division.

Roger, Bob. (September 2007). Patented hand-held ice reducing tools: Part I -- ice picks and chippers.*The Chronicle*. 60(3). pg. 114-125.

Roger, Bob. (March 2008). Patented hand-held ice-reducing tools: Part III -- ice crushers.*The Chronicle*. 61(1). pg. 30-34.

Siegel, Bob Jr. (April 1971). Ice from nature to consumer -- tools and methods. *The Chronicle*. 24(1). pg. 1-5.

Inventors and Inventions

Doster, Alexis, III, Goodwin, Joe and Ross, Jane M. Eds. (1978). *The Smithsonian Book of Invention*. Smithsonian Exposition Books, Smithsonian Institution, Washington, DC.

Eco, Umberto and Zorzoli, G.B. (1962). *The picture history of inventions: From plough to polaris*. Translated from Italian by Anthony Lawrence. The Macmillan Company, NY, NY.

Newcomen Society. (1928). *Martin Triewald's short description of the atmospheric engine*. Translated from the edition of 1734. Courier, London.

Usher, A.P. (1954). *A history of mechanical inventions*. Harvard University Press, Cambridge, MA.

Knives and Swords
Also see cutlers, crooked knives, and Vol. 11: *Handbook for Ironmongers*.

Hrisoulas, James. (1987). *The complete bladesmith: Forging your way to perfection.* Paladin Press, Boulder, CO.

Hrisoulas, James. (1994). *The pattern-welded blade: Artistry in iron.* Paladin Press, Boulder, CO.

Hrisoulas, James. (2005). *The master bladesmith: Advanced studies in steel.* Paladin Press, Boulder, CO.

Kapp, Leon, Kapp, Hiroko and Yoshihara, Yoshindo. (1987). *The craft of the Japanese sword.* Kodansha International, Tokyo, Japan.

Kapp, Leon, Kapp, Hiroko and Yoshihara, Yoshindo. (2002). *Modern Japanese swords and swordsmiths: From 1868 to the present.* Kodansha International, Tokyo, Japan.

Irvine, Gregory. (2000). *The Japanese sword: The soul of the Samurai.* V&A Publications, London.

Nagayama, Kokan. (1997). *The connoisseur's book of Japanese swords.* Kodansha International, Tokyo, Japan.

Neumann, George C. (1973). *Swords and blades of the American Revolution.* David & Charles, Newton Abbott, UK. Reprinted in 1991 by Rebel Pub. Co., Texarkana, T

Reibold, M., Paufler, P., Levin, A. A., Kochmann, W., Pätzke, M. and Meyer, D. C. (November 2006). Carbon nanotubules in an ancient Damascus sabre. *Nature.*

Stone, George Cameron. (1999). *A glossary of the construction, decoration, and use of arms and armor in all countries in all times.* Dover Publications, Mineola, NY.

Sukhanov, I. P. (2004). *Masterpieces and rarities of edged weaponry from the funds of St. Petersburg museums, art manufacturers and private collections.*

Turnbull, Stephen. (2004). *Samurai: The story of Japan's great warriors.* Sterling Publishing Co., New York**Error! Bookmark not defined.**, NY.

Y. al-Hassan, Ahmad and Hill, Donald R. (1992). *Islamic technology: An illustrated history.* Cambridge University Press, Cambridge, England.

Lighting Devices

Thuro, Catherine. (1976). *Oil lamps: The kerosene era in North America.* Collector Books, A Division of Schroeder Publishing Co., Inc., Paducah, KY.

Loggers and Timber Harvesting (Lumbering)
Also see sawyers and the Davistown Museum online special topics bibliography on the Mast Trade.

Adams, Peter. (1981). *Early loggers and the sawmill*. The Early Settler Life Series. Crabtree Publishing Company, New York, NY.

Andrews, Ralph Warren. (date unknown). *This was logging*. Astragal Press, Mendham, NJ.

Andrews, Ralph Warren. (1956). *Glory days of logging*. Bonanza Books, NY, NY.

Andrews, Ralph Warren. (1968). *Timber: Toil and trouble in the big woods*. Bonanza Books, NY, NY.

Bawden, Frank G. (September 1976). Cutting and hewing timber. *The Chronicle*. 29(3). pg. 35-9.

Beaudry, Michael. (2002). *The axe wielder's handbook*. Horizon Publishers, Bountiful, UT.

Blodgett, Wentworth P. (March 1964). Thoroughshot and boom auger. *The Chronicle*. 17(1). pg. 9.

Bryant, Ralph Clement. (1913). *Logging: The principles and general methods of operation in the United States*. John Wiley & Sons, Inc., New York, NY.

Candee, Richard M. (1969 - 1970). Merchant and millwright: The water-powered sawmills of the Piscatauqua. *Old-Time New England*. 60. pg. 131 - 149.

Delson, Barnet. (September 1968). Colonial felling axes. *The Chronicle*. 21(3). pg. 37-38.

Foley, V. and Moyer, R.H. (June 1977). The American axe. Was it better? *CEAIA*. 30. pg. 28-32.

Forman, Benno M. (1969 - 1970). Mill sawing in seventeenth century Massachusetts. *Old-Time New England*. 60. pg. 110 - 130.

Georgia-Pacific. (1974). *Special edition for the American bicentennial: Forest products industry museums, displays and exhibits in the United States*. Georgia-Pacific, Portland, OR.

Hamilton, Edward P. (1964). *The village mill in early New England*. Old Sturbridge Village Booklet Series, Old Sturbridge, Inc., Sturbridge, CT.

Heavrin, Charles A. (September 1982). The felling axe in America. *CEAIA*. 35. pg. 43-53.

Hempstead, Alfred Geer. (1975). *The Penobscot boom and the development of the West Branch of the Penobscot River for log driving 1825 - 1931*. Down East, Camden, ME.

Henry Disston & Sons, Inc. (1902). *Handbook for lumbermen with a treatise on the construction of saws and how to keep them in order*. The Astragal Press, Mendham, NJ. Reprinted in 1994.

Hiton, C. Max . (1942). Rough pulpwood operating in northwestern Maine 1935-1940. *The Maine Bulletin*. 45(1).

Holbrook, Stewart H. (1961). *Yankee loggers: A recollection of woodsmen, cooks, and river drivers.* International Paper Company, NY, NY.

Holbrook, Stewart H. (1962). *The American lumberjack.* Collier Books, NY, NY.

Hughson, John W. and Bond, Courtney, C.J. (1965). *Hurling down the pine: The story of the Wright, Gilmour and Hughson families, timber and lumber manufacturers in the Hull and Ottawa region and on the Gatineau River, 1800-1920.* The Historical Society of the Gatineau, Old Chelsea, Quebec, Canada.

Johnson, Kevin. (2007). *Early logging tools.* Schiffer Publishing, Atglen, PA.

Penn, Theodore Z. and Parks, Roger. (1975). The Nichols-Colby sawmill in Bow, New Hampshire. *IA, Journal of the Society for Industrial Archeology.* 1. pg. 1-12.

Peterson, Charles E. (1975). Early lumbering: A pictorial essay. In: Hindle, Brooke, Ed. *America's wooden age: Aspects of its early technology.* Sleepy Hollow Restorations, Tarrytown, NY.

Phelps, Hermann. (1982). *The craft of log building.* Lee Valley Tools, Ottawa, Ontario, Canada.

Poirer, Noel. (June 2001). The colonial timberyards in America. *The Chronicle.* 54(2). pg. 54-59.

Rivard, Paul E. (1990). *Maine sawmills: A history.* Maine State Museum, Augusta, ME.

Simmons, Fred C. (1951). *Northeastern loggers' handbook.* Agriculture Handbook No. 6. U. S. Department of Agriculture, Washington, DC.

Smith, David C. (1972). *A history of lumbering in Maine 1861-1960.* Maine Studies No. 93. University of Maine Press, Orono, ME.

Smith, Joseph Coburn. (December 1972, June 1973, September 1973). The Maine woodsmen. *The Chronicle.* In: Pollak, Emil and Pollak, Martyl, Eds. (1991). *Selections from The Chronicle: The fascinating world of early tools and trades.* The Astragal Press, Mendham, NJ.

Williams, Richard L. (1976). *The loggers.* Time-Life Books, NY, NY.

Wood, Richard G. (1935). *A history of lumbering in Maine, 1820-1861.* University Press [of Maine], Orono, ME. Reprinted on April 10, 1961 in *The Maine Bulletin, Maine Studies No. 33.*

- We have excerpted so many quotes from this book that we have placed them into the Davistown Museum online information files: lumbering in Maine and potash.

Machinists and Machinery
(also see Measuring and Drafting, Metallurgy, and Engineering)

Battison, Edwin A. (1966). Eli Whitney and the milling machine. *Smithsonian Journal of History.* 1. pg. 9-34.

Battison, Edwin A. (October 1973). The cover design: A new look at the 'Whitney' milling machine. *Technology and Culture.* 14(4). pg. 592-598.

Blanchard, Clarence, Ed. (Spring 2006). Gerstner machinist boxes. *The Fine Tool Journal.* 55(4). pg. 20-1.

Burlingame, Luther D. (August 6, 1914). Pioneer steps toward the attainment of accuracy. *American Machinist.* 41. pg. 237 - 243.

Cope, Kenneth L. (1993). *American machinist's tools: An illustrated directory of patents.* Astragal Press, Mendham, NJ.

Cope, Kenneth L. (1994). *Makers of American machinist's tools: A historical directory of makers and their tools.* Astragal Press, Mendham, NJ.

- The classic period of American machinists' tools.
- A most useful and essential reference.

Cope, Kenneth L. (1998). *More makers of American machinist's tools: Part two of a historical directory of makers and their tools.* Astragal Press, Mendham, NJ.

- These two volumes constitute a comprehensive survey of machinist toolmakers.

Drepperd, Carl W. (July 1948). Tools and standards. *The Chronicle.* 3(16). pg. 137-138.

Foley, Vernard. (July 1983). Leonardo, the wheel lock, and the milling process. *Technology and Culture.* 24. pg. 399-427.

Gordon, Robert B. (1991). Machine archeology: The John Gage planer. *IA: The Journal of the Society for Industrial Archeology.* 17(2). pg. 3-14.

Greenfield Tap & Die. (1969). *Facts about taps & tapping.* Greenfield Tap & Die, Greenfield, MA.

Holtzapffel, Charles and Holtzapffel, John Jacob. (1846-1847). *Turning and mechanical manipulation.* 5 vols. Holtzapffel & Co., London.

- See annotations in the European Precedents and the Early Industrial Revolution bibliography.

Kearney & Trecker Corp. (1957). *Milling practice series: Book one: Right and wrong in milling practice.* Kearney & Trecker Corp., Milwaukee, WI.

Kearney & Trecker Corp. (1957). *Milling practice series: Book two: The milling machine and its attachments*. Kearney & Trecker Corp., Milwaukee, WI.

Kennametal Inc. (1965). *Kennametal tool application handbook*. No. 9. Latrobe, PA.

Lukin, James. (n.d.). *Turning lathes: A guide to turning, screw cutting, metal spinning & ornamental turning*. Astragal Press, Mendham, NJ.

Roe, Joseph Wickham. (1916). *English and American tool builders: The men who created machine tools*. Yale University Press, New Haven, CT. Reprinted in 1987 by Lindsay Publications, Bradley, IL.

Rose, Joshua. (1887-1888). *Modern machine-shop practice*. 2 vols. C. Scribner's Sons, NY, NY.

South Bend Lathe Works. (1958). *How to run a lathe: Revised edition 55: The care and operation of a screw-cutting lathe*. South Bend Lathe Works, South Bend, IN.

Sparey, Lawrence H. (undated). *The amateur's lathe*. Pitman Publishing Corporation, London.

L.S. Starret Company. (1968). *The Starret Book for Student Machinists*. L.S. Starret Company, Athol, MA.

Turner, Frederick W. (1941). *Machine shop work: A comprehensive treatise on approved shop methods including construction and use of tools and machines, details of their efficient operation and a discussion of modern production methods*. American Technical Society, Chicago, IL.

Turner, Gerard L'e. (2000). Elizabethan instrument makers. The origins of the London trade in precision instrument making. Oxford University Press, Oxford, UK.

Wagener, Albert M. and Arthur, Harlan R. (1941). *Machine shop theory and practice*. D. Van Nostrand Co., Inc., NY, NY.

Woodbury, Robert S. (1958). *History of the gear-cutting machine; a historical study in geometry and machines*. Technology Press, Massachusetts Institute of Technology, Cambridge, MA.

Woodbury, Robert S. (1960). *History of the milling machine; a study in technical development*. Technology Press, Massachusetts Institute of Technology, Cambridge, MA.

Woodbury, Robert S. (1961). *History of the lathe to 1850: A study in the growth of a technical element of an industrial economy*. M.I.T. Press, Cambridge, MA.

Woodbury, Robert S. (1972). *Studies in the history of machine tools*. M.I.T. Press, Cambridge, MA.

Measuring and Drafting Tools

Aiken, Ken. (Fall 2000). From gauges to scales. *Fine Tool Journal.* 50(2). pg. 6-9.

Butterworth, Dale and Blanchard, Clarence. (Fall 2005). Central Maine log rule makers and their rules. *The Fine Tool Journal.* pg. 11-14.

Butterworth, Dale and Whalen, Tom. (2007). *From logs to lumber: A history of people & rule making in New England.* Agicook Press, Marshfield, MA.

Cajori, Florian. (1910). *A history of the logarithmic slide rule.* Reprinted by the Astragal Press, Mendham, NJ.

Cannon, Phil. (Fall 2001). Features of some 19th century carpenter rules. *The Fine Tool Journal.* 51(2). pg. 9.

Eugene Dietzgen Co. (no date). *Use and care of drawing instruments with instructive exercises.* Eugene Dietzgen Co., Chicago, IL.

Gordon, Robert B. (1988). Gaging, measurement and the control of artificer's work in manufacturing. *Polhem.* 6. pg. 159-172.

Gordon, Robert B. (1997). Who turned the mechanical ideal into mechanical reality? In: Cutcliffe, Stephen H. and Reynolds, Terry S., Eds. *Technology & American history: A historical anthology from Technology & Culture.* The University of Chicago Press, Chicago, IL.

- "Reliable scales or rules were not available to American artificers before 1850, when J. R. Brown made a linear dividing engine suitable for graduating steel rules for shop use. In 1851 Brown began manufacture of a vernier caliper that made it possible for mechanical artificers to measure to 0.001 inch. But manufacture of the micrometer caliper, the instrument most useful in precision shop work, began in America only in 1868." (pg. 157).

Hallam, Douglas J. (1984). *The first 200 years. A short history of Rabone Chesterman Limited.* Birmingham, UK.

Hambly, Maya. (1988). *Drawing instruments: 1580-1980.* Sotheby's Publications, London.

Hodgson, Fred T. (1890). *Steel squares and their uses.* Industrial Publication Co., NY, NY.

- Available as a loan from the E.A.I.A. Library.

Hopp, Peter M. (date unknown). *Slide rules: Their history, models, and makers.* Astragal Press, Mendham, NJ.

Hoppus, E. (1820). *Hoppus's tables for measuring or practical measuring made easy, by a new set of tables: Which shew at sight the solid content of any piece of timber, stone, &c. either square, round, or unequal-sided, and the value at any price per foot cube; also, the superficial content of boards, glass, painting, plastering, &c. with copious explanations of the uses and applications of the tables. Contrived to answer all the occasions of gentlemen and artificers, the contents being given in feet, inches, and twelfth parts of an inch. With some very curious observations concerning measuring of timber by several dimensions.* 17th Edition. London.

International Correspondence Schools. (1921). *How to use the steel square.* David McKay Company, Philadelphia, PA.

Kebabian, Paul B. (June 1988). The English carpenter's rule: Notes on its origin. *The Chronicle.* 41(2). pg. 24-27.

Kebabian, Paul B. (March 1989). Further notes on the early English three fold ship carpenter's rule. *The Chronicle.* 42(1). pg. 3.

Kisch, Bruno. (1965). *Scales and weights: A historical outline.* Yale University Press, New Haven, CT.

Klein, Herbert Arthur. (1989). *The science of measurement: A historical survey.* Dover, NY, NY.

McConnell, Don. (Fall 2005). The Carpenter's rule in London: 1537-1602. *The Fine Tool Journal.* pg. 7-10.

More, Richard. (1602). *The carpenter's rule.* Felix Kingston, London, UK.

Rabone, John & Sons. (1880). *The carpenter's slide rule: Its history and use.* Third Edition. Reprinted in 1982 by Ken Roberts Publishing Co., Fitzwilliam, NH.

von Jezierski, Dieter. (1977). *Slide rules: A journey through three centuries.* Astragal Press, Mendham, NJ.

Zupko, Ronald Edward. (1977). *British weights & measures: A history from antiquity to the seventeenth century.* The University of Wisconsin Press, Madison, WI.

Mechanics

Dugas, Rene. (1955). *A history of mechanics.* Dover Publications, NY, NY. Reprinted in 1988.

Graham, Frank D. and Emery, Thomas J. (1925). *Audels plumbers and steam fitters guide #3: A practical illustrated trade assistant and ready reference for master plumbers, journeymen and apprentices steam fitters, gas fitters and helpers, sheet metal workers and draughtsmen master builders and engineers explaining in practical concise language and by well done illustrations, diagrams, charts graphs and pictures the principles of modern plumbing practice.* Theo. Audel & Co., NY, NY. Reprinted in 1946.

Henry Ford Trade School. (1942). *Shop theory: Revised edition.* McGraw-Hill Book Co., Inc., NY, NY.

Hubbard, Howard G. (January 1938). The first internal combustion engine. *The Chronicle.* 2(3). pg. 17, 19-20.

Overman, Frederick. (1851). *Mechanics for the millwright, machinist, engineer, civil engineer, architect and student, containing a clear elementary exposition of the principles and practice of building machines.* Lippincott, Grambo, Philadelphia, PA.

Young, F. (1882). *Every man his own mechanic.* London.

Milliners
(also see Flax dressing, and Textiles)

Walker, Niki. (2001). *The milliner.* Crabtree Publishing Company, NY, NY.

Mills and Milling (food)

McGuire, Barton. (1973). *A mill primer.* Early Trades and Crafts Society, NY.

Storck, John and Teague, Walter D. (1952). *Flour for man's bread: A history of milling.* Publisher unknown, Minneapolis, MN.

Nails and Nailmaking

Cooke, Lawrence S. (March 1961). Nail rod and some of its by-products. *The Chronicle.* 14(1). pg. 6-7.

DeValinger, Leon, Jr. (June 1960). Nail-making device at the Delaware State Museum. *The Chronicle.* 13(2). pg. 17.

Didsbury, J. (December 1959). The French method of nail-making. *The Chronicle.* 12(4). pg. 47-48.

Wilson, Kenneth. (June 1960). Nailers' anvils at Old Sturbridge Village. *The Chronicle.* 13(2). pg. 17-19, 25, 27, 28.

Papermaking

Foster, Charles I. (July 1947). Paper Industry I, 1810 - 1860. *The Chronicle.* 3(12). pg. 103, 105-109.

Foster, Charles I. (September 1947). Paper Industry II, 1810 - 1860. *The Chronicle.* 3(13). pg. 113-114, 117-118.

Hunter, Dard. (1957). *Papermaking: The history and technique of an ancient craft*. Alfred A. Knopf, NY, NY.

Osborne, William C. (1974). *The paper plantation: Ralph Nader's study group report on the pulp and paper industry in Maine*. Viking, NY, NY.

Smith, David C. (1970). *History of papermaking in the United States, 1691-1969*. Lockwood Pub. Co., NY, NY.

Patternmaking
Also see foundry operations.

(1927). *Patternmaking: Methods, materials and equipment*. International Textbook Company, Scranton, PA.

Barrows, Frank Wilson. (1906). *Practical pattern making: Fully illustrated by engravings made from special drawings for this work by the author*. The N.W. Henley Publishing Company, NY, NY.

Ritchey, James, Monroe, Walter M. and Beese, Charles W. (1940). *Pattern making: A practical treatise for the pattern maker on woodworking and wood turning, tools and equipment, construction of simple and complicated patterns, modern molding machines and molding practice*. American Technical Society, Chicago, IL.

Rose, Joshua. (1889). *The pattern maker's assistant: Embracing lathe work, branch work, core work, sweep work; and practical gear construction; and the preparation and use of tools*. Sixth edition. Reprinted in 1995 by The Astragal Press, Mendham, NJ.

Willard, G.H. (1910). *Pattern-making*. Popular Mechanics Company, Chicago, IL.

Pewter

Laughlin, Ledlie Irwin. (1981). *Pewter in America*. American Legacy Press, New York, NY.

Montgomery, Charles F. (1973). *A history of American pewter*. Praeger Publishers, New York, NY.

Planemaking

Armour, W.J. (January 15, 1898). Practical plane making. *Work, The Illustrated Journal for Mechanics*. XV(461). Reprinted in 1985 in *Newsletter No. 11*, Tool and Trades History Society.

Bates, Alan G. (1986). Plane making styles. *Newsletter Nos. 14 & 15*. Tool and Trades History Society.

Carter, Matthew. (June 1983). British planemakers before 1700. *The Chronicle*. 36(2). pg. 27-30.

Chemeng County Historical Society. (1983). *Shavings from the past. The wooden plane collection of the Chemeng County Historical Society and the DeWitt Historical Society of Tompkins County.* The Chemeng County Historical Society and the DeWitt Historical Society of Tompkins County, in cooperation with the Early American Industries Association, Ithaca, NY.

Greber, Josef M. (1987). *Die geshichtes des hobels.* Seth Burchard, trans. *The history of the woodworking plane.* Th. Scäfer, Germany.

Goodman, W.L. (1968). *British planemakers from 1700.* Third Edition enlarged and revised by Jane & Mark Rees, published by Roy Arnold in 1993, Astragal Press, Mendham, NJ.

- See annotations in the collector's guides bibliography.

Goodman, William L. (November 1972). Woodworking apprentices and their tools in Bristol, Norwich, Great Yarmouth and Southampton, 1535-1650. *Industrial Archaeology.* 9. pg. 391-392.

Hack, Garrett. (1997). *The handplane book.* Taunton Press, Newtown, CT.

Heckel, Dave. (Fall 2001). Sargent Craftsman planes. *The Fine Tool Journal.* 51(2). pg. 24.

- "Sargent manufactured planes for Sears under the brand names Fulton (low price), Dunlap (middle price), and Craftsman (high price). Sargent manufactured planes have the Sears code letters <u>BL</u> either on the cutter or the body of the plane." (pg. 24).

Hilton, William B. (1977). *Index of plane and spoke shave patents 1812 to 1910.* Published by William B. Hilton, Lynn, MA.

Humphrey, Michael R. (1991 - 1998). *The catalog of American wooden planes.* Bacon Street Press, Sherborn, MA.

Ingraham, Ted. (June 2008). Plane chatter: The mother lode. *The Chrionicle.* 61(2). pg. 83-85.

- An important article on the use of the counter or backing out plane to make complex molding planes.

Kauffmann, Henry J. (April 1953). Some notes on American plane makers. *The Chronicle.* 6(2). pg. 30-31.

Lampert, Nigel. *Through much tribulation: Stewart Spiers and the planemakers of Ayr.* Astragal Press, Mendham, NJ.

- An important English planemaker.

Lasswell, Pat. (Fall 2000). The patterns in imprints. *Sign of the Jointer.* 2(3). pg. 76-80.

Moody, John A. (1981). *American cabinetmaker's plane, its design and development: 1700-1900.* The Tool Box, Evansville, IN.

Moody, John A. (1981). *The American cabinetmakers Plow plane. Its design and improvement 1700-1900.* The Tool Box, Evansville, IN.

Parke, David L. (1981). *Wooden planes at the Farmers' Museum.* Farmers' Museum, Cooperstown, NY.

Pollak, Emil and Pollak, Martyl. (1994). *A guide to the makers of American wooden planes, third edition.* Astragal Press, Mendham, NJ.

- See annotations in the collector's guides bibliography.

Pollak, Emil and Pollak, Martyl. (2001). *A guide to the makers of American wooden planes, fourth edition.* Revised by Thomas L. Elliott. Astragal Press, Mendham, NJ.

- See annotations in the collector's guides bibliography.

Powell, John. (1986). Plane making styles. *Newsletter No. 15.* Tool and Trades History Society.

Roberts, Kenneth D. (1975). *Wooden planes in 19th century America.* Ken Roberts Publishing Co., Fitzwilliam, NH.

- See annotations in the collector's guides bibliography.

Roberts, Kenneth D. (1978). *Wooden planes in 19th century America, volume II: Planemaking by the Chapins at Union Factory, 1826 - 1929.* Ken Roberts Publishing Co., Fitzwilliam, NH.

- See annotations in the collector's guides bibliography.

Roberts, Warren E. (1986). Planemaking in the United States: The cartography of a craft. *Material Culture.* 18(3). pg. 167 - 185.

Rosebrook, Donald and Fisher, Dennis. (2003). *Wooden plow planes: A celebration of the planemakers' art.* Astragal Press, Mendham, NJ.

Sellens, Alvin. (1978). *Woodworking planes: A descriptive register of wooden planes.* Self-published, 234 Clark St., Augusta, KS.

Smith, Roger K. (1981-1992). *Patented transitional & metallic planes in America 1827 - 1927.* 2 vols. North Village Publishing Co., Lancaster, MA.

- See annotations in the collector's guides bibliography.

Smith, Roger K. (Spring 1989). Transitional and metal planes: Stanley no. 18 and no. 19 knuckle-joint block planes, general information and type study. *Plane Talk*. 13(1). pg. 155-162.

Van der Sterre, Gerrit. (2001). *Four centuries of Dutch planes.* Primavera Pers, Netherlands.

Van der Sterre, Gerrit. (2001). *Four centuries of Dutch planes and planemakers*. Astragal Press, Mendham, NJ.

Welsh, Peter C. (Winter 1966). The metallic woodworking plane. *Technology and Culture*. 7(1). pg. 38-47.

West, Phillip. (September 1998). Pioneers in planemaking. *The Chronicle*. 51(3). pg. 69-71.

Whelan, John M. (date unknown). *Making traditional wooden planes*. Astragal Press, Mendham, NJ.

Whelan, John M. (1993). *The wooden plane: Its history, form & function*. The Astragal Press, Mendham, NJ.

- See annotations in the collector's guides bibliography.

Wildung, Frank H. (April 1955). Making wood planes in America. *The Chronicle*. 8(2). pg. 19-21.

Wildung, Frank H. (July 1955). Making wood planes in America (continued). *The Chronicle*. 8(3). pg. 28-30.

Wing, Anne and Wing, Donald. (date unknown). *The case for Francis Purdew or granfurdeus disputatus*. Self-published?

- See annotations in the European precedents bibliography.

Wing, Anne and Wing, Donald. (2005). *Early planemakers of London: Recent discoveries in the Tallow Chandlers and the Joiners Companies*. The Mechanik's Workbench, Marion, MA.

- Top ten among the Davistown Museum favorites.

Potters

Branin, M. Lelyn. (1978). *The early potters and potteries of Maine*. Maine Heritage Series No. 3, Maine State Museum, Augusta, ME.

- The only guide to Maine potters of the 19th and 20th centuries.

Guilland, Harold F. (1971). *Early American folk pottery*. Chilton Book Company, NY, NY.

Watkins, Lura Woodside. (April 1945). Early New England redware potters. *The Chronicle*. 3(3). pg. 21, 30, 32-33, 36.

Quarrying

Armstrong, Army. (2002). Film: *Granite by the sea: The history of granite quarrying on Vinalhaven Island*. Vinalhaven Historical Society, Vinalhaven, ME.

Bunting, William H. (March 2003). Hallowell Granite Works Stinchfield Quarry. *The Chronicle*. 56(1). pg. 40-41.

Gage, Mary E. and Gage, James. (2002). *The art of splitting stone: Early rock quarrying methods in pre-industrial New England 1630-1825*. Powwow River Books, Amesbury, MA.

Gage, Mary E. and Gage, James. (2003). *Stories carved in stone*. Powwow River Books, Amesbury, MA.

Grindle, Roger L. (1977). *Tombstones and paving blocks: The history of the Maine granite industry*. Courier-Gazette, Rockland, ME.

- See annotations in the Maine tool manufacturers bibliography.

Wood, Paul. (June 2006). Tools and machinery of the granite industry. *The Chronicle*. 59(2). pg. 37-52.

Wood, Paul. (September 2006). Tools and machinery of the granite industry, part II. *The Chronicle*. 59(3). pg. 81-96.

Wood, Paul. (March 2007). Tools and machinery of the granite industry, part IV. *The Chronicle*. 60(1). pg. 10-32.

Railroads

Botkin, B. A. and Harlow, Alvin F. (1989). *A treasury of railroad folklore; the stories, tall tales, traditions, ballads, and songs of the American railroad*. Bonanza Books, New York, NY.

Clarke, Thomas Curtis. (1988). *The American railway; its construction, development, management, and appliances*. Castle, Secaucus, NJ.

Hastings, Paul. (1972). *Railroads; an international history*. Praeger Publishers Inc, New York, NY.

Penney, A. R. (1988). *A history of the Newfoundland railway; volume I (1881-1923)*. Harry Cuff Publications Ltd, St. John's, Canada.

Poor, Henry Varnum. (1860). *History of the railroads and canals of the United States of America.* Augustus M. Kelley Publishers, New York, NY.

Stephens, Carlene. (1989). Most reliable time: William Bond, the New England railroads, and time awareness in 19th-century America. *Technology and Culture.* 30. pg. 1-24.

Raw Material Preparation

Kemper, Jackson, III. (n.d.) *American charcoal making.* Eastern National Park and Monument Association, Hopewell Village, PA.

Van Wagenen, Jared. (July 1947). Potash manufacture. *The Chronicle.* 3(12). pg. 106, 109.

Van Wagenen, Jared. (September 1947). The charcoal burner. *The Chronicle.* 3(13). pg. 113, 118.

Van Wagenen, Jared. (April 1949). The old-time tanner. *The Chronicle.* 3(19). pg. 161, 164.

Rope Making and Sail Making

Ashley, Clifford W. (1944). *The Ashley book of knots.* Doubleday & Company, Inc., Garden City, NY.

Beecher, Mark H. (December 1972). Hand rope making. *The Chronicle.* 25(4). pg. 58-62.

Brewington, M. V. (June 1950). The sailmaker's gear. *The Chronicle.* 3(23). pg. 205-213.

Brewington, M. V. (September 1950). The sailmaker's gear. *The Chronicle.* 3(24). pg. 217-221, 223.

Hasluck, Paul N., Ed. (no date). *Knotting and splicing ropes and cordage.* David McKay Company, Philadelphia, PA.

Mikelson, Bob. (September 1971). Sailmaking and sailmaker's tools. *The Chronicle.* 24(3). pg. 29-33.

Sprague, William B. (February 1940). The rope maker. *The Chronicle.* 2(13). pg. 97, 100-101.

Sawyers (Saws)

Andrews, Ralph Warren. (1957). *This was sawmilling.* Superior Pub. Co., Seattle, WA.

Baader, William. (September 1968). Early saws. *The Chronicle.* 21(3). pg. 38-39.

Baker, Phil. (June 2005). The 19th century American backsaw. *The Gristmill.* 9. pg. 24-29.

Baker, Phil. (Spring 2005). The 19th-century American back saw. *The Fine Tool Journal*. pg. 6.

Baker, Phil. (Fall 2005). Oh!! If a saw could talk. *The Fine Tool Journal*. pg. 15.

Baker, Phil. (2006). The nineteenth-century American backsaw. *The Chronicle*, 59(1). pg. 13.

Baker, Phil. (Spring 2006). The backsaw blade: Tapered & parallel. *The Fine Tool Journal*. 55(4). pg. 26.

Baker, Phil. (June 2009). Nickel-plated backsaws. *The Chronicle*. 62(2). pg. 74-5.

Bale, M. Powis. (1880). *Woodworking machinery its rise, progress, and construction with hints on the management of saw mills and the economical conversion of timber*. Crosby Lockwood and Co., London. Reprinted in 1992 by Glen Moor Press, Lakewood, CO.

Disston, Henry & Sons. (1916). *The saw in history*. Philadelphia, PA. Reprinted in April 1978 by the Midwest Tool Collector's Association and the Early American Industries Association.

Disston, Henry & Sons. (1922). *Saw, tool and file book*. Philadelphia, PA.

- See annotations in the toolmakers and manufacturers bibliography.

Drabble & Sanderson. (1925). *The saw doctor's handbook*.

Ewan, N.R. (May 1941). Up-and-down saw mills.*The Chronicle*. 2(17). pg. 137, 144.

Friberg, Todd L. (2000). *Patented American saw sets, an illustrated patent directory 1812 - 1925*. Early American Industry Association, Murphy, NC.

Grimshaw, Robert. (1881). *Saws; The history, and development etc.* Philadelphia, PA.

Grimshaw, Robert. (1901). *Saw-filing and management of saws: A practical treatise on filing, gumming, swaging, hammering, and the brazing of band saws, the speed, work, and power to run circular saws, etc., etc.* H.W. Henley & Co., NY, NY.

- The Astragal Press has published a reprint of the 1880 original titled *Grimshaw on Saws*. It includes saw advertisements from the period.

(November 30, 1870). *Ironmonger*. 12. pg. 1009-1011.

(October 25, 1871). *Ironmonger*. 13. pg. 912.

James Leffel & Co. (1874, 1881). *Leffel's construction of mill dams and Bookwalter's millwright and mechanic*. Reprinted in 2001 by the Early American Industry Association, Murphy, NC.

- Mill dams, equipment for grist mills and saw mills.

Jones, P. d'A and Simons, E.N. (1960/1). *The story of the saw*. Spear & Jackson, Ltd., Sheffield, UK.

Kebabian, John S. (March 1973). An early Philadelphia pit-saw. *CEAIA*. 26. pg. 13-14.

Kebabian, John S. (September 1973). Sawmills -- early and not so early. *The Chronicle*. 26(3). pg. 41-45.

Kellogg, Elijah. (1867-1883). *Uncle Seth builds a sawmill*. Reprinted in 1973. Prepared for April 24, 1973 meeting of the Early Trades & Crafts Society.

Lamond, Tom. (June 2005). When does a nib become a nub? *The Gristmill*. 9. pg. 12.

Nicholson File Company. (1959). *Sawology: A Nicholson handbook*. Nicholson File Company, Providence, RI.

- "---being a brief account of the History, Manufacture, Variety and Uses of saws for the cutting of ferrous and non-ferrous metals, hard plastics and rubber, wood, and other dense materials . . . A useful handbook and guide for the shop superintendent, production foreman, mechanic, or home craftsman." (table of contents page).

Roberts, Robert W. (December 1970). The up and down sawmill. *The Chronicle*. 23(4). pg. 49-51.

Schaffer, Erwin L. (1999). *Hand-saw makers of North America*. Osage Press, Rockford, IL.

Simonds Saw and Steel Company. (1929). *The cross-cut saw*. Simonds Saw and Steel Company, Fitchburg, MA.

Simonds Saw and Steel Company. (1937). *The circular saw: A guide book for filers, sawyers and woodworkers*. Simonds Saw and Steel Company, Fitchburg, MA.

Simonds Saw and Steel Company. (1937). *How to file a cross-cut saw*. Simonds Saw and Steel Company, Fitchburg, MA.

Simonds Saw and Steel Company. (1937). *Woodworking saws and planer knives: Their care and use*. Simonds Saw and Steel Company, Fitchburg, MA.

Stubbs, Graham. (June 2006). American bucksaws. *The Chronicle*. 59(2). pg. 59-69.

Taran, Pete. (Winter 2001). Disston type study: The medallions: Part one. *Fine Tool Journal*. 50(3). pg. 10-13.

- A very important article on the history of the Disston Saw Company with excellent photographs; and an essential reference for any collector.
- The medallions on the earliest Disston saws (1840 - 1865 -- four variants noted) have an eagle on it. The earliest variant has the most detailed and esthetically pleasant drawing. After 1865, the medallion has the more well known scales. Taran illustrates 12 variations of the medallion with scales, which include numerous ways of spelling Philadelphia and the company name. After 1942, the word Philadelphia is dropped from the medallion and H. Disston and Sons becomes Disston.

The Tool Chest. (November 1999). Mr. Disston's view of saw nibs. *The Tool Chest*. 54. pg. 32.

Scientific Instruments

Bedini, Silvio. (1964). *Early American scientific instruments*. Washington, DC.

Bion, N. (1758). *The construction and principal uses of mathematical instruments, N. Bion: Translated and supplemented by Edmund Stone, 2nd Ed., 1758*. Reprinted by Astragal Press, Mendham, NJ.

Cohen, I. Bernard. (1950). *Some early tools of science*. Cambridge, MA.

Turner, Gerard L'E. (date unknown). *Scientific instruments 1500 - 1900: An introduction*. Astragal Press, Mendham, NJ.

Screwdrivers

Robinson, Trevor. (June 1996). The ratchet screwdriver. *The Chronicle*. 49(2). pg. 54-57.

- See annotations in the Maine toolmakers bibliography.

Rybczynski, Witold. (2000). *One good turn: A natural history of the screwdriver and the screw*. Scribner, NY, NY.

Sheet Metal Working
(also see Machinists)

Dyer, Herbert J. (1951). *How to work sheet metal: A practical man's description of metal working practice "straight from the bench"*. Model & Allied Publications Ltd., Hertfordshire, England.

Graham, Frank D. and Anderson, Edwin P. (1942). *Audels sheet metal workers handy book*. Theodore Audel & Co., Division of Howard W. Sams & Co., Inc., New York, NY. 1968 printing.

Woodworth, J.V. (1910). *Dies, their construction and use for the modern working of sheet metals*. Norman W. Henley Publishing, NY.

Ships, Shipbuilding, and Shipwrights
Also see Vol. 7, *Art of the Edge Tool.*

Abell, Sir Westcott. (1948). *The shipwright's trade.* The University Press, Cambridge, England, UK.

American Bureau of Shipping. (1982). *Rules for building and classing steel vessels.* American Bureau of Shipping, New York, NY.

Anderson, J. W. (1911). *Shipmasters' business companion.* James Brown & Son, Glasgow, UK.

Anderson, Romola and Anderson, R.C. (1947). *The sailing-ship: Six thousand years of history.* Robert M. McBride & Company, NY, NY.

Bass, George F., Ed. (1972). *A history of seafaring based on underwater archaeology.* Walker and Company, New York, NY.

Bradford, Gershom. (1927). *A glossary of sea terms.* Rumford Press, Concord, NH.

Bradford, Gershom. (1972). *The mariner's dictionary.* Barre Publishers, Barre, MA.

Brewington, M.V. (June 1962). The log canoe builder and his canoe. *The Chronicle.* 15(2). pg. 13-15.

Butler, Joyce. (1993). *William E. Barry's sketch of an old river with an illustrated essay shipbuilding on the Kennebunk.* The Brick Store Museum, West Kennebunk, ME.

Chapelle, Howard I. (1930). *The Baltimore Clipper, its origin and development.* Marine Research Society, Salem, MA.

Chapelle, Howard I. (1935). *The history of American sailing ships.* Bonanza Books, New York, NY.

Chapelle, Howard I. (1941). *Boatbuilding: A complete handbook of wooden boat construction.* Reprinted in 1969 by W. W. Norton & Co., NY, NY.

Chapelle, Howard I. (1960). *The national watercraft collection.* United States National Museum Bulletin 219. Smithsonian Institution, Washington, DC.

Culler, R.D. (1974). *Skiffs and schooners.* International Marine Publishing Company, Camden, ME.

Dodds, James. (2001). *Rudyard Kipling's the shipwrights' trade.* Mystic Seaport Museum, Mystic, CT.

Dodge, Ernest S., Ed. (1972). *Thirty years of the American Neptune.* Harvard University Press, Cambridge, MA.

Dow, George Francis and Edmonds, John Henry. (1996). *The pirates of the New England coast 1630 - 1730*. Dover Publications, New York, NY.

Duncan, Roger F. (2000). *Dorothy Elizabeth: Building a traditional wooden schooner*. W.W. Norton, NY, NY.

Gardner, John. (September 1969). Shipwrights' tools. *Log of Mystic Seaport*. 21(3). pg. 91-95.

Gillingham, Walter P. (March 1950). Ships' tackle blocks. *The Chronicle*. 3(22). pg. 194-195.

Goldenberg, Joseph. (1976). *Shipbuilding in colonial America*. University of Virginia Press for the Mariner's Museum, Charlottesville, VA.

- In the top ten on our most important list.

Hall, Christopher and Lee, Lance. (1975). *Apprenticing revived*. The Apprenticeshop of the Bath Maritime Museum, Bath, ME.

Hall, Christopher and Lee, Lance. (1975). *The Crotch Island pinky*. The Apprenticeshop of the Bath Maritime Museum, Bath, ME.

Hammond, Robert R. (1967). *An era to remember: A historical sketch of the shipbuilding industry in west Washington county*. Self published, Harrington, ME.

- An invaluable and irreplaceable record of the ships built in Addison, Columbia Falls, Cherryfield, Harrington and Milbridge, Maine.

Horsley, John E. (1978). *Tools of the maritime trades*. International Marine Publishing, Camden, ME.

Leavitt, John F. (1971). *Wake of the Coasters*. Wesleyan University Press, Middletown, CT.

Madsen, Betsy Ridge and Burnham, Maria. (1981). *Dubbing, hooping, and lofting: Shipbuilding skills*. The Cricket Press, Inc., Manchester, MA.

Martin, Kenneth R. (1975). *Whalemen and whaleships of Maine*. Harpswell Press, Brunswick, ME.

Morison, Samuel Eliot. (1965). *Spring tides*. Houghton Mifflin Company, Boston, MA.

Morris, Paul C. (1979). *American sailing coasters of the North Atlantic*. Bonanza Books, New York, NY.

Pickett, Gertrude M. (1979). *Portsmouth's heyday in shipbuilding*. Published by Joseph G. Sawtelle.

Robinson, Edward. (December 31, 1817). *Account book for sloop Betsy, Warren, Maine.* Handwritten.

- This is an original accounting book for the sloop Betsy, owned by Edward Robinson.
- It is accompanied by three of his Journals of the individual accounts of customers? These also contain miscellaneous newspaper clippings.

Rumsey, Barbara. (1995). *Hodgdon shipbuilding and mills: A documentary history of the first hundred years: 1816-1916.* The East Boothbay Series, #1, Winnegance House and Boothbay Region Historical Society, Boothbay, ME.

Smith, Hervey Garrett. (1990). *The arts of the sailor: knotting, splicing, and ropework.* Dover, NY, NY.

Stackpole, Edouard A. (1967). *The Charles W. Morgan: The last wooden whaleship.* Meredith Press, New York, NY.

Steel, David. (1794). *The Elements and practice of rigging and seamanship.* Vol. 1. David Steel, London, UK.

Story, Dana. (1964). *Frame-up! The story of Essex, its shipyards and its people.* Barre Publishers, Barre, MA.

Story, Dana. (1971). *The building of a wooden ship "sawn frames and trunnel fastened".* Barre Publishers, Barre, MA.

Story, Dana. (1995). *The shipbuilders of Essex: A chronicle of Yankee endeavor.* Ten Pound Island Book Company, Gloucester, MA.

Whittemore, Edwin. (January 1955). Ship building in early New England. *The Chronicle.* 8(1). pg. 4-7, 9.

Shoemakers
(see cobblers)

Silversmiths and Jewelers

Bovin, Murray. (1970). *Centrifugal or lost wax jewelry casting for schools, tradesmen, craftsmen.* Self published, 68-36 108th St., Forest Hills, NY.

De Matteo, William. (MCMLXVI). *The silversmith in eighteenth-century Williamsburg: An account of his life & times, & of his craft.* Williamsburg Craft Series, Colonial Williamsburg, VA.

Finegold, Rupert and Seitz, William. (1983). *Silversmithing.* Chilton Book Co., Radnor, PA.

Kane, Patricia E. et. al. (1998). *Colonial Massachusetts silversmiths and jewelers.* Yale University Art Gallery/ University Press of New England.

- Two hundred and ninety six biographies, essays on tools and styles of silversmithing, and a glossary.

Kauffman, Henry J. (1969). *The colonial silversmith: His techniques & his products.* T. Nelson, Camden, NJ.

Martin, Charles and D'Amico, Victor. (1949). *How to make modern jewelry.* Art for Beginners Series, The Museum of Modern Art, Simon and Schuster, NY, NY.

Steam Engines

Hills, Richard. (1989). *Power from steam: A history of the stationary steam engine.* Cambridge University Press, Cambridge, MA.

Spratt, H.P. (1950). *Outline history of transatlantic steam navigation.*

Thurston, Robert Henry. (1891). *A manual of the steam-engine. Part 1, structure and theory, Part 2, design, construction, and operation.* 2 vols. Publisher unknown, NY.

Westbury, Edgar T. (1970). *Building a steam engine from castings.* Model & Allied Publications Ltd., Hertfordshire, England.

Surveying

Bedini, Silvio A. (1975). Artisans in wood: The mathematical instrument-makers. In: Hindle, Brooke, Ed. *America's wooden age: Aspects of its early technology.* Sleepy Hollow Restorations, Tarrytown, NY.

- Surveying instruments, compasses, octants, and scales.

Buff & Buff Mfg. Co. (1938). *Surveying instruments: For civil and mining engineers.* Buff & Buff Mfg. Co., Boston, MA.

- A nice Buff & Buff is on display in the Davistown Museum permanent collection in the main hall.

Finch, J. K. (1918). *Plane surveying: A practical treatise on the art of plane surveying, including chaining, leveling, compass and transit measurements, land and construction surveying, topographic surveying, and mapping.* American Technical Society, Chicago, IL.

Gurley, W. & L. E. (1874). *A manual of American engineer & surveyor's instruments.* Twenty-first edition. Reprinted in 1993 by Astragal Press, Mendham, NJ.

320

Smart, Charles E. (1962). *The makers of surveying instruments in America since 1700*. Regal Art Press, Troy, NY.

Tanning

Didsbury, J. (March 1964). Oak and hemlock bark for tanning. *The Chronicle*. 17(1). pg. 10-12.

Didsbury, J. (June 1964). Oak and hemlock bark for tanning (continued). *The Chronicle*. 17(2). pg. 23-24.

Textiles

Albright, Frank P. (September 1972). Notes on spinning wheels. *The Chronicle*. 25(3). pg. 36-39.

Anonymous. (December 1958). Some notes on the cotton gin. *The Chronicle*. 11(4). pg. 60-64.

Catling, Harold. (1970). *The spinning mule*. David & Charles, Newton Abbot, UK.

Chase, William H. (1950). *Five generations of loom builders.* Draper, Hopedale, MA.

Cole, A.H. (1926). *The American wool manufacture*. Vol. 1. Harvard University Press, Cambridge, MA.

Fannin, Allen. (1976). *Handspinning: Art & technique.* Van Nostrand Reinhold Company, New York, NY.

Gaines, Ruth. (May 1941). Homespun. *The Chronicle*. 2(17). pg. 137, 139-141.

Gemming, Elizabeth. (1979). *Wool gathering: Sheep raising in old New England.* Coward, McCann & Geoghegan, New York, NY.

Gross, Laurence. (1987). Wool carding: A study of skill and technology. *Technology and Culture* 28. pg. 807-827.

Huff, F. and Didsbury, J. (June 1959). The spinning wheel. *The Chronicle*. 12(2). pg. 13-16.

Johnson, Laurence A. (March 1958). The niddy-noddy. *The Chronicle*. 11(1). pg. 4-7.

MacFarlane, Janet and Parslow, Virginia. (October 1954). Hand processing wool in America. *The Chronicle*. 7(4). pg. 37 - 40.

McGouldrick, Paul F. (1968). *New England textiles in the nineteenth century: Profits and investments*. Harvard University Press, Cambridge, MA.

Ralph, William. (June 1972). The spinning wheel: A neglected tool. *The Chronicle*. 25(2). pg. 22-27.

Ralph, William. (September 1972). Collecting spinning wheels. *The Chronicle*. 25(3). pg. 40-42.

Robinson, Harriet H. (1898). *Loom and spindle*. Thomas Crowell & Co., NY, NY.

Robinson, Stella. (1983). *Textiles*. Endeavour Books. Wayland Ltd., Brighton, England.

Starbuck, David R. (1986). The Shaker mills in Caterbury, New Hampshire. *IA, Journal of the Society for Industrial Archeology*. 12. pg. 11-37.

Swain, Frank K. (May 1941). Two unusual flax wheels. *The Chronicle*. 2(17). pg. 173-174.

van Wagenen, Jared, Jr. (1953). *The golden age of homespun*. Hill and Wang, NY.

Timber Framing and Housebuilding

Benjamin, Asher. (1827). *The American builder's companion; or, a system of architecture, particularly adapted to the present style of building*. R. P. & C. Williams, Boston, MA. Reprinted in 1969 by Dover Publications, NY.

Beaudry, Michael. (2009). *Crafting frames of timber*. Mud Pond Hewing and Framing, Montville, ME.

Benson, Tedd. (1980). *Building the timber frame house: The revival of a forgotten craft*. Scribner, NY, NY.

Berg, Donald J. (1986). *How to build in the country: Good advice from the past on how to choose a site, plan, design, build, landscape & furnish your home in the country*. Ten Speed Press, Berkeley, CA.

Blackburn, Graham. (1974). *Illustrated housebuilding*. Overlook Press, Woodstock, NY.

Chamberlain, Samuel. (1937). *Beyond New England thresholds*. Hastings House, NY, NY.

Chambers, Robert. (August/September 1986). Scribe-fitting a log house. *Fine Homebuilding*. 34.

Chappell, Steve. (1998). *A timber framer's workshop*. Fox Maple Press, West Brownfield, ME.

Condit, Carl W. (1968). *American building: Materials and techniques from the first colonial settlements to the present*. Chicago, IL.

Early Trades and Crafts Society. (1971). *Barn builder's words*. Prepared by the Early Trades and Crafts Society for the members of the Early American Industries Association.

Elliot, Stewart. (1978). *The timber frame planning book*. Contemporary Books, Chicago, IL.

Elliot, Stewart and Wallas, Eugene. (1997). *The timber framing book*. Housesmiths Press, York, ME.

Fickes, Clyde P. and Groben, W. Ellis. (1945). *Building with logs*. Publication No. 579. U. S. Department of Agriculture, Washington, DC.

Fine Homebuilding Editors. (1996). *Timber-framed houses*. The Taunton Press, Newtown, CT.

Gauthier-Larouche, Georges. (1974). *Evolution de la maison rurale traditionelle dans la region de Quebec*. Les Presses de L'Universite Laval Quebec.

Harris, Richard. (2001). *Discovering timber-framed buildings*. Shire Publications, Ltd., Essex, UK.

Hewett, Cecil A. (1997). *English historic carpentry*. Linden Publishing, Fresno, CA.

Hunt, W. Ben. (1974). *How to build and furnish a log cabin*. Collier Books, NY.

Isham, Norman Morrison. (1968). *Early American houses: The seventeenth century*. Classic Guide books to the Visual Arts, American Life Foundation, Watkins Glen, NY.

Kahn, Lloyd. (1973). *Shelter*. Shelter Publicatioins, Bolinas, CA.

Kahn, Lloyd. (2004). *Home work: Handmade shelter*. Bolinas, CA.

Kahn, Lloyd. (2008). *Builders of the Pacific coast*. Bolinas, CA.

Keith, Wilbur C. (1992). *Homebuilding and woodworking in colonial America*. The Globe Pequot Press, Old Saybrook, CT.

Kellogg, Elijah. (1867-1883). *Uncle Seth builds a windmill*. Reprinted in 1973. Prepared for Oct. 23, 1973 meeting of the Early Trades & Crafts Society.

Langsner, Drew. (1982). *The logbuilder's handbook*. Rodale Press, Emmaus, PA.

Mackie, Allan B. (1997). *Building with logs*. Firefly Books, Buffalo, NY.

McRaven, Charles. (1994). *Building and restoring the hewn log house*. Betterway Books, Cincinnati, OH.

Newman, Rupert. (2005). *Oak-framed buildings*. Guild of Master Craftsman Publications, East Sussex, UK.

Petesen, David. (July/August 1985). Building the traditional hewn-log home: A mini manual. *Mother Earth News*. 94.

Phleps, Hermann. (1982). *The craft of log building*. Lee Valley Tools, Ltd., Ottawa, Ontario, CA.

Roberts, Warren E. (July 1977). The tools used in building log houses in Indiana. *Pioneer America*. Reprinted in September, 1978 by the Mid-West Tool Collectors Association and the Early American Industries Association.

Shurtleff, Harold R. (1939). *The log cabin myth*. Cambridge, MA.

Sobon, Jack. (1997). *Historic American timber joinery - a graphic guide*. Timber Framers Guild, Becket, MA.

Sobon, Jack and Schroeder, Roger. (1984). *Timber frame construction: All about post and beam building*. Storey Communications, Inc., Pownal, VT.

Wilbur, C. Keith. (1992). *Homebuilding and woodworking in colonial America*. The Globe Pequot Press, Old Saybrook, CT.

Tinsmithing

Bailey, Gillian W. B. (April 1951). Tin plate and the tinker. *The Chronicle*. 4(2). pg. 13-15.

Butts, Isaac R. (1863). *The tinman's manual, and builder's and mechanic's handbook, designed for tinmen, japanners, coppersmiths, engineers...* I.R. Butts & Co., Boston, MA.

- Available as a loan from the E.A.I.A. Library.

Demer, John H. (December 1973). How tinsmiths used their tools. *The Chronicle*. In: Pollak, Emil and Pollak, Martyl, Eds. (1991). *Selections from The Chronicle: The fascinating world of early tools and trades*. The Astragal Press, Mendham, NJ.

Demer, John H. (1978). *Jedediah North's tinner's tool business*. The Early American Industries Association, South Burlington, VT.

- Available as a loan from the E.A.I.A. Library.

Lansansky, Jeannette. (1982). *To cut, piece & solder: The work of the rural Pennsylvania tinsmith 1778-1908*. Oral Traditions Project of the Union County Historical Society, Courthouse, Lewisburg, PA.

Shagena, Jack L. (2006). *An illustrated history of tinware in America: How the tinsmith, peddler, tinker, and toolmaker built an industry*. Self published by Jack L. Shagena, Bel Air, MD.

Smith, Elmer L. (1976). *Tinware: Yesterday and today*. Applied Arts Publishers, Lebanon, PA.

Vosburgh, H. K. (1879). *The tinsmith's helper and pattern book: With useful rules, diagrams and tables*. Reprinted 1910 revised edition in 1994 by Astragal Press, Mendham, NJ.

Toolmaking

Also see the extensive citations in volumes 6 and 11 of the *Hand Tools in History* publication series.

Burch, Monte. (2004). *Making Native American hunting, fighting and survival tools.* Lyons Press, Guilford, CT.

Comte, Hubert. (1997). *Tools: Making things around the world.* Translated from French by Molly Stevens and David Marinelli. Harry N. Abrams, Inc., NY, NY.

Durell, Edward. (September 1965). Dating a tool. *The Chronicle.* 18(3). pg. 38-40.

Hawley, Ken. (Autumn 1991). Edge tool makers. *The Tool and Trades History Society Newsletter.* 35. pg. 48.

Horsley, John E. (1978). *Tools of the maritime trades.* International Marine Publishing, Camden, ME.

Sarpolus, Dick. (2001). *Collectible blowtorches.* Astragal Press, Mendham, NJ.

Townsend, Raymond R. (December 1956). Hand tools for making fishhooks. *The Chronicle.* 9(4). pg. 37-41.

Waldorf, D. C. (1993). *The art of flint knapping.* Fourth Edition. Self published.

- There is also an accompanying video: *The art of flint knapping: Video companion.*

Wolcott, S.C. (October 1958). Classification of certain American tools of certain trades. *The Chronicle.* 11(3). pg. 54-55.

Woodworth, Joseph V. (1910). *American tool making and interchangeable manufacturing.* Norman W. Henley Publishing, NY, NY.

Whaling

Brewington, M.V. (September 1968). The tools of a whaler's cooper. *The Chronicle.* 21(3). pg. 40-42.

Foreman, Henry Chandlee. (1966). *Early Nantucket and its whale houses.* Hastings House, New York, NY.

Horsley, John E. (1978). *Tools of the maritime trades.* International Marine Publishing, Camden, ME.

Lytle, Thomas G. (1984). *Harpoons and other whalecraft.* The Old Dartmouth Historical Society Whaling Museum, New Bedford, MA.

- The word whalecraft in Lytle's title refers to tools used in the hunting and killing of whales rather than sailing craft utilized to capture the whales.
- Appendix A is currently considered the definitive listing of New Bedford area whaling tool (whalecraft) manufacturers. A number of these whalecraft manufacturers were practicing blacksmiths who also made edge tools for ship building. This appendix is summarized in a *Registry of Maine Toolmakers* appendix.
- For an interesting listing of the construction location of the New Bedford whaling ships, see "Ship Registers of New Bedford, Massachusetts." A copy is located in the Davistown Museum files as is a copy of Lytle's Appendix A "Whalecraft Manufacturers of New Bedford and Fairhaven, Massachusetts."

Verrill, A. Hyatt. (1916). *The real story of the whaler: Whaling, past and present.* D. Appleton and Company, New York, NY.

Wheelwrights

Peloubet, Don, Ed. (date unknown). *Wheelmaking: Wooden wheel design and construction.* Astragal Press, Mendham, NJ.

Sturt, George. (2000). *The wheelwright's shop.* Cambridge University Press.

Woodworking
Also see planemaking, timber framing, coopers, and shipbuilding

Aber, R. James. (April 13, 1980). *A glossary of woodworking joints.* Written for a meeting of Crafts of NJ.

Anonymous. (September 1967). Tool study -- hatchets. *The Chronicle.* 20(3). pg. 46-47.

Anonymous. 1992. *The cutting edge, an exhibition of Sheffield tools.* Exhibition Catalog. The Ruskin Gallery, Sheffield, UK.

Ball, John E. (1991). *Audel: Carpenters and builders Library: Volume One: Tools, steel square, joinery.* Macmillan Publishing Company, NY, NY.

Bjerkoe, Ethel Hall. (1962). *The cabinetmakers of America.* Schiffer Publishing Ltd., Exton, PA.

Blackburn, Graham. *Traditional woodworking handtools: A manual for the woodworker, a guide for the enthusiast.* The Astragal Press, Mendham, NJ.

Bramwell, Martyn, Ed. (1976). *The international book of wood.* Simon And Schuster, NY, NY.

Chippendale, Thomas. (1762). *The gentleman and cabinet-maker's director: Being a large collection of the moft elegant and useful designs of household furniture, in the moft fashionable taste. The third edition.* London. Reprinted in 1966 by Dover Publications Inc., NY, NY.

326

Carlson, Robert. (1975). *Auger points.* Early Trades and Crafts Society.

Carlson, Robert H. and Stevens, Thomas A. (December 1967). On the origin of the spiral auger. *The Chronicle.* 20(4). pg. 49-53.

Christensen, Erwin O. (1952). *Early American wood carving.* Dover Publications Inc., NY, NY.

Dunbar, Michael. (1989). *Restoring, tuning & using classic woodworking tools.* Sterling Publishing Co., Inc., NY, NY.

Early Trades and Crafts Society. (undated). *A sampling from the active scrapbook.* Prepared for Oct. 15 meeting of the Early Trades and Crafts Society.

- Includes: *The tool handle or stock maker's clamp* by Frank Bawden, *The manufacture of a wooden bucket,* by Richard A. Martin, *The vanishing auto tool kit* by Gene Kosche, *The shrink rule* by Herman Friedman, *Stair builder's slip stick* by William B. Hilton, *Cooper's croze* by Miner J. Cooper and others.

Eason, Julie Anne. (Summer 1998). The hand tools you use with your feet. *The Fine Tool Journal.* 48(1). pg. 8-10.

- The history, design and function of the shaving horse.

Edlin, Herbert L. (1969). *What wood is that? A manual of wood identification.* The Viking Press, New York, NY.

Feirer, John L. (1960). *Industrial arts: Woodworking.* Chas. A. Bennett Co., Inc., Peoria, IL.

Feirer, John L. (1988). *Cabinetmaking and millwork.* Fifth Edition. Glencoe Publishing Company, Mission Hills, CA.

Fisher, Leonard Everett. (1966). *Colonial American craftsmen: The cabinetmakers.* Franklin Watts, Inc., NY, NY.

Foley, Vernard and Moyer, Richard H. (June 1977). The American axe: Was it better? *The Chronicle.* 30(2). pg. 28-32.

The Forest Products Laboratory. (1955). *Wood handbook: Basic information on wood as a material of construction with data for its use in design and specification.* Forest Service, U. S. Department of Agriculture, Washington, DC.

G. & D. Cook & Co. (1860). *Illustrated catalogue of carriages and special business advertiser.* Reprinted in 1970 by Dover Publications, Inc., New York, NY.

Gaynor, James M., Ed. (1997). *Eighteenth-century woodworking tools: Papers presented at a tool symposium: May 19-22, 1994.* Colonial Williamsburg Historic Trades. Volume III. The Colonial Williamsburg Foundation, Williamsburg, VA.

- This volume is among the most important information sources about early toolmaking in America.
- The following papers in this text are particularly relevant for an understanding of the evolution of toolmaking in New England and Maine.
    - Walker, Philip. *Woodworking tools before 1700.*
    - Hey, David. *The development of the English toolmaking industry during the seventeenth and eighteenth centuries.*
    - Kebabian, Paul B. *Eighteenth-century American toolmaking.*
    - Hagedorn, Nancy L. *Tools for sale: The marketing and distribution of English woodworking tools in England and America.*
    - Wing, Donald and Anne. *Planemaking in eighteenth-century America.*
    - Ingraham, Ted. *The joiner's trade and the wooden plane in eighteenth-century New England.*
    - Hummel, Charles F. *Using tools to earn a living and the Dominy family of East Hampton, Long Island.*
    - Underhill, Roy. *"The debate of the carpenter's tools."*
- Check the index of Vol. 8 to find additional comments on this text.

Goodman, W.L. (1964). *The history of woodworking tools.* David McKay Company, Inc., NY, NY.

- See annotations in the European Precedents and Early Industrial Revolution bibliography.

Graham, Frank D. and Emery, Thomas J. (1923). *Audels carpenters and builders guide #1: A practical illustrated trade assistant on modern construction for carpenters - joiners, builders - mechanics and all wood workers.* 4 vols. Theo. Audel & Co., NY, NY.

- Known as Audels, this is the classic and most sought after carpenter's reference. It is still very useful and is a perennial best seller at the Liberty Tool Co. There are never enough copies to satisfy the demand.
- Numerous editions exist, the earlier ones are the most interesting.

Green, Harvey. (2007). *Wood: Craft, culture, history.* Penguin, NY, NY.

Hampton, C.W. and Clifford, E. (1934). *Planecraft: Hand planing by modern methods.* Reprinted in 1972 by Woodcraft Supply Co., Woburn, MA.

Hansen, John. (1970). *A cabinet maker's and joiner's miscellaney.* Unpublished, prepared for the Nov. 17, 1970 meeting of the Early Trades and Crafts Society.

Hasluck, Paul N., Ed. (1905). *The handyman's book of tools, materials, and processes employed in working wood.* Cassell and Company, Limited, NY, NY. Reprinted as *Wood working* in November, 1987, by North Village Publishing Co., Lancaster, MA.

Heine, Gunther. (June 1995). Toolmakers of Hamburg in the 19th century: The manufacture of tools of the woodworking trades. *The Chronicle.* 48(2). pg. 47-53.

- Hamburg, Germany is important because many of their tools were copied in the United States by German immigrants.

Heuvel, Johannes. (1963). *The cabinetmaker in eighteenth-century Williamsburg.* Williamsburg Craft Series, Williamsburg, VA.

Hewett, Cecil A. (date unknown). *English historic carpentry.* Astragal Press, Mendham, NJ.

Hibben, T. (1933). *The carpenter's tool chest.* London.

Hoadley, R. Bruce. (2000). *Understanding wood: A craftsman's guide to wood technology.* Taunton, Newtown, CT.

Hodgson, Fred T. (1883). *The carpenters' steel square, and its uses. Being a description of the square, and its uses in obtaining the lengths and bevels of all kinds of rafters, hips, groins, braces, brackets, purlins, collar-beams, and jack-rafters; also, its application in obtaining the bevels and cuts for hoppers, spring mouldings, octagons, stairs, diminished stiles, etc., etc., etc.* Palliser, Palliser & Co., Bridgeport, CT.

Jones, Bernard E., Ed. (1980). *The complete woodworker.* Ten Speed Press, Berkeley, CA.

Kebabian, Paul B. (1978). *American woodworking tools.* New York Graphic Society, Boston, MA.

Kebabian, Paul B. and Lipke, William C., Eds. (1979). *Tools and technologies: America's wooden age.* Robert H. Fleming Museum, University of Vermont, Burlington, VT.

Klingler, Eugene L. (June 1988). American braces in the 19th century: Pre-Barber and post-Barber -- Part I. *The Chronicle.* 41(2). pg. 21-23.

Lamond, Thomas C. (1997). *The spokeshave book: Manufactured and patented spokeshaves & similar tools.* Tom Lamond, 30 Kelsey Place, Lynbrook, NY 11563.

Lamond, Thomas C. (Summer 1998). An updated look at Windsor Beader. *The Fine Tool Journal.* 48(1). pg. 11-12.

Lamond, Thomas C. (September 1999). Spokeshaves and similar tools. *The Chronicle.* 52(3). pg. 108-113.

Lane, Joshua W. (2003). *Woodworkers of Windsor: A Connecticut community of craftsmen and their world, 1635-1715*. Historic Deerfield, Deerfield, MA.

Lanz, Henry. (1985). *Japanese woodworking tools: Selection, care and use*. Sterling Publishing Co., New York, NY.

Lasswell, Pat. (Winter 2001). Carpenter's inventory: North Providence, Rhode Island: 1796. *Sign of the Jointer*. 2(4). pg. 85-91.

- A reproduction of a listing of used carpenter's tools done by Joseph Fuller upon the death of Henry Whipple and includes their value.

Mercer, Henry C. (January 1926). Ancient carpenter's tools, part IV: Tools for surfacing, chopping and paring (continued). *Old Time New England The Bulletin of the Society for the Preservation of New England Antiquities*. 16(3). pg. 118-137.

- This series replicates ancient carpenter's tools in installments.
- Axe hatchet, adze, cooper's adze, Korean adze, draw knife, round shave or scorper, peg cutter, witchet or rounding plane, spoke shave, plane, jack plane or fore plane, trying plane, floor plane, cooper's long jointer and smoothing plane.

Mercer, Henry C. (April 1927). Ancient carpenter's tools, part VI: Tools for shaping and fitting (continued). *Old Time New England The Bulletin of the Society for the Preservation of New England Antiquities*. 17(4). pg. 179-191.

- Chisel, forming chisel or firmer, skew forming chisel, Dutch paring chisel, paring chisel, gouge, mortise chisel, axe mortise chisel, carpenter's mallet, commander, mortising axe or post axe, post boring machine and twibil.

Mercer, Henry C. (July 1928). Ancient carpenter's tools, part VIII: Tools for shaping and fitting (continued). *Old Time New England The Bulletin of the Society for the Preservation of New England Antiquities*. 19(1). pg. 28-43.

- Wheelwright's reamer, centre bit, button bit, plug centre bit, Japanese annular auger, spiral auger, Cooke's auger, gimlet or wimble, carpenter's brace and bit, Dutch brace and bit, wheelwright's burning iron and Chinese wood punch or reamer.

Mercer, Henry C. (1929). *Ancient carpenters' tools together with lumbermen's, joiners' and cabinet makers' tools in use in the eighteenth century*. Horizon Press. Fifth edition reprinted in 1975 by the Bucks County Historical Society.

- See annotations in the Industrial Revolution bibliography.

Nagyszalanczy, Sandor. (1998). *The art of fine tools*. Taunton Press, Newtown, CT.

- Sumptuous photos.

Neary, John. (December 1980?). Book reviews: Books on tools that evoke nostalgia and explain how to check the set of an adze. *Americana*. pg. 81-84.

Northcott, W. Henry. (1868). *A treatise on lathes and turning*. Longmans, Green and Co., London. Second edition published in 1876 by Linden Publishing Co, Inc., Fresno, CA.

Odate, Toshio. (1998). *Japanese woodworking tools: Their tradition, spirit and use*. Linden Publishing, Inc., Fresno, CA.

Peters, Rick. (2000). *Woodworker's guide to wood: Softwoods, hardwoods, plywoods, composites, veneers*. Sterling Pub. Co., NY, NY.

Price, James E. (Spring 1996). The olde tool column: Obscure patented hollow augers. *The Fine Tool Journal*. 45(4). pg. 11.

Price, Jim. (Winter 2005). Early handforged spoke pointer. *Fine Tool Journal*. 54(3). pg. 23.

Pye, David. (1971). *The nature and art of workmanship*. Van Nostrand Reinhold, NY, NY.

Rees, Mark. (1987). Airtight case making - the planes and their uses. *Tools & Trades*. 4.

Roberts, Kenneth D. (1980). *Some 19th century English woodworking tools: Edge and joiner tools and bit braces*. K. Roberts Publishing Company, Fitzwilliam, NH.

Romaine, Lawrence B. (April 1953). A Yankee carpenter and his tools. *The Chronicle*. 6(2). pg. 33-34.

Romaine, Lawrence B. (April 1955). The American carpenter. *The Chronicle*. 8(2). pg. 16.

Schwarz, Christopher. (Spring 2007). Rotary lapping machines. *The Fine Tool Journal*. 56(4). pg. 7-10.

Shea, John G. (1975). *Making authentic country furniture: With measured drawings of museum classics: With 794 construction plans and illustrations*. Dover Publications Inc., NY, NY.

Singleton, Esther. (1913). *The furniture of our forefathers*. Garden City, NY.

Smitten, Stanley L. (1870). *The woodworking lathe to 1850*. No publisher.

Stokes, J. (1829). *Complete cabinetmaker's and upholsterer's guide*. Dean and Munday, London.

Straffin, Dean. (December 1989). A treatise on the chisel axe. *The Chronicle*. 42(4). pg. 89 - 90.

Thwing, L. L. (November, 1937). Clapboards. *The Chronicle*. 2(2). pg. 15 - 16.

Thwing, L. L. (December 1949). The woodworker in France and England. *The Chronicle*. 3(21). pg. 181-182.

Thwing, L. L. (June 1950). The woodworker in England. *The Chronicle*. 3(23). pg. 205, 213.

Walker, Philip. (1997). Woodworking tools before 1700. In: *Eighteenth century woodworking tools, Papers presented at a tool symposium, May 19-22, 1994.* Gaynor, Jay, Ed. Colonial Williamsburg Foundation, Williamsburg, VA.

Welsh, Peter C. (1965). Woodworking tools, 1600-1900. *Contributions from the Museum of History and Technology, Paper 51.* Bulletin 241. United States National Museum, Washington, DC, pg. 179-227.

Welsh, Peter C. (1966). *Woodworking tools 1600 – 1900.* Smithsonian Institute, Washington, D.C.

Wildung, Frank H. (1957). *Woodworking tools at Shelburne Museum.* Museum Pamplet Series, No. 3. The Shelburne Museum, Shelburne, VT. Facsimile.

Wolcott, Stephen C. (July 1934). The Frow -- a useful tool. *The Chronicle*. 1(6). pg. 1, 3.

- Froe: "A cleaving tool for splitting cask staves and shingles from the block. Etymology: perhaps alteration of obsolete *froward* turned away, from Middle English; from the position of the handle." (Merriam-Webster Dictionary online.)

Wood Turning Center and Yale University Art Gallery. (2001). *Wood turning in North America since 1930: Exhibition catalog.*

Wyatt, E. M. (1936). *Common woodworking tools: Their history.* Milwaukee, WI.

Wyllie, Robin H. (December 1986). The Swedish axe in North America. *The Chronicle*. 39(4). pg. 53-56.

Wrenches

Gaier, Dan. (Winter 2005). Reach for the wrench. *Fine Tool Journal*. 54(3). pg. 16-18.

Page, Herb (Mr. Oldwrench). (Spring 2001). "No name" wrenches. *The Fine Tool Journal*. 50(4). pg. 23-24.

- Also see an extensive listing of Herb Page's wrench articles in the US and New England Toolmakers bibliography.

Page, Herb (Mr. Oldwrench). (Fall 2002). Reach for the wrench: The song of the monkey-wrench. *The Fine Tool Journal*. 52(2). pg. 17-18.

- This article is reprinted in the Davistown Museum online essay on the Boston wrenches.

Page, Herb. (Fall 2005). Reach for the wrench: Vintage auto wrenches. *The Fine Tool Journal*. pg. 16-18.

Page, Herb. (2006). Reach for the wrench. *Fine Tool Journal*. 56(2). pg. 17.

Staten, Vince. (1996). *Did monkeys invent the monkey wrench? Hardware stores and hardware stories*. Simon & Schuster, NY, NY.

Collector's Guides, Handbooks, and Dictionaries

American Society for Metals. (1964). *Metals handbook.* Volume 2, 8th Ed. American Society for Metals, Materials Park, OH.

Appleton. (1866). *Appleton's dictionary of machines, mechanics, engine-work, and engineering.* 2 vols. D. Appleton and Company, New York, NY.

Arbor, Marilyn. (1994). *Tools and Trades of America's Past: The Mercer Museum collection.* The Mercer Museum, Doylestown, PA.

Astragal Press. (n.d.). *Books on early tools, trades and technology.* PO Box 239, Mendham, NJ 07945-0239. www.astragalpress.com.

- This catalog of Astragal publications allows quick access to most of the important contemporary publications on tools and technology. Many of them are also listed within these bibliographies.

Babcock & Wilcox Company. (1923). *Steam; its generation and use.* Bartlett Orr Press, New York, NY.

Bacheller, Milton H., Jr. (2000). *American marking gages, patented and manufactured.* Self-published, 185 South St., Plainville, MA 02762.

Baird, Ron and Comerford, Dan. (1989). *The hammer: The king of tools: A collectors handbook.* Ron Baird and Dan Comerford, Publishers, Fair Grove, MO.

Barlow, Ronald S. (1991). *The antique tool collector's guide to value.* Third edition. Windmill Publishing Company, El Cajon, CA 92020.

- One of the more frequently utilized references for checking values and identifications.

Barnwell, George W. Ed. (1941). *The new encyclopedia of machine shop practice: a guide to the principles and practice of machine shop procedure.* Wm. H. Wise & Co., Inc., NY, NY.

Batory, Dana M. (1997). *Vintage woodworking machinery: An illustrated guide to four manufacturers.* Vol 1. Astragal Press, Mendham, NJ.

Batory, Dana M. (2004). *Vintage woodworking machinery.* Vol. 2. Astragal Press, Mendham, NJ.

Bettesworth, A. and Hitch, C. (1981). *The builder's dictionary; or, gentleman and architect's companion, Vol. I.* The Association for Preservation Technology, Ottawa, Canada.

Bettesworth, A. and Hitch, C. (1981). *The builder's dictionary; or, gentleman and architect's companion, Vol. II.* The Association for Preservation Technology, Ottawa, Canada.

Blackburn, Graham. (1974). *The illustrated encyclopedia of woodworking handtools, instruments & devices.* Simon and Schuster, NY, NY.

Blanchard, Clarence. (2006). A trip to the D'Elia. *Fine Tool Journal.* 56(1). pg. 10.

Blanchard, Clarence. (2006). *The Stanley little big book: A comprehensive pocket price guide for planes: 2006.* Antique & Collectible Tools Inc., 27 Fickett Rd, Pownal, ME 04069.

Blanchard, Clarence. (2008). *The Stanley little big book: A comprehensive pocket price guide for rules, levels, & other Stanley tools: 2007-2008.* Antique & Collectible Tools Inc., 27 Fickett Rd, Pownal, ME 04069.

Blase, Francis Jr. (1984). *Heebner & Sons, pioneers of farm machinery in America.* Hatfield Publishing Company, Hatfield, PA.

Bureau of Naval Personnel. (1963). *Basic hand tools.* Navy Training Course NAVPERS 10085-A. United States Navy.

Burke, James. (1978). *Connections.* Little, Brown and Company, Boston, MA.

Carr, Ronald, Smith, Charles and Stubbs, Graham. (2007). *Vintage blowtorches: An identification and rarity guide.* The Blow Torch Collectors Association, Las Vegas, NV.

Colvin, Fred H. and Stanley, Frank A. (1926). *American machinists' handbook and dictionary of shop terms: A reference book of machine shop and drawing room data, methods and definitions.* Fourth Edition.McGraw-Hill Book Company, Inc., New York, NY.

- One of the more important of the many types and editions of machinist's handbooks.

Commissioner of Patents. (1859). *Report of the Commissioner of Patents for the year 1859: Arts and manufactures.* George W. Bowman, Washington D.C.

Cope, Kenneth L. (1993). *American machinist's tools: An illustrated directory of patents.* Astragal Press, Mendham, NJ.

- See annotations in the Toolmaking Trades bibliography under machinists.

Cope, Kenneth L. (1994). *Makers of American machinist's tools: A historical directory of makers and their tools.* Astragal Press, Mendham, NJ.

- See annotations in the Toolmaking Trades bibliography under machinists.

Cope, Kenneth L. (1998). *More makers of American machinist's tools: Part two of a historical directory of makers and their tools.* Astragal Press, Mendham, NJ.

- See annotations in the Toolmaking Trades bibliography under machinists.

Cope, Kenneth L. (1999). *American wrench makers 1830 - 1915*. Astragal Press, Mendham, NJ.

- Another of Cope's indispensable references.
- Note the enlarged edtion listed below.

Cope, Kenneth L. (2001). *American foot power & hand power machinery*. Astragal Press, Mendham, NJ.

Cope, Kenneth L. (2001). *American foot power & hand power machinery*. M.J. Donnelly Antique Tools, Avoca, NY.

Cope, Kenneth L. (2001). *American lathe builders: 1810 - 1910*. Astragal Press, Mendham, NJ.

Cope, Kenneth L. (2002). *American wrench makers 1830-1930*. 2nd edition. Astragal Press, Mendham, NJ.

Cope, Kenneth L. (2002). *American planer, shaper and slotter builders*. Astragal Press, Mendham, NJ.

Cope, Kenneth L. (2003). *American cooperage machinery and tools*. Astragal Press, Mendham, NJ.

- See annotations in the Toolmaking Trades bibliography under coopers.

D'Allemagne, Henry Rene. (1968). *Decorative antique ironwork - a pictoral treasury*. Dover Publications, NY, NY.

- A catalog of the collection of ironwork (wrought iron, steel, and cast iron objects, hardware, and tools) in the Musee Le Secq des Tournelles in Rouen, France.

Diagram Group, The. (1981). *Handtools of arts and crafts: The encyclopedia of the fine, decorative and applied arts*. St. Martin's Press, NY, NY.

- This wonderful text has drawings of the tools used for writing and drawing, printing, making books, painting, pictures without paint, clay and pottery, modeled and cast sculpture, carved sculpture, cabinetmaking, wood decoration, working with glass, fine metalwork, lapidary and beadmaking, thread preparation, weaving, knitting and knotting, needlework and leatherwork.

Donnelly, Martin J., Antique Tools. *The catalogue of antique tools: The world's premier antique hand tool value guide*. Martin J. Donnelly Antique Tools, PO Box 281, Bath, New York. (800) 869-0695. www.mjdtools.com.

- Martin Donnelly's catalogs, florid language notwithstanding, provide a wealth of information about tools and toolmakers to his subscribers.
- The Davistown Museum's Center for the Study of Early Tools library contains a nearly complete collection of the many auction catalogs he has issued.

Early Trades and Crafts Society. (1972). *A tool collector's picture book.* Unpublished, prepared for the Sept. 26, 1972 meeting of the Early Trades and Crafts Society.

Early Trades and Crafts Society. (1989). *A tool collector's picture book.* Unpublished, prepared for the Feb. 1989 meeting of the Early Trades & Crafts Society.

Elliott, Thomas L. (2003). *A field guide to the makers of American wooden planes.* Astragal Press, Mendham, NJ.

Evans, Oliver. (1850). *The young mill-wright and miller's guide.* 13th Ed. Lea & Blanchard, Philadelphia, PA. Reprinted by Arno Press, New York, NY.

- Among the most important of all antiquarian references.

Farnham, Alexandar. (1970). *Tool collector's handbook: Prices paid at auction for early American tools.* Alexandar Farnham, Stockton, NJ.

Feintuch, Burt and Watters, David H., Eds. (2005). *The Encyclopedia of New England.* Yale University Press, New Haven, CT and London.

Ferguson, Eugene S. (1968). *Bibliography of the history of technology.* Society for the History of Technology, Cambridge, MA.

Garrett Wade Company. (2001). *Tools, a complete illustrated encyclopedia.* The Garrett Wade Co., Inc., NY, NY.

Goodman, W.L. (1968). *British planemakers from 1700.* Third Edition enlarged and revised by Jane & Mark Rees, published by Roy Arnold in 1993, Astragal Press, Mendham, NJ.

- The most important reference on English planemakers.

Graham, Frank D. and Emery, Thomas J. (1923). *Audels carpenters and builders guide #1: A practical illustrated trade assistant on modern construction for carpenters - joiners, builders - mechanics and all wood workers.* 4 vols. Theo. Audel & Co., NY, NY.

- See annotations in the Toolmaking Trades bibliography under woodworkers.

Hack, Garrett. (1999). *Classic hand tools.* Taunton Press, Newtown, CT.

Hall, Elton W. (2006). The D'Elia antique tool museum in Scotland, Connecticut. *The Chronicle*. 59(1). pg. 7.

Henderson, John Goulding and Bates, Jack M. (1953). *Metallurgical dictionary*. Reinhold, New York, NY.

Heuring, Jerry and Heuring, Elaine. (1990). *E.C. Simmons Keen Kutter collectibles: An illustrated price guide*. Second edition. Collector Books, P. O. Box 3009, Paducah, KY 42001.

- The only guide to Keen Kutter tools.

Hinckley, F. Lewis. (1960). *Directory of the historic cabinet woods*. Bonanza Books, NY, NY.

Houghton, William M. (1971). *Selected topics of rural historical interest and explanatory drawings*. Madison County Historical Society, Morrisville, NY.

Hume, Ivor Noel. (1969). *A guide to artifacts of colonial America*. Alfred A. Knopf, Inc., NY, NY.

Hume, Ivor Noel. (2001). *If these pots could talk: Collecting 2,000 years of British household pottery*. Chipstone Foundation, Milwaukee, WI.

Husfloen, Kyle, Ed. Blanchard, Clarence, contributing Ed. (2003). *Antique Trader: Tools price guide: Tools from the 1700s through the 20th century*. Krause Publications, Iola, WI.

Johnson, Jacob. (1800-1807). *The book of trades, or library of the useful arts*. Parts I, II, and III. Whitehall, Philadelphia, PA.

- These have been issued as modern reprints by Peter Stockham.

Kean, Herbert P. (date unknown). *Restoring antique tools*. Astragal Press, Mendham, NJ.

Kean, Herbert P. (2002). *Tool tales*. Astragal Press, Mendham, NJ.

Kean, Herbert P. and Pollak, Emil S. (1990). *Collecting antique tools*. Astragal Press, Mendham, NJ.

Kemp, Peter. (1976). *The Oxford companion to ships and the sea*. Oxford University Press, London.

Kijowski, Gene W. (1990). *Directory of American tool makers: Colonial times to 1899: Working draft edition*. The Early American Industries Association.

- This essential reference is utilized throughout our inventory of tools on exhibit using the abbreviation **DATM**.
- A new version is now published, see Nelson below.

Klenman, Allen. (1990). *Axe makers of North America.* Whistle Punk Books, Currie's Forestgraphics Ltd., Victoria, B.C.

Kling, Peter M. (1906). *Why a boy should learn a trade.* Press of Percy F. Smith, Pittsburgh, PA.

Knight, Charles. (1845). *The pictorial gallery of arts: Useful arts.* C. Knight and Co., London, UK.

- A Xerox of parts of this text is in stock.

Knight, Charles. (1851). *Cyclopaedia of the industry of all nations.* George P. Putnam, NY, NY.

Knight, Edward H. (undated c. 1875). *The practical dictionary of mechanics: being a description of tools, instruments, machines, processes, and engineering; history of inventions; general technological vocabulary; and digest of mechanical appliances in science and the arts.* 4 vols. Cassell Petter & Galpin, London.

- The definitive reference on the technological history and the tools and inventions of the 19th century, with over 5,000 engravings. This is the counterpart to the Encyclopedia Britannica when it comes to tools. If you want to know what a cliseometer is, look here.

Knight, Edward H. (1877). *American mechanical dictionary.* 3 vols. Cambridge, University Press, Hurd & Houghton, NY.

- Another edition of the above reference.

Lardner, D. (1832). *The cabinet cyclopedia of the useful arts.* Vol. 1. London, UK.

Larson, Lars and Blanchard, Clarence. (2001). *Patented American planes for wood, leather, and the allied trades, 1795-1934.* Vol 1 - 3. Astragal Press, Mendham, NJ.

Levine, Bernard. (1985). *Levine's guide to knives and their values.* DBI Books, Northbrook, IL.

Martin, Richard A. (1977). *The wooden plane.* Early American Industries Association, South Burlington, VT.

McCreight, Tim. (1997). *Jewelry: Fundamentals of metalsmithing.* Hand Books Press, Rockport, MA.

McCreight, Tim. (2004). *Complete metalsmith.* Brynmorgen Press, Inc., Portland, ME.

Moore, R. (1888). *The universal assistant, and complete mechanic, containing over one million industrial facts, calculations, receipts, processes, trade secrets, rules, business forms, legal items, etc., in every occupation, from the household to the manufactory.* J. S. Ogilvie, NY, NY.

Nelson, Robert E., Ed. (1999). ***Directory of American Toolmakers***: *A listing of identified makers of tools who worked in Canada and the United States before 1900*. Early American Industries Association.

- This essential reference is referred to as **DATM (1999)** when we have used it in our tool inventory listings and the Registry of Maine Toolmakers.
- The most important and useful reference in this bibliography.

Neumann, George C. (1991). *Swords & blades of the American Revolution*. Rebel Pub. Co, Texarkana, T

Neumann, George C. and Kravic, Frank J. (1997). *Collector's illustrated encyclopedia of the American Revolution.* Scurlock Publishing Co., Texarkana, Texas.

- The best guide to Revolutionary era artifacts.

Nicholson, John. (1826). *The operative mechanic and practical machinist.* 2 Vols. H.C. Carey & I. Lea, Philadelphia, PA.

Noel-Hume, Ivor. (1972). *A guide to artifacts of colonial America*. Alfred A. Knopf, New York, NY.

Overstreet, Robert M. (2003). *Official Overstreet identification and price guide to Indian arrowheads: The ultimate reference to United States point types*. Eighth edition. House of Collectibles, NY, NY.

- This text is useful for identifying Native American lithics and other tools, including those in our museum collections.

Park, Edwards. (1983). *Treasures of the Smithsonian*. Smithsonian Books, Washington, DC.

Parker, John Henry. (1896). *A concise glossary of architectural terms*. James Parker & Co., Oxford. Reprinted in 1992 by Studio Editions, London.

Pearson, Ronald W. (1994). *The American patented brace 1829-1924: An illustrated directory of patents*. Astragal Press, Mendham, NJ.

Peterson, Harold. (1956). *Arms & armor in colonial America 1526 - 1763*. Bramhall House, NY.

Peterson, Harold L. (1958). *American knives: The first history and collectors' guide*. C. Scribner's Sons, New York, NY.

Pollak, Emil and Pollak, Martyl. (1994). *A guide to the makers of American wooden planes, third edition.* Astragal Press, Mendham, NJ.

- The single most important reference on American planemakers. The first edition was published in 1983; the fourth edition in 2001.

Pollak, Emil and Pollak, Martyl. (2001). *A guide to the makers of American wooden planes, fourth edition.* Revised by Thomas L. Elliott. Astragal Press, Mendham, NJ.

- This, the latest edition is the most up-to-date.

Proudfoot, Christopher and Walker, Philip. (1984). *Christie's collectors guides; woodworking tools.* Phaidon, Oxford, United Kingdom.

Price, James E. (1992). *A sourcebook of United States patents for bitstock tools and the machines that made them.* Published by the author, Naylor, MO.

Rees, Jane and Rees, Mark. (1999). *Tools: A guide for collectors.* 2nd Edition. Sean Arnold.

Reichman, Charles. (March 1980). Tool museum at Troyes. *The Chronicle.* 33(1). pg.1-3.

- Maison de l'Outil et de la Pensee Ouvriere, Troyes, France.
- Vol. 6 of the *Hand Tools in History* publication series contains some photos taken at this museum.

Reichman, Charles. (June 1990). Gleanings: The Pollaks and the Astragal Press. *The Chronicle.* 43(2). pg. 46-47, 55.

- Interesting background information about the evolution of the Astragal Press, the most important source of information on American planemakers and their maker's stamps as well as publisher of many important books on tools.

Roberts, Kenneth D. (1975). *Wooden planes in 19th century America.* Ken Roberts Publishing Co., Fitzwilliam, NH.

- The first of the guides to American planemakers.

Roberts, Kenneth D. (1978). *Wooden planes in 19th century America, volume II: Planemaking by the Chapins at Union Factory, 1826 - 1929.* Ken Roberts Publishing Co., Fitzwilliam, NH.

Roberts, Kenneth D. (1979). *Scottish & English metal planes by Spiers & Norris.* Ken Roberts Publishing Co., Fitzwilliam, NH.

Roberts, Kenneth D. (1980). *Some 19th century English woodworking tools: edge and joiner tools and bit braces.* Ken Roberts Publishing Co., Fitzwilliam, NH.

Rogers, William. (1913). *Rogers machinists guide: A practical illustrated treatise on modern machine shop practice.* Theo. Audel & Company, NY, NY.

Rose, Joshua. (1995). *The Pattern maker's assistant:1889 edition with 250 illustrations.* Astragal Press, Mendham, NJ.

Rosebrook, Donald. (1999). *American level patents: Illustrated and explained.* Volume I. Astragal Press, Mendham, NJ.

Salaman, R.A. (1975). *Dictionary of tools used in the woodworking and allied trades, C. 1700-1975.* Charles Scribner's Sons, New York, NY.

- The first comprehensive guide to woodworking tools. A copy of this reference is available for visitors to browse in the main hall of the Davistown Museum.
- Note that there is now a revised edition published in 1997.

Salaman, R.A. (1978). *A bibliography of tools.* The Early American Industries Association.

Salaman, R.A. (1986). *Dictionary of leather-working tools, c. 1700-1950 and the tools of allied trades.* Macmillan Publishing Company, NY, NY.

- The most comprehensive reference on the subject.

Schulz, Alfred and Schulz, Lucille. (1989). *Antique and unusual wrenches.* Published by Alfred & Lucille Schulz, R.1 Box 151, Malcolm, NE 68402.

- The classic guide to wrench makers and long a standard reference for collectors. A copy of this reference is available for visitors to browse in the main hall of The Davistown Museum.

Sellens, Alvin. (1990). *Dictionary of American hand tools: A pictorial synopsis.* Alvin Sellens, Publisher, Augusta, KS.

- The best general dictionary of American hand tools. A copy of this reference is available for visitors to browse in the main hall of The Davistown Museum.

Shumway, George. (1980). *Rifles of colonial America, Vol. II.* George Shumway Publisher, PA.

Sloane, Eric. (1964). *A museum of early American tools.* Funk & Wagnalls, Inc. Reprinted in 1973 by Ballantine Books, NY, NY.

- See annotations in the Davistown Museum online Children's bibliography.

Smith, Joseph. (1816). *Explanation or key, to the various manufactories of Sheffield, with engravings of each article.* J. Smith, Sheffield, England. Reprinted in 1975 by the Early American Industries Association, South Burlington, VT.

- See annotations in our European Precedents and the Early Industrial Revolution bibliography.

Smith, Roger K. (1981-1992). *Patented transitional & metallic planes in America 1827 - 1927.* 2 vols. North Village Publishing Co., Lancaster, MA.

- These two volumes are the definitive reference for tracing the development of the steel hand plane in the United States and includes detailed descriptions, background histories, and excellent photographs. These volumes represent a lifetime of work by the most knowledgeable scholar on transitional and metallic planes in America and are the most essential of all references for the collector of planes.
- Smith begins the first volume with a discussion of the history of cast iron planes and includes a picture of the earliest known Roman plane found at Pompeii, dating to 79 A.D., a smooth plane from the 14th or 15th century and a picture of the first American cast iron plane made in 1827 by Hazard Knowles. Excellent photographs throughout both volumes.
- Smith has a brief introduction which provides the context for placing the development of the metal plane in America as occurring after craftsmen had relied on traditional wooden planes for many generations: "The 'American System' of manufacture had its beginning in 1813 when Simeon North received the first government contract to specify interchangeable parts in the manufacture of guns at Middletown, Connecticut." (pg. 8).

Standards and Curriculum Division, Bureau of Naval Personnel. (1944). *Hand tools.* U. S. Government Printing Office, Washington, DC.

Stanley, Philip E. (1984). *Boxwood & ivory: Stanley traditional rules, 1855 - 1975.* The Stanley Publishing Co., Westborough, MA.

- The only comprehensive reference on Stanley rules.

Stanley, Philip E. (1985). *A concordance of major American rule makers.* Philip Stanley, Westborough, MA.

Taylor, Lyman, Ed. (1961). *Metals handbook.* 8th Edition. 3 Volumes. American Society for Metals.

- The three volumes are titled: *Properties and Selection of Metals, Heat Treating, Cleaning and Finishing, and Machinery.*
- A definitive reference for the foundryman and patternmaker. This set was formerly owned by the Mineloa Pattern Works, Inc., Garden City Park, Long Island, NY and bought from the head patternmaker, Eric R. Swedberg, who had retired to Dover, NH.

Tiemann, H.P. (1933). *Iron and steel: A pocket encyclopaedia.* New York, NY.

Timmins, R. & Sons. (no date). *Tools for the trades and crafts: An eighteenth century pattern book.* R. Timmins & Sons, Birmingham Reprinted in 1976 by K. Roberts Pub. Co., Fitzwilliam, NH.

Tomlinson, Charles. (1854). *Cyclopedia of useful arts, mechanical and chemical, manufactures, mining, and engineering.* George Virture, London, UK.

Tomlinson, Charles. (1972). *Illustrations of the trades.* The Early American Industries Association.

Ure, Andrew. (1845). *A dictionary of arts, manufactures and mines: Containing a clear exposition of their principles and practice.* D. Appleton & Co., NY, NY.

Walter, John. (1988). *Antique and collectable Stanley planes: 1988 price guide.* The Tool Merchant, Akron, Ohio.

Walter, John. (1989). *Antique and collectable Keen Kutter hand tools: 1989 price guide.* The Tool Merchant, Akron, Ohio.

Walter, John. (1990). *Antique & Collectible Stanley tools: A guide to identity & value.* The Tool Merchant, Marietta, OH.

- The most comprehensive guide to Stanley tools. This text represents a lifetime of study of Stanley tools and is a major contribution to the literature on Stanley tools. The Walters publications are the tool references most frequently used by the Liberty Tool Co.

Walter, John. (2000). *Antique & collectible Stanley tools: 2000 pocket price guide.* The Tool Merchant, Marietta, OH.

Westley, Robert. (1993). *Guide to imprints of Canadian plane makers and hardware dealers.* MacLachlan Woodworking Museum, Kingston, Ontario, Canada.

- A second edition was published in 1997 without the wedge profiles but with more biographical information on 95 makers/dealers and notes on how to determine which stamps on a plane are for makers and dealers.

Wendel, Charles H. (1997). *Encyclopedia of American farm implements & antiques.* Krause Publications, Iola, Wisc.

Wendel, Charles H. (2001). *The encyclopedia of tools & machinery.* Krause Publications, Iola, WI.

Whelan, John M. (1993). *The wooden plane: Its history, form & function.* The Astragal Press, Mendham, NJ.

- This text contains the most extensive descriptions and illustrations of plane profiles in the literature. It is indispensable for identifying plane types. A copy of this reference is available for visitors to browse in the main hall of the Davistown Museum.

Wilbur, C. Keith. (1987). *Antique medical instruments: Price guide included.* Schiffer Publishing Ltd., West Chester, PA.

344

Tool Catalogs

Some English company catalogs are included below and the individual company listings may contain citations for the companies catalogs.

Abercrombie & Fitch. (1965). *Antique guns 1965.* Abercrombie & Fitch, NY, NY.

American Steel & Wire Company. (undated). *Catalogue of American nails, wire, barbed wire, staples tacks, poultry netting, etc.* Form 5860. United States Steel Corporation Subsidiary.

Arthur, Henry. (Sept. 1874). *Price list for September 1st, 1874, of leather and findings and boot & shoe uppers.* Henry Arthur, 84 & 86 Gold Street, Corner Ferry, NY. Reprinted by Alexander Farnham, RD 2, Stockton, NJ 08559.

Arnold, Roy. (late 1860's). *The mid Victorian Elwell catalog of forged tools.* A facsimile with an introduction by Richard Filmer. Reprinted in 2000 by Astragal Press, Mendham, NJ.

Arnold & Walker. (1975 - 1977). *The traditional tools of the carpenter and other craftsmen.* Catalogue 2, 3, 4, 5. 77 High St., Needham Market, Suffolk, UK.

Astragal Press. (1989). *The Stanley catalog collection: 1855 - 1898: Four decades of rules, levels, try-squares, planes, and other Stanley tools and hardware.* Astragal Press, Mendham, NJ.

Astragal Press. (1994). *The handsaw catalog collection: A select compilation of the four leading manufacturers (1910-1919) : E.C. Atkins & Co., Henry Disston & Sons, Simonds Manufacturing Co., and Spear & Jackson.* Astragal Press, Mendham, NJ.

Auburn Tool Company. (1869). *Price list of planes, plane irons, rules, gauges, hand screws, &c., manufactured and sold by Auburn Tool Company, (successors to Casey, Clark & Co.).* Wm. J. Moses' Printing and Publishing House. Reprinted by Ken Roberts Publishing Co., Fitzwilliam, NH.

Bailey, Leonard & Co. (1876). *Illustrated catalogue and price list of patent adjustable iron bench planes, try squares, bevels, rules, levels, hammers, &c., &c.* Leonard Bailey & Co., Hartford, CT. Reprinted by Ken Roberts Publishing Co., Fitzwilliam, NH.

Bailey, Leonard & Co. (January 1883). Catalog: *Leonard Bailey & Co.'s patent adjustable iron bench planes, try squares, bevels, spoke shaves, box scrapers, &c.* Leonard Bailey & Co., Hartford, CT. Reprinted in May 1975 by Ken Roberts Publishing Co., Fitzwilliam, NH.

Bailey Wringing Machine Company. (February 15, 1876). *Price list: Defiance metallic bench planes.* Bailey Wringing Machine Company, 99 Chambers St., NY, NY. Reprinted by Kendall Bassett.

Barbour, J. & E. R. (no date). Catalog: *J. & E. R. Barbour, dealers in mechanical rubber goods, engineers specialties, steamboat, railroad, and mill supplies, contractors for steam machinery &*

appliances. Nos. 8 and 10 Exchange Street, Portland, Maine. Poole Bros. Printers and Electrotypers, Mechanic Falls, ME.

Bartholomew, H. S. (ca. 1889). *Price-list of breast drills, braces, ferrules, etc., etc. 190.* H. S. Bartholomew, Bristol, CT. Reprinted in September 1991 by ATTIC.

Barton, D. R. & Co. (1873). *Catalogue and revised standard list of mechanics' tools and machine knives, manufactured by D. R. Barton & Co., 136 Mill Street, Rochester, N.Y.* Evening Express Printing House, Rochester, NY. Reprinted in April 1983 by Ken Roberts Publishing Co., Fitzwilliam, NH.

Barnes, W. F. & John Co. (1903). *Catalogue No. 59. Foot power lathes manufactured by W. F. & John Barnes Co. Rockford, Illinois, U.S.A.* Reprinted in 1982 by The Mid-West Tool Collectors Association, Columbia, MO.

Barnes, W. F. & John Co. (1907). *Catalogue No. 67. Patent foot and hand power wood working machinery manufactured by W. F. & John Barnes Co. Rockford, Illinois, U.S.A. The "Original Barnes".* Reprinted in 1978 by The Mid-West Tool Collectors Association, Columbia, MO.

Belcher Brothers & Co. (1860). *Price list of boxwood & ivory rules, for sale only by WM. Belcher, 233 Pearl St., New York. Measuring tapes, thermometers, sandpaper, steel squares, braces and bits, gauges, spokeshaves, try-squares, Bemis' cast steel goods, &c.* Reprinted in 1982 by Ken Roberts Publishing Co., Fitzwilliam, NH.

Brown & Sharpe Mfg. Co. (1902). *1902: Catalogue: Brown & Sharpe Mfg. Co.: Machinery and tools.* Brown & Sharpe Mfg. Co., Providence, RI.

Brown & Sharpe Mfg. Co. (1941). *Brown & Sharpe small tools: Catalog no. 34.* Brown & Sharpe Mfg. Co., Providence, RI.

Buck Brothers. (1890). *Price list of chisels, plane irons, gouges, carving tools, nail sets, screw drivers, handles, &c. manufactured by Buck Brothers.* Riverlin Works, Millbury, MA. Reprinted in 1976 by Ken Roberts Publishing Co., Fitzwilliam, NH.

Carr, WM. H. & Co. (1838). *American manufactured hardware, &c, for sale by WM. H. Carr & Co.* William Brown, Philadelphia, PA.

Chandler & Farquhar Co. (1924). *Chandler & Farquhar Company machinists' tools and supplies, mill supplies, general hardware.* R. R. Donnelley & Sons Co., Chicago, IL.

Chapin, Hermon. (1853). *Cataloque and invoice prices of rules, planes, gauges, &c. manufactured by Hermon Chapin. Union Factory, Pine Meadow, Conn.* Reprinted in 1976 by Ken Roberts Publishing Co., Fitzwilliam, NH.

Chapin, Philip E. (1878). *The Foss patent adjustable iron planes. Manufactured by Philip E. Chapin, Pine Meadow, Conn.* Reprinted in Sept. 1981 by Ken Roberts Publishing Comapny.

The Chapin-Stephens Co. (no date). Catalog: *A plane statement. Read it. 86 years experience.* Pine Meadow, CT.

The Chapin-Stephens Co. (ca. 1914). *Catalog No. 114: The Chapin-Stephens Co. Union Factory: Rules planes gauges plumbs and levels hand screws handles spoke shaves box scrapers, etc.* Pine Meadow, CT. Reprinted in March 1975 by Ken Roberts Publishing Co., Fitzwilliam, NH. (4)

Chapple, William. (1876) *Revised list, William Chapple (late Peter Wilcock), plane manufacturer.* Reprinted in Jan. 1890 by Ken Roberts Publishing Company.

Cheney, Henry Hammer Co. (1904). *Illustrated catalogue of the Henry Cheney Hammer Co.* Reprinted September 23, 2003 by The Special Publications Committee M-WTCA.

Collins & Co. (1921). *Illustrated catalogue of axes, hatchets, adzes, picks, sledges, hoes, wrenches, bush hooks, etc., etc. manufactured by Collins & Co. established in 1826.* The Collins Company, Collinsville, CT. Reprinted in 1974 by the Early Trades & Crafts Society, Long Island, NY.

Collins Company, The. (1935). *A brief account of the development of The Collins Company in the manufacture of axes, machetes and edge tools.* The Case, Lockwood & Brainard Co. Reprinted in 1985 by Ken Roberts Publishing Co., Fitzwilliam, NH.

Colwell Cooperage Company. (no date). *This trade mark Colco N.Y. stands for efficient tools, stock and supplies for shipping containers: Barrels, boxes, pails, tubs, crates, baskets. General catalogue no. 26.* Colwell Cooperage Company, Foot Jersey Avenue, Jersey City, NJ.

Colwell, E. D. (no date). *Everything the cooper needs: Catalogue of bungs.* E. D. Colwell, 412-418 Greewich St., NY, NY.

Connecticut Valley Manufacturing Co., Inc. (1939). *Connecticut Valley Manufacturing Co. Inc. wood boring tools.* Walker-Rackliff Co., New Haven, CT.

Cope, Kenneth L. (no date available). *A Brown & Sharpe catalogue collection, 1868, 1887, 1899.* Astragal Press, Mendham, NJ.

Crucible Steel Company of America. (1919). *Catalog of products of the Black Diamond (Park) steel works, Pittsburgh, Pa.* Crucible Steel Company of America, Pittsburgh, PA.

Davis Level & Tool Co. (ca. 1880). *Price list of the Davis Level & Tool Co. Springfield, Mass. manufacturers of hardware tools, adjustable spirit plumbs, levels, and inclinometers, iron pocket levels, builders' levels and level glasses, saw clamps, improved iron bench planes, calipers and dividers, surface gauges, machinists' screw drivers, etc. hack saws and breast drills, thread gauges, &c., &c.* Reprinted in May 1975, by Roger K. Smith, Lancaster, MA.

- Reproductions of some of the pages in this price list can be seen in the Davistown Museum online information file on Davis Level & Tool Co.

Delta Manufacturing Division. (Jan. 1932 - March 1941). *The Deltagram books I, II, III*. Milwaukee, WI.

C. Drew & Company. *Catalog No. 34: C. Drew & Co. established 1837: Factory at Kingston, Mass. Plymouth county*. Topping Bros., New York Agents office and warehouse, 122 Chambers Street. Re-issued by the Marine Historical Association, Inc. and Antique Trades and Tools of Connecticut in October 1972.

- Reproductions of a couple of the catalog pages can be seen in the Davistown Museum online information file on C. Drew & Co.

Eagle Square Manufacturing Co. (no date). *Complete line of steel squares made by the Eagle Square Manufacturing Co., South Shaftsbury, Vermont. What the scales and tables are and how to use them.*

- This appears to be a reprint, but has no information other than a stamp stating "ACTIVE Antique Crafts and Tools in Vermont" on the back.

Earle M. Jorgensen Co. (1984). *Steel and aluminum stock list and reference book*. Earle M. Jorgensen Co., Los Angeles, CA.

Farm Tools, Inc. (no date). *Special Fordson implement catalog no. 20*. Farm Tools, Inc., Mansfield, OH.

Farrington, I.B. (1879). *Price list of I.B. Farrington's ornamental designs for scroll sawing and all kinds of scroll saw machines*. Henry H. Price, Book and Job Printer. Reprinted 1976 by Early Trades & Crafts Society.

Folding Sawing Machine Co. (1897). *Folding sawing machine*. Reprinted in February 1981 by The Mid-West Tool Collectors Association.

Ford Motor Company. (1939). *Johansson gage blocks and accessories, catalog no. 14*. Ford Motor Company, Dearborn, MI.

Forman, Benno M. (1988). *American seating furniture 1630-1730: An interpretive catalogue*. W. W. Norton & Co., NY, NY.

Garrett Wade Company. (1984). Catalog: *Woodworking tools*. With September 1983 Price list. Garrett Wade, 161 Ave. of the Americas, NY, NY.

General Electric Co. (May 1910). Bulletin No. 4737: *Electric hardening furnace*. Power and Mining Department, Schenectady, NY.

Goodell-Pratt Company. (1905). *Catalogue no. 7: Tools manufactured by Goodell-Pratt Company, Greenfield, Mass., U.S.A.* Reprinted in June 1977 by Roger K Smith, Lancaster, MA.

Goodnow & Wightman. (1882). *Price list of Goodnow & Wightman, importers, manufacturers and dealers in tools of all kinds for machinists, pattern-makers, carvers, model makers, amateurs, cabninet makers, jewelers, etc. Boston, Mass.* Reprinted as a joint project by the Early Trades & Crafts Society and the Mid-West Tool Collector's Association.

Gransfors Bruks AB. (2002) *The Axe Book.* Gransford Bruks AB, Bergsjo, Sweden.

Greenfield Tool Co. (1854). *Price list of joiners' bench planes and moulding tools manufactured by the Greenfield Tool Company.* Reprinted in 1981 by Ken Roberts Publishing Co., Fitzwilliam, NH.

Greenfield Tool Co. (1872). *Illustrated catalogue and invoice price list of joiners' bench planes, moulding tools, handles, plane irons, &c., manufactured by the Greenfield Tool Company.* Reprinted in January 1978 by Ken Roberts Publishing Co. Fitzwilliam, NH.

Greenlee Bros. & Co. (1940). *Greenlee mortising and boring tools.* Greenlee Bros. & Co., Rockford, IL.

Hall, Elton W. (March 2008). The development of the illustrated tool catalog. *The Chronicle.* 61(1). pg. 1-16.

Hammacher, Schlemmer & Co. (no date). *Piano, organ and violin tools, catalog no. 142.* Reprinted by Martin J. Donnelly Antique Tools, Bath, NY.

Hammacher, Schlemmer & Co. (1937). *Williams superior carbon steel wrenches.* Hammacher, Schlemmer & Co, New York, NY.

Harbor Tool Supply Co., Inc. *Starrett saw data, second edition.* Harbor Tool Supply Co., Inc, Needham Heights, MA.

Harvey, H.H. (1896). *H. H. Harvey's special illustrated catalogue and price list for 1896 – 7. Granite, marble, and soft stone workers', blacksmiths' and contractors', hammers and tools, manufactured by him at Augusta, Maine.* Publisher unknown.

Henry Disston & Sons. (1876). *Price list.* Reprinted in 1994 by Astragal Press.

Higganum Mfg. Co. (1873). Brochure: *The superior hay spreader.* The Higganum Mfg. Co., Higganum, CT.

Higganum Mfg. Co. *Illustrated catalogue of Higganum specialties, cider mills, wine presses, hay cutters, Clark's harrow and seeder, corn shellers, &c.* The Higganum Mfg. Co., Higganum, CT.

Hill & Co., James R. *Harness makers' and dealers' supply catalogue*. Page Belting Co., Concord, NH.

Hirth & Krause. (1890). *Hirth & Krause, dealers in all kinds of leather & findings. Shoe store supplies, etc. Grand Rapids, Michigan*. Reprinted by The Midwest Tool Collectors Association and The Early American Industries Association in 1980.

Hoe, R., & Co. (1855). *R. Hoe & Co., manufacturers of patent ground warranted cast-steel saws*. Reprinted in November 1976 by Early Trades & Crafts Society, Long Island, NY.

Holtzapffell & Co. (1851). *Holtzapffell & Co., No. 64, Charing Cross, London, engine, lathe, & tool manufacturers, and general machinists*.

Holyoke Supply Co. (1918). *Condensed Catalogue of Steam and Plumbing Supplies, First Edition pocket Size*. The Howland Publishing Co., New York City, NY.

Hoole Machine and Engraving Works. (1911). *Hoole Machine and Engraving Works*. Reprinted in May 1985 by The Special Publications Committee, Midwest Tool Collectors Association.

Hynson Tool & Supply Company. (1903). *52 annual catalogue: Hynson Tool & Supply Company*. St. Louis, MO. Reprinted in April 1980 by the Mid-West Tool Collectors Association and the Early American Industries Association.

- Coopers' tools.

The Irwin Auger Bit Company. *How to select, use, and care for wood bits*. O.T. Printing, Cols., OH.

Jackson, S. Robert . (no date). *The level you need.* Hungerford-Holerook Co. Watertown, NY. Reprinted by Martin J. Donnelly Antique Tools.

Jackson & Tyler. (no date). *Price list, Jackson & Tyler tools and supplies of all kinds.* Reprinted in September 1993 by The Special Publications Committee, Mid-West Tool Collectors Assocation.

Jennings, C. E., & Co., Jennings, Charles E. and Griffin, Francis B. (1985). *Price list. no. 13: C. E. Jennings & Co., manufacturers of C. E. Jennings' Arrowhead high grade tools.* The Stanley Publishing Co., P. O. Box 689, Westborough, Mass. 01581.

Jennings, Russell Mfg. Co. (1981). *Reprint of price list of Russell Jennings Mfg. Co., c. 1899 with supplementary data.* Ken Roberts Publishing Co., P. O. Box 151, Fitzwilliam, NH.

- The most renown manufacturer of augers and auger bits.

Johnson, Helen Louise.(1898) *The enterprising housekeeper*. The Enterprise Manufacturing Company of PA, Philadelphia, PA.

Knutsson, Johan and Kylsberg, Bengt. (1985). *Verktyg och verkstader pa Skoklosters slott.* Skoklosters slott.

Lang & Jacobs. (1884). *Catalogue and price list of Lang & Jacobs' head quarters for coopers' supplies & cooperage stock, Boston, Mass. Coopers' tools & truss hoops a specialty.* Reprinted by the Early Trades and Crafts Society, Long Island, NY.

The Lufkin Rule Co. (1888). *Lufkin measuring instruments, exerpts from trade catalogues, 1888 to 1940, documentary and arrangement by Kenneth D. Roberts.* Reprinted in April 1983 by Clark-Briton Printing Co., Cleveland, OH.

Marble Arms & Mfg. Company. (1932-33). *Marble's outing equipment.* Catalog No. 21. Marble Arms & Mfg. Company, Gladstone, MI. Reprinted in 1992 by Philip J. Whitby.

Marble Safety Axe Company. (1905). *Marble's specialties for sportsmen.* Marble Safety Axe Company, Gladstone, MI. Reprinted in 1992 by Centennial Press, Littleton, CO.

Marples, William & Sons. (February 1979). *1909 edition price list: William Marples & Sons, Limited, Sheffield.* Reprinted by The Mid-West Tool Collectors Association, The Early American Industries Association and Arnold and Walker.

Marshall-Wells Hdwe. Co. (circa 1910). *Zenith tools and cutlery.* Reprinted in April 1987 by The special Publications Committee, Mid-West Tool Collectors Association.

Martin, Glenn L. (1953). *Catalog of exhibit of antique tools at the University of Maryland, from the collection of Herbert T. Shannon.* University of Maryland, MD.

Mathieson, Alex & Sons, Ltd. (1899). *Saracen Tool Works, East Campbell Street, Glasgow. Established, 1792. Warehouses, Edinburgh, 23 Cockburn Street. Liverpool, 41 Byrom Street. Selections from the illustrated price list of wood working tools manufactured by Alex. Mathieson & Sons, Ltd. Glasgow.* Eighth Edition. Reprinted in March 1975 by Ken Roberts Publishing Co., Fitzwilliam, NH.

Millers Falls Co. (April 15, 1878). Catalog: *Millers Falls Co., Millers Falls, Mass., No. 74 Chambers Street, NY.* Reprinted in 1992 by Philip J. Whitby.

Millers Falls Co. (1887). *Catalog.* Reprinted in 1981 by Astragal Press.

Morse Twist Drill & Machine Co. (1935). *Machinist's practical guide: The Morse Twist Drill and Machine Co.: Incorporated 1864: Makers of twist drills, reamers, milling cutters, taps, dies, sockets, gauges, chucks, machinery and machinists' tools.* Morse Twist Drill & Machine Co., New Bedford, MA.

Moseley & Stoddard Manufacturing Co. (1896). (cover) *For sale by H. P. Lucas, Pittsfield, Mass.* (inside) *Special fair catalogue, number 60, of improved machinery for the farm, dairy and creamery.*

Moseley & Stoddard Manufacturing Co., Rutland, Vermont. Reproduced by the Early Trades & Crafts Society in 1975.

Niagara Machine & Tool Works. (no date available). *Catalog no. 50: Tools and machines for tinsmiths and sheet metal workers, presses, shearing machines, punches, forming rolls, tinsmiths' tool, etc.* Reprinted by Lindsay Publications, Bradley, IL.

North Bros. Manufacturing Co. (1908). *"Yankee" Tool book describing, with illustrations, some up-to-date labor saving tools. More especially: ratchet screw drivers, spiral screw drivers, automatic or push drills, breast and hand drills etc., etc.* Philadelphia, PA. reprinted in 1985 by O.M. Ramsey and Philip Whitby of the Midwest Tool Collectors Association.

- Yankee Tool is now part of the Stanley Tool Company.
- To find more information on "Yankee" screwdrivers, see the Davistown Museum online Zachary T. Furbish information file.

North Bros. Manufacturing Co. (1912). *Catalogue of "Yankee" tools ice cream freezers etc., etc.* North Bros. Mfg. Co., Philadelphia, PA. Reprinted in November 1988 by the Mid-West Tool Collectors Association.

Ohio Tool Company. (1900). *Ohio adjustable planes.* Ohio Tool Company, Charleston, VA.

Ohio Tool Company. (ca. 1900). Catalog: *High-grade mechanics' tools.* Ohio Tool Co., Columbus, Ohio, Auburn, NY. Reprinted in 1981 by Ken Roberts Publishing Co., Fitzwilliam, NH.

Ohio Tool Company. (1910). *Ohio Tool Company established 1823: Catalogue no 23.* Ohio Tool Co., Columbus, Ohio, Auburn, NY. Reprinted in May 1976 by Roger K. Smith, Lancaster, MA.

Oldham, Joshua. (1887). Catalogue and price list: *Joshua Oldham, manufacturer of saws, machine knives, &c. of every description.* Joshua Oldham, NY, NY. Reprinted in 1976 by the Early American Industries Association, So. Burlington, VT.

Osborne, C. S. & Co. (1911). Catalog: *Established 1826: C. S. Osborne & Co., Inc.: Standard tools.* C. S. Osborne & Co., Inc., Newark, NJ. Reprinted in 1976 as a joint project by the Early American Industries Association, the Early Trades and Crafts Society and the Mid West Tool Collectors.

Parton, James. (1917). *A captain of industry: The story of David Maydole: Inventor of the adz-eye hammer: To which is added a catalog of the principal varieties of hammers made by The David Maydole Hammer Company.* The David Maydole Hammer Co., Norwich, NY.

The Peavey Manufacturing Co. (1978). *Celebrated logging tools since 1857.* Box 371, Brewer, ME. (207) 843-7861.

Pomeroy, A.H. (1886). *Illustrate catalogue of scroll saws, lathes, fancy woods, clock movements, pocket cutlery, mechanics' tools, &c. &c.* Pilgrim Publishers, MA.

Pratt & Whitney Co. (1950). *Pratt & Whitney Co. small tools.* Niles-Bement-Pond Co., West Hartford, CT.

Preston, Edward & Sons, Ltd. (1991). *The "PRESTON" catalogue: rules, levels, planes, braces and hammers, thermometers, saws, mechanic's tools, &c., Edward Preston & Sons, Ltd., Birmingham, England, Catalogue No. 18, May 1909.* Reprinted by Astragal Press, Mendham, NJ.

Rabone, John & Sons. (1982). *John Rabone & Sons 1892 catalogue of rules, tapes, spirit levels, etc.* Reprinted by Ken Roberts Publishing Co., P. O. Box 151, Fitzwilliam, NH 03447.

Rees, Mark. (1991). *Edward Preston & Sons of Birmingham, an outline history.* Astragal Press, Mendham, NJ. Reprint of the *Preston 1909 catalogue.*

Reo Motor Car Company. (1923). *Standard bodies on the Speed Wagon chass* Reo Motor Car Company, Lansing, MI.

Richmond Cedar Works. (Feb. 1, 1926). *Richmond Cedar Works: Richmond, VA.U.S.A. Manufacturers of Virginia white cedar wooden ware, ice cream freezers, washing machines etc. Price list.* Richmond Cedar Works, Richmond, VA.

Rixford, O. S. (ca. 1887). *O. S. Rixford's scythes and axes East Highgate, Vermont.* T. O. Metcalf & Co., Printers, 48 Oliver St., Boston, MA.

Roberts, Ken. (1979). *Scotch and English metal planes manufactured by Stewart Spiers and manufactured by T. Norris & Son.* Ken Roberts Publishing Co., Fitzwilliam, NH.

Roberts, Ken. (1989). *The Stanley Rule & Level Company's combination planes featuring the development and use of the Miller, Traut, and Stanley 45 and 55 planes. Miller's patent combined plow, filletster and matching plane.* Astragal Press, Mendham, NJ.

Rogers. (no date). *Roger's tool & book catalogue.* Hitchin.

Romaine, Lawrence B. (1960). *A guide to American trade catalogs, 1744-1900.* Bowker, NY, NY.

Rugg Mfg. Co. *1884-5 Price list of hand and drag rakes, snow shovels and handles.* Rugg Mfg. Co., Montague, MA.

Russell & Tremain. (1862). Advertising circular. *Russell's screw power mower and reaper, combined. Without cog gearing. Manufactured by Russell & Tremain, Manlius, Onondaga Co., N.Y.* Reprinted by the Early Trades & Crafts Society, Levittown, NY.

Sandusky Tool Co. (1877). *Illustrated list of planes, plane irons, etc. The Sandusky Tool Co.* Register Team Printing Establishment. Reprinted in January 1978 by Ken Roberts Publishing Co., Fitzwilliam, NH.

Sandusky Tool Co. (September 1st, 1925). *The Sandusky Tool Company, established 1868, catalog no. 25*. Sentinel Printing Co., Keene, NH. Reprinted in January 1978 by Ken Roberts Publishing Co., Fitzwilliam, NH.

Sargent & Co. *Wood bottom and iron planes*. Reprinted in May 1975 by Roger K. Smith, Lancaster, MA.

Sargent & Co. (1993). *The Sargent tool catalog collection: A reprint of the Sargent tools illustrated in the Company's 1894, 1910, and 1922 catalogs*. Forward by Paul Weidenschilling,The Astragal Press, Mendham, New Jersey.

Sargent & Co. *Duralumin and steel carpenter squares*. New Haven, CT.

Sawyer Tool Mfg. Co. (1904). *Catalog "G": Illustrated price list of fine tools and hardware specialties manufactured by the Sawyer Tool Mfg. Co. (incorporated)*. Fitchburg, MA. Reprinted by Martin J. Donnelly Antique Tools, Bath, NY.

Sawyer Tool Mfg. Co. *Fine tools, a good place to buy good tools*. Babcock, Hinds & Underwood, Binghampton, N.Y. Reprinted by Martin J. Donnelly Antique Tools, Bath, NY.

Scribner, J.M. (1880). *Scribner's lumber & log book for ship builders, boat builders, lumber merchants, farmers, and mechanics*. Geo. W. Fisher, Rochester, NY.

Sears, Roebuck and Co. *How to select and maintain Craftsman circular saw blades, s pecial purpose blades, dado sets, and saw blade stabilizers*.

Shannon, J.B. (1873). *Illustrated catalogue and price list of carpenters' tools*. McCalla & Stavely, Printers, Philadelphia, PA.

Shapleigh Hardware Co. (1927). *Diamond Edge planes*. Shapleigh National Series No. 1354, Shapleigh Hardware Co., St. Louis, MO. Reprinted in 1992 by Philip J. Whitby.

Shelburne Museum. (1957). *The blacksmith's and farriers' tools at Shelburne Museum*. Museum Pamplet Series, No. 7. The Shelburne Museum, Shelburne, VT.

Simmons Hardware Company. (1930). *E.C. Simmons Keen Kutter cutlery and tools*. Simonds Steel and Saw Co., Fitchburg, MA.

Simmons Hardware Company. (no date). *E. C. Simmons Keen Kutter cutlery and tools*. 21st catalog. Reprinted in 1971 by American Reprints, St. Louis, MO.

Simonds Saw and Steel Company. (1937). *Woodworking saws and planer knives: Their care and use*. Simonds Steel and Saw Co., Fitchburg, MA.

Simonds Saw and Steel Company. (1937). *The circular saw: A guide book for filers, sawyers and *woodworkers*. Simonds Steel and Saw Co., Fitchburg, MA.

Simonds Saw and Steel Company. (1937). *How to file a cross-cut saw.* Simonds Steel and Saw Co., Fitchburg, MA.

Simonds Saw and Steel Company. (1937). *The cross-cut saw.* Simonds Steel and Saw Co., Fitchburg, MA.

The H.D. Smith & Co. (1920). *Catalog: Makers of drop forged fine tools: Originators - not imitators.* The H. D. Smith & Co. Plantsville, CT.

- Established 1850, incorporated 1892, trademarks: PERFECT HANDLE, SharpenEzy, Gittatit, ENCHASED, ULTIMATE.

Snow & Neely Co. *"Our Best" lumbering tools.* Snow & Neely Co., Bangor, ME.

The Standard Rule Co. (ca. 1883). Catalog: *The Standard iron and wood planes.* Standard Rule Co., Unionville, CT. Reprinted by Ken Roberts Publishing Company.

The Standard-Simmons Hardware Co. *Simmons mail order want book: E. C. Simmons Keen Kutter cutlery and tools.* The Standard-Simmons Hardware Co., Toledo, OH. Reprinted in 1989 by R. L. Deckeback, Royal Oak, MI.

Stanley Rule & Level Co. (no date). Catalog: *Stanley improved labor saving carpenters' tools including "Bailey" adjustable plane.* Stanley Rule & Level Co., New Britain, CT.

Stanley Rule & Level Co. (1859). *1859 Price list of boxwood and ivory rules, levels, try squares, sliding T bevels, gauges, &c., manufactured by the Stanley Rule and Level Company, also including the price list of boxwood and ivory rules manufactured by A. Stanley & Co., New Britain, Conn. Jan. 1855.* Press of Case, Lockwood & Company, Hartford, CT. Reprinted in May 1975 by Ken Roberts Publishing Co., Fitzwilliam, NH.

Stanley Rule & Level Co. (January 1, 1867). *Price list of U. S. standard boxwood and ivory rules, levels, try squares, gauges, handles, mallets, hand screws, &c. manufactured by the Stanley Rule and Level Company, New Britain, Conn., and Brattleboro', VT.* Press of Case, Lockwood & Company, Hartford, CT. Reprinted by Ken Roberts Publishing Co, Bristol, CT.

Stanley Rule & Level Co. (January 1872). *Price list of U. S. standard boxwood and ivory rules: Levels, try squares, gauges, iron and wood bench planes, mallets, hand screws, spoke shaves, srew drivers, etc. manufactured by the Stanley Rule and Level Co., New Britain, Conn.* Reprinted in February 1981 by Ken Roberts Publishing Company, Fitzwilliam, NH.

Stanley Rule & Level Co. (January 1879). *Price list of U. S. standard boxwood and ivory rules, plumbs and levels, try squares, bevels, gauges, mallets, iron and wood adjustable planes, spoke* shaves, screw drivers, awl hafts, handles, etc. manufactured by the Stanley Rule and Level Company, New Britain, Conn., U.S.A. Reprinted in 1973 by Ken Roberts Publishing Co., Bristol, CT.

Stanley Rule & Level Co. (Jan. 1, 188?). *Bailey's patent adjustable bench planes and other improved carpenters' tools manufactured by the Stanley Rule and Level Company, New Britain, CT.* Reprinted in April 1975 by Ken Roberts Publishing Company, Fitzwilliam, NH.

Stanley Rule & Level Co. (January 1888). *Price list: Improved labor-saving carpenters' tools manufactured by the Stanley Rule and Level Co.* Reprinted in April 1975 by Ken Roberts Publishing Company, Fitzwilliam, NH.

Stanley Rule & Level Co. (January 1892). *Price list: Improved labor-saving carpenter's tools manufactured by the Stanley Rule and Level Co.* Reprinted in July 1972 by H. C. Maddocks, Jr., West Boylston, MA.

Stanley Rule & Level Co. (January 1898). *Price list of U. S. standard boxwood and ivory rules, plumbs and levels, try squares, bevels, gauges, mallets, iron and wood adjustable planes, spoke shaves, screw drivers, awl hafts, handles, etc. manufactured by the Stanley Rule and Level Co. New Britain, Conn., U.S.A.* Reprinted in May 1975 by Ken Roberts Publishing Company, Fitzwilliam, NH.

Stanley Rule & Level Co. (1900). *The Stanley bed rock: A new plane.* Stanley R. & L. Co., New Britain, CT. Reprinted in January 1983 by Bob Kaune, Port Angeles, WA.

The Stanley Rule & Level Co. (1909). Catalog: *Carpenters & mechanics tools: No. 102.* Reprinted in May 1975 by Roger K. Smith, Lancaster, MA.

The Stanley Rule and Level Plant. (1921). *"55" plane and how to use it.* The Stanley Rule and Level Plant, New Britain, CT.

Stanley Tools. (no date). *Read this before you use Stanley planes: A plane is no better than its cutter.* Stanley Tools, New Britain, CT.

Stanley Tools. (no date). *The Stanley catalog collection, 1855 to 1898: Four decades of rules, levels, try-squares, planes, and other Stanley tools and hardware.* The Astragal Press, Morristown, NJ.

The Stanley Works. Catalog: *Stanley tools ~ in sets.* The Stanley Rule & Level Plant, New Britain, CT. Reprinted in September 2002 by the Midwest Tool Collector's Association.

The Stanley Works. (1927). *Stanley tools.* The Stanley Works, New Britain, CT.

The Stanley Works. (1929). *Stanley tools for carpenters and mechanics: Catalog no. 129.* The Stanley Rule & Level Plant, New Britain, CT. Reprinted in June 1977 by Roger K. Smith, Lancaster, MA.

Stanley. (1953). *Do it better with Stanley tools.* Hanus Sicher, New York City, NY.

Stanley. *Stanley rafter and framing squares.* Stanley, New Britain, CT.

The Stanley Works. (1955). Catalog: *45 plane: Seven planes in one.* Reprint from Manual. Stanley Tools, New Britain, CT.

Stanley. Catalog: *Stanley tools.* The Stanley Rule & Level Plant, New Britain, CT. Reprinted in September 2002 by the Midwest Tool Collector's Association.

Stanley. (1994). *Tool traditions catalog.* Stanley Mail Media, Inc., Phoenix, AZ.

Stanley. (1995). *Tool traditions catalog.* Two different editions. Stanley Mail Media, Inc., Phoenix, AZ.

Stanley. Insert: *Read this before you use: Combination plane no. 46.* The Stanley Rule & Level Plant, New Britain, CT. Reprinted in September 2002 by the Midwest Tool Collector's Association.

Starrett, L. S. (July 1895). *No. 13 catalogue and price list of fine mechanical tools, manufactured by L. S. Starrett Athol, Mass. U.S.A.* Boston Eng. & McIndoe Printing Co., Boston, MA. Reprinted in October 1989 by Bud Brown Publishing Co., Reading, PA.

The L.S. Starrett Co. (1924). *Catalog No. 23: Fine mechanical tools.* The L.S. Starrett Co., Athol, MA.

The L.S. Starrett Co. (1924). *Catalog No. 26: Fine mechanical tools.* The L.S. Starrett Co., Athol, MA.

Stearns, E. C. & Co. (1924). *Catalogue of hardware manufactured by E. C. Stearns & Co.* Syracuse, NY. Reprinted in June 1977 by Roger K. Smith, College Press, S. Lancaster, MA.

Steel Shot & Grit Co. (no date). Catalog: *Certified steel abrasives . . . Samson steel shot . . . for sawing and polishing.* Steel Shot & Grit Co., Pittsburgh, PA.

J. Stevens Arms & Tool Co. (1898). *Shop-pointers & all steel tools.* J. Stevens Arms & Tool Co., Chicopee Falls, MA.

John Stortz & Son, Inc. (1970). *Price list for catalog no. C-69.* John Stortz & Son, Inc., Philadelphia, PA.

Swan, James Company. (1911). *Illustrated catalog and price list of premium mechanics' tools manufactured by The James Swan Co.* Seymour, CT. Reprinted in November 1981 by Ken Roberts Publishing Company, Fitzwilliam, NH.

Tools for Working Wood. (Summer 2003). Catalogue: *Tools for working wood: Hand tools, books, videos and accessories for the serious woodworker.* Tools for Working Wood. 3(1).

Underhill Edge Tool Co. (1859). *Wholesale Prices of Chopping Axes, Carpenter's, Cooper's, Butcher's, and Many Other Kinds of Mechanics' Tools, Manufactured by the Underhill Edge Tool Company.* Underhill Edge Tool Co., Nashua, NH.

Union Mfg. Co. (no date). *Union iron and wood planes.* Reprinted in June 1981 by Ohio Tool Collectors Association.

Union Tool Company. (1969). *Union Tool Company: Machinist and carpenter's tools: Combined 1969 catalog.* Orange, MA.

Union Twist Drill Co. (1912). *Catalog No. 100. milling and high power cutters: High speed steel.*

United Hardware & Tool Corporation. (1925). *United Hardware & Tool Corporation manufacturers' distributors and importers guaranteed hardware and tools of quality: Fulton catalog no. 40.* 74 Reade Street, NY. Reprinted in November 1983 by the Mid-West Tool Collectors Association.

Universal-Cyclops Steel Corporation. *Tool steel catalog.*

Vose & Co. (April 1853). *Illustrated book of stoves manufactured by Vose & Co., Albany, N.Y.* Vose & Co., Albany, NY. Reprinted in 1983 by the Early American Industries Association.

Vulcan Crucible Steel Company. (1929). *Catalogue No. 7: Vulcan tool steels: High grade tool steels and special steels.* Vulcan Crucible Steel Company, Aliquippa, PA.

Walter's, Wm. P. Sons. (1888). *Illustrated catalogue of wood workers' tools and foot power machinery.* Market Street, Philadelphia. Reprinted October 1981, by The North Village Publishing Co., Lancaster, MA. Available from Roger K. Smith, 1444 N. Main Street, Lancaster, MA 01523.

Walker-Turner Co. (1938). *Walker Turner power tools.* Walker-Turner Co., Plainfield, NJ.

The Western Tool and Manufacturing Co. *Fourteenth edition catalog and price list.* Winters Company, Springfield, OH.

Westervelt, A. B. & W. T. (1883). *No. 6. Illustrated catalogue and price list of copper weather vanes, bannerets and finials, manufactured by A. B. & W. T. Westervelt, office and warerooms: 102 Chambers Street, Corner Church Street, New York.* Reprinted in 1982 as *American antique weather vanes* by Dover Publications, Inc.

Weston Electrical Instrument Co. (1911). *Construction: Weston switchboard wattmeters synchroscopes and power-factor meters: manufacture, construction and design of Weston A. C.*

switchboard indicating instruments: Sections 1 and 2 of Catalog 16. Weston Electrical Instrument Co., Newark, NJ.

Weston Electrical Instrument Co. (1911). *Weston switchboard indicating wattmeters: Direct-current, single-phase and polyphase wattmeters. Section 3 of Catalog 16*. Weston Electrical Instrument Co., Newark, NJ.

Weston Electrical Instrument Co. (1911). *Weston switchboard synchroscopes: Section 4 of Catalog 16*. Weston Electrical Instrument Co., Newark, NJ.

Weston Electrical Instrument Co. (1911). *Weston switchboard power-factor and frequency meters: Sections 5 and 6 of Catalog 16*. Weston Electrical Instrument Co., Newark, NJ.

Weston Electrical Instrument Co. (1911). *Weston switchboard alternating-current ammeters and voltmeters: Section 7 of Catalog 16*. Weston Electrical Instrument Co., Newark, NJ.

Wheeler, Madden & Bakewell. (1860). *Monhagen Saw Works, Middletown, Orange County, N.Y. illustrated price list.* The "Press" Printing Establishment, Exchange Building. Reprinted in 1976 by Early American Industries Association, South Burlington, VT.

The L. & I.J. White Company. *Catalogue of coopers' tools including turpentine tools, edge tools and machine knives.* L. & I.J. White Co., Buffalo, NY.

Whitton, Blair, Ed. (1979). *Bliss toys and dollhouses: 89 illustrations, including the complete 1911 catalogue.* Dover Publications Inc., NY, NY.

Wiley & Russell Mfg. Co. (1888). *Patent screw-cutting machinery and tools, etc.* Printed by McIndoe Bros., Boston, MA, under C.H. Tiebout & Sons, Agents, Brooklyn, NY.

Wilkinson, A. J. & Co. (1867). *Wilkinson, A. J. & Co.'s illustrated catalogue of hardware and tools.* No. 2 Washington Street, Boston, MA. Reprinted in September, 2001, by the Mid-West Tool Collectors Association.

Wilkinson, John Co. (no date). *Catalogue No. 88: Price list of tools and machines for metal and wood workers.* The John Wilkinson Co., 77 State St., Chicago, IL.

Williams, J.H. & Co. (1937). *Superior Drop-Forgings and Drop-Forged Tools.* J.H. Williams & Co., Buffalo, NY.

Winchester Repeating Arms Co. (1923). *Winchester: Trade mark: Pocket catalog of tools: 1923: For sale at the Winchester Store.* Winchester Repeating Arms Co., New Haven, CT.

Winchester Repeating Arms Co. (1926). *Winchester: 1926 - 27 product catalog: The Winchester store.* Winchester Repeating Arms Co., New Haven, CT. Reprinted in 1985 by R. L. Deckebach, 413 Walnut St., Royal Oak, MI.

Winchester Repeating Arms Co. (1931). *Winchester, world standard guns and ammunition, Winchester Repeating Arms Co., New Haven, Conn., U.S.A.* Reprinted by R. L. Deckebach, Royal Oak, Michigan in 1989.

Winnie Machine Works. *18th annual catalogue.* Winnie Machine Works, Chicago, IL.

Winsted Edge Tool Works. (ca. 1900). *Catalog of the Winsted Edge Tool Works, West Winsted, Conn. chisels, drawing knives, gouges, etc.* Reprinted in 1989 by the Early Trades & Crafts Society, Long Island, NY.

Witherby, T.H. *Price list of Winsted Edge Tool Works, chisels, drawing-knives, gouges, etc.* The Case, Lockwood, & Brainard Co. Print, Hartford, CT.

Wood, Wm. T. & Co. (ca. 1894). *Price list, Wm. T. Wood & Co. manufacturers of finest quality ice tools, Arlington, Mass.* Reprinted in 1974 by the Early Trades & Crafts Society, Long Island, NY.

Young, Otto &Co. (1893). *Otto Young & Co. Chicago, ills. tool and material catalogue.* W.S. Conkey Co., Printers and Ginders, Chicago, IL. Reprinted in July 1998 by The Special Publications Committee, Mid-West Tool Collectors Assocation.

Walter, John. (1988). *Antique & collectable Stanley planes 1988 price guide.* The Tool Merchant, Akron, OH.

Walter, John. (1988). *Antique & collectable Keen Kutter hand tools 1989 price guide.* The Tool Merchant, Akron, OH.

Woodcraft Supply Corp., 313 Montvale Ave., Woburn, MA.

- The Davistown Museum has in the library the following Woodcraft catalogs:
    - 1974, cover missing with format very similar to 1977
    - Spring-Summer 1977
    - Fall-Winter 1978
    - Winter 1981
    - 1982
    - 1983 Spring Supplement
    - 1987-88 Supplement
    - March 1997

Wyke, John. (1978). *A catalogue of tools for watch and clock makers.* University Press of Virginia, Charlottesville, VA.

Zutphen, D. Stolp. (1919). *D. Stolp, Zutphen. Gereedschappen. Catalog January 1, 1919.* Reprinted and translated from Dutch in 1982 by The Mid-West Tool Collectors Association and The Early American Industries Association.

Tool Journals, Newsletters, and Auction Listings

Allen, Hank, Ed. (1996). *Tool shed treasury: An anthology of articles and stories 1978 - 1995: CRAFTS (collectors of rare and familiar tools society)*. Astragal Press, Mendham, NJ.

American Woodworker.
www.rd.com/americanwoodworker/action.do?categoryId=7000&siteId=2222

APT Bulletin: The Journal of Preservation Technology. The Association for Preservation Technology International, 4513 Lincoln Ave., Suite 213, Lisle, IL 60532-1290. www.apti.org

- The Davistown Museum has in the library the following copies of: *Bulletin: The Association for Preservation of Technology.*
  - Volume X, nos. 2 and 4, 1978.

The Anvil's Ring. ABANA, PO Box 816, Farmington, GA 30638-0816. www.abana.org

Carpenter. United Brotherhood of Carpenters and Joiners of America.
www.carpenters.org/carpentermag/

- The Davistown Museum has in the library the following copies.
- June, 1928, XLVII(6).

The Chronicle. **Early American Industries Association (EAIA)**, Elton Hall, 167 Bakerville Rd., South Dartmouth, MA 02748-4198. www.eaiainfo.org

- The EAIA's *Chronicle* is one of the most important contemporary sources of information about the tools and trades of pre-industrial North America.
- *The Chronicle* began publication in 1933 and is still issued monthly. The Davistown Museum has two volumes of bound issues at the Hulls Cove office and more recent single issues in the library stacks at the Museum.
- **Please help the museum by donating back copies of this journal that we do not have in stock.**
- The Davistown Museum has in the library the following copies:
  - Volumes 1 through 11, November, 1933 to December 1958.
  - Volumes 12 through 26, March, 1959 to December 1973.
  - Vol. 30, Nos. 1 – 4, March, June, September and December 1977.
  - Vol. 31, Nos. 2 and 3, June and September 1978.
  - Vol. 32, Nos. 1 – 4, March – December 1979.
  - Vol. 33, No. 1, March 1980.
  - Vol. 36, No. 3, September 1983.
  - Vol. 37, Nos. 2 – 4, June – December 1984.
  - Vol. 38, Nos. 1 – 3, March – September 1985.
  - Vol. 39, Nos. 1- 4, March - December 1986.
  - Vol. 40, Nos. 1 and 3, March and September 1987.

- o Vol. 41, Nos. 2 – 4, June – December 1988.
- o Vol. 42, Nos. 1 and 2, March and June 1989.
- o Vol. 43, Nos. 1 and 2, March and June 1990.
- o Vol. 44, No. 4, December 1991.
- o Vol. 45, Nos. 2 and 3, June and September 1992.
- o Vol. 46, Nos. 1, 3 and 4, March, September and December 1993.
- o Vol. 48, Nos. 1 – 4, March – December 1995.
- o Vol. 49, Nos. 2 - 4, June - December 1996.
- o Vol. 50, Nos. 1, 3 and 4, March, September and December 1997.
- o Vol. 51, Nos. 1 – 4 plus May Supplement, June – December 1998.
- o Vol. 52, Nos. 1 – 3, March – December 1999.
- o Vol. 53, Nos. 1 – 4, March – December 2000.
- o Vol. 54, Nos. 1, 2 and 4, March, June and December 2001.
- o Vol. 55, Nos. 1 – 4, March – December 2002.
- o Vol. 56, Nos. 1 - 4, March - December 2003.
- o Vol. 57, Nos. 1, 2 and 4, March, June and December 2004.
- o Vol. 58. Nos. 1 - 4, March - December 2005.
- o Vol. 59. No. 1, March 2006.
- o Vol. 60. No. 1, March 2007.
- o Vol. 62. No. 2, June 2009.
- o Index to Volumes 41 through 46, 1988 – 1993

The Cultivator. (1852). Vol. IX. Luther Tucker, Albany, NY.

- A monthly journal devoted to agriculture, horticulture, floriculture, and to domestic and rural economy. Illustrated with engravings of farm houses and farm buildings, improved breeds of cattle, horses, sheep, swine and poultry, farm implements, domestic utensils, &c. An index is included.

Blacksmith's Journal. PO Box 1699, Washington, MO 63090. www.blacksmithsjournal.com

Diecutting Diemaking Intelligence Newsletter. DDIN International.

- A quarterly 72 page publication devoted to all elements of the diecutting process.

Fine Tool Journal. Antique & Collectible Tools, Inc., 27 Fickett Rd., Pownal, ME 04069. www.finetoolj.com

- An excellent source of information on antiquarian tools, with specific articles on tool related topics in every issue. For example, vol. 49, no. 2, fall 1999, has an important article on Henry Mercer, his life and his museum in Pennsylvania.
- The Davistown Museum is seeking back issues of this important information resource. Numerous articles from the *Fine Tool Journal* are being added to these bibliographies. Thank you to Clarence Blanchard for his donation of many back issues to the Museum.
- Most back issues of the journal are in stock at the Museum's Davistown Library in Liberty.

- The Davistown Museum has in the library the following copies:
    - Vol. 41, No. 4, Spring 1992.
    - Vol. 43, No. 3, Winter 1994.
    - Vol. 44, No. 4, Spring 1995.
    - Vol. 45, Nos. 2 and 4, Fall 1995 and Spring 1996.
    - Vol. 46, Nos. 1, 2 and 4, Summer, Fall 1996 and Spring 1997.
    - Vol. 47, Nos. 1, 2 and 4, Summer, Fall 1997 and Spring 1998.
    - Vol. 48, Nos. 1 - 4, Summer, Fall 1998 and Winter, Spring 1999.
    - Vol. 49, Nos. 1 - 3, Summer, Fall 1999 and Winter 2000.
    - Vol. 50, Nos. 1 – 4, Summer, Fall 2000, Winter and Spring 2001.
    - Vol. 51, Nos. 1 – 4, Summer and Fall 2001, Winter and Spring 2002.
    - Vol. 52, Nos. 1 – 4, Summer and Fall 2002, Winter and Spring 2003.
    - Vol. 53, Nos. 1 - 4, Summer and Fall 2003, Winter and Spring 2004.
    - Vol. 54, Nos. 1 - 4, Summer and Fall 2004, Winter and Spring 2005.
    - Vol. 55, Nos. 1 - 3, Summer and Fall 2005, Winter 2006.
    - Vol. 56, Nos. 1, 2 and 4, Summer and Fall 2006, Spring 2007.
    - Vol. 57, Nos. 2 - 4, Fall 2007, Winter and Spring 2008.

Fine WoodWorking. The Taunton Press, Inc., Taunton Lake Road, Newtown, CT 06470. www.taunton.com/finewoodworking/index.asp

- The Davistown Museum has in the library the following copies:
    - Vol. 1, No. 3, Summer 1976
    - No. 29, July/August 1981
    - No. 32, Jan/Feb 1982 (2 copies)
    - No. 37, Nov/Dec 1982

The Gristmill. Midwest Tool Collector's Association. www.mwtca.org/gristm.htm

- Another excellent source of information about tools and trades. The Davistown Museum is seeking back copies of this publication.
- The Davistown Museum has in the library the following copies:
    - No. 119, June 2005

Hammer's Blow. ABANA, PO Box 816, Farmington, GA 30638-0816. www.abana.org/membership/publications/hb/hammers_blow.html

MVWC Newsletter. Missouri Valley Wrench Club, Newsletter Editor, 659 E. 9th, York, NE 68467-3109.

Muzzle Blasts. Maxine Moss Drive, Friendship, IN 47021. nmlra.org/merchandise.htm#Muzzle%20Blasts%20Back%20Issues

- This journal is published monthly by the National Muzzle Loading Rifle Association. It focuses on black powder firearms and often contains articles on gunmaking, reproductions of early American tools and the history of the muzzleloading era.

Muzzleloading and Traditional Living Magazine. 82 Little Houston Brook Rd., Concord Twp., Maine 04920. www.muzzleloadingandtraditionalliving.com

Nautical Research Journal. Published Quarterly by the Nautical Research Guild, 19 Pleasant Street, Everett, MA 02149.

- The Davistown Museum has in the library the following copies:
    - Vol. 10, Nos. 1, 3 and 4, 1959.
    - Vol. 12, Nos. 2 and 3. (Partial copies only).
    - Vol. 13, Nos. 2 – 4, Autumn and Winter 1965.
    - Vol. 15, Nos. 2 and 4, Summer and Winter 1968.
    - Vol. 16, Nos. 1, 2 and 4, Spring, Summer and Winter 1969.
    - Vol. 17, Nos. 1, 3 and 4, Spring, Fall and Winter 1970.
    - Vol. 18, Nos. 1, 2 and 4, Spring, Summer and Winter 1971.
    - Vol. 19, Nos. 2 and 4, Summer and Winter 1972.
    - Vol. 20, Nos. 1, 2 and 4, October 1973, January and October 1974.
    - Vol. 21. Nos. 2 and 3, June and September 1975.
    - Vol. 22, No. 4, December 1976.
    - Vol. 23, Nos. 1 – 4, March – December 1977.
    - Vol. 24, Nos. 1 and 4, March and December 1978.
    - Vol. 25, Nos. 1 – 3, March – September 1979.
    - Vol. 26, Nos. 1 – 3, March – September 1980.
    - Vol. 27, Nos. 1, 2 and 4, March, June and December 1981.
    - Vol. 28, Nos. 1, 3 and 4, March, September and December 1982.
    - Vol. 29, Nos. 1 and 2, March and June 1983.
    - Vol. 30, No. 2, June 1984.
    - Vol. 31, Nos. 1, 2 and 4, March, June and December 1985.
    - Vol. 32, Nos. 1 – 4, March – December 1987 (no issues published in 1986).
    - Vol. 33, No. 2, June 1988.
    - Vol. 34, Nos. 2 – 4, June, September and December 1989.
    - Vol. 35, Nos. 1 and 2, March and June 1990.
    - Vol. 36, Nos. 1 – 3, March – September 1991.
    - Vol. 42, No. 4, December 1997.

The Pine Tree Shilling: Opening a Window on Life in the American Colonies - 1650-1780. 507 Meany Rd., PO Box 1005, Charlestown, NH 03603-1005. ljmiller@turbont.net.

- Published 4 times a year by living history re-enactors. This newsletter includes articles on how to make and use historically accurate tools. It focuses on the every day life in the American colonies.

Plane Talk: The Quarterly Journal of Plane Collecting and Research. PO Box 338, Morristown, NJ 07963-0338. Published by the Astragal Press.

- This journal is no longer in publication. The Astragal Press offers a text containing several of the most recent volumes.

Popular Woodworking. www.popularwoodworking.com

Rural Heritage. 281 Dean Ridge Lane, Gainesboro, TN 38562. www.ruralheritage.com/index.htm

- "A bimonthly journal in print since 1976, online since 1997, in support of small farmers and loggers who use draft horse, mule and ox power."

Shavings. Newsletter of the Early American Industries Association, Inc. 1402 Hickman Bluff Rd., Lavaca, Arkansas 72941. www.eaiainfo.org

- Published six times a year.

Sign of the JOINTER: The Quarterly Journal of Antique Wooden Planes. 6211 Elmgrove Rd., Spring, TX 77389. www.woodenplane.org/index.htm

- This newsletter began in the summer of 1999 and is published four times a year. The Davistown Museum library has a complete collection through _____.

The Spinning Wheel Sleuth. PO Box 422, Andover, MA 01810, 978 475-8790. www.spwhsl.com

- Began publication in 1993; currently published four times a year.

Stanley Tool Collector News. Published 3 times a year by The Tool Merchant, 208 Front St., Marietta, OH 45750.

- The first issue was printed in the fall of 1990 and has now been discontinued.

Technology and Culture. The International Quarterly of the Society for the History of Technology, The University of Chicago Press, Chicago, IL. shot.press.jhu.edu/tc.html

- The Davistown Museum received a donation of 25 issues from the Columbia University Libraries.
  - Current Bibliography (1990) published 1992. 33.
  - January, April, July, October 1992. 33(1-4).
  - January, April, July 1993. 34(1-3).
  - Current Bibliography (1992) published 1994 supplement. 35.
  - January, April, July, October 1994. 35(1-4).
  - Current Bibliography (1993) published 1995 supplement. 36.
  - January, April, April Supplement, July, October 1995. 36(1-4).

- January, April 1996. 37(1-2).
- January, April, July, October 1997. 38(1-4).

Tool Chest. Hand Tool Preservation Association of Australia, PO Box 1163, Carlton, Victoria, Australia 3053. home.vicnet.net.au/~toolclub/pub.htm

- Published four times a year along with *The Sharp Edge* newsletter.

Tool Collectors' Picture Book. Early Trades & Crafts Society, Long Island, NY.

- The Davistown Museum has in the library the following copies:
  - 14th: *The handy Hammer*. May 1982.
  - 15th: *The spoke-shave and related tools*. April 1983.

Tool Talk. PAST (Preserving Arts and Skills of the Trades), 1445 Fourth St., Berkeley, CA 94710-1335. pasttools.org/

- Published four times a year.

Tool Talks. Stanley Tools, New Britain, CT.

Tools & Technology: The Newsletter of the American Precision Museum. Windsor, VT 05089. www.americanprecision.org

- Published quarterly.
- Copies of this reference are available for visitors to browse in the Museum Library.

Tools and Trades. The Journal of the Tool and Trades History Society, The Membership Secretary, Woodbine Cottage, Budleigh Hill, East Budleigh, Devon EX9 7DT, United Kingdom. www.taths.org.uk

- TATHS is the quarterly illustrated newsletter for the society.

The Tool Shed. Journal of the Collectors of Rare And Familiar Tools Society (CRAFTS), 38 Colony Court, Murray Hill, NJ 07974. www.craftsofnj.org/toolshed/tool_shed.html

Woodworker: The Magazine for the Craftsman in Wood.

- The Davistown Museum has in the library the following copies:
  - August 1976, 80(993).

Woodworker's Journal: The Voice of the Woodworking Community. www.woodworkersjournal.com

Woodworkers Library. Linden Publishing, 2006 South Mary, Fresno, CA 93721. www.lindenpub.com

Woodworking Magazine. www.popularwoodworking.com/wwmhomepage/

Working Wood: The Quarterly Journal for the Craftsman Working in Wood. Quailcraft Publishing Co., Ltd., Aldershot, Hants, England.

- The Davistown Museum has in the library the following copies:
    o Summer 1982, 4(1).

Ye Olde Tool Chest. Newsletter of the Pacific Northwest Tool Collectors. 24575 Butler Rd., Elmira, Oregon 97437.

- The Davistown Museum has in the library the following copies:
    o March 1988, 5(2).

Auction Listing Catalogs

Auction catalogs are a particularly useful source of information on antiquarian tools and their value, especially if illustrated. The following listing is under construction and includes only a few of the catalogs in the Davistown Museum collection. One of the best sources of tools on the web are auctions.

Baxter Auction Gallery, Indianapolis, IN:

- Saturday, Oct. 28 2000.
- Saturday, Oct. 26, 2002.

David Stanley Auctions, Stordon Grange, Osgathorpe, Loughborough, UK. *Special collective sale by auction of quality antique woodworking & allied trades tools.*

- October 4, 1983.

Fine Tool Journal: FTJ Absentee Auction and Brown Auction Services, http://www.finetoolj.com/auctions/index.html.

Horst Auction Center, Ephrata, PA:

- Saturday, Feb. 5, 2000.

Johnny King Auctioneers, Ware Shoals, SC:

- Sept. 7, 9, 16, 20.

Martin J. Donnelly Antique Tools. *The Catalogue of Antique Tools: The World's Premier Antique Hand Tool Value Guide*. Martin J. Donnelly Antique Tools, PO Box 281, Bath, New York. (800) 869-0695. www.mjdtools.com.

- The Davistown Museum has a nearly complete collection of the many auction catalogs he has issued.
- See annotations in the collector's guides bibliography.

Richard A. Bourne, Co., Inc., Corporation Street, Hyannis, MA:

- Tuesday, September 16, 1980.

Tony Murland: International Tool Auction catalog of items: England:

- July 25th, 1997, Lot 2209.

Toolshop International Auction Catalogues.78 High Street, Needham Market, Suffolk, IP6 8AW, England. www.antiquetools.co.uk/auction.html

- "Our catalogues are the finest tool catalogues in the world and have become a benchmark reference publication for all those individuals who have an interest in antique tools."

Your Country Auctioneer:

- Sept. 24-25, 1999.

Index